Philosophy of Science and Mathematics

Philosophy of Science and Mathematics

Y N Thakur

MD Publications Pvt Ltd
New Delhi
www.mdppl.com

Published by :

MD Publications Pvt Ltd
"MD House", 11, Darya Ganj,
New Delhi - 110 002
Phone : +91-11- 45355555
E-mail : contact@mdppl.com
Website: www.mdppl.com

ISBN : 978-81-7533-334-5

Published and Printed by Mr. Pranav Gupta on behalf of **MD Publications Pvt Ltd** at Repro India Limited.

PREFACE

Many students of philosophy of science and mathematics, as well as many teachers of it, have complained in my presence of the scarcity of philosophy of science and mathematics texts that are at one and the same time thorough enough not to be superficial, yet written to be accessible to undergraduate readers. I have written this text in the hope of having done something to remove that scarcity. Along the way I have also tried to make what contribution I could to the growing movement within philosophy of science and mathematics away from the exclusive use of the physical sciences as a source domain for illustrations and examples. Most undergraduate level philosophy of science and mathematics texts, let us face it, are really philosophy of physics texts. I believe that this focus on one of the premier life sciences of our time has produced a happier and more fruitful result with respect to making the philosophy of science interesting and accessible to a larger audience of undergraduates.

This text presents a thorough survey of contemporary philosophy of science and mathematics in the sense that it covers both the positivist model of science and the currently popular alternatives to the positivist model that flow from Thomas Kuhn's watershed work. Many current texts cover one or the other of these models of science rather well but condemn the remaining one to a superficial treatment or to no treatment at all. Several aspects of the text call for special comment. The philosophy of science as an invitation to participate in wild-eyed intellectual revolution, as it is so often assumed to be by other writers. While I do not agree with much of what the social constructivists have to say, social constructivism is here to stay, and any survey text in philosophy of science that seeks even minimal adequacy ought to contain a treatment of it at least as thorough as the one contained in this text.

Throughout the text I make no attempt to hide from the reader my sympathies for scientific realism. I am frankly skeptical of any philosopher of science who claims that it is possible to write an effective philosophy of science and mathematics text while remaining absolutely neutral on the central issue of realism and antirealism. Nothing of benefit to students is accomplished by trying to maintain such a false pose of neutrality about something it is so obviously absurd to be neutral about. It is possible to present fairly all sides to an issue while taking a position oneself.

I have learned much from philosophical discussions with my friends over the years, and their philosophical influence on me has made this a far better book than it would have been otherwise. My relatives and brother encouragement was also very helpful and appreciated.

I am very grateful to Mr. Pranav Gupta, Director of MD Publications Pvt Ltd, New Delhi, for his kind cooperation and support for this book.

Y N Thakur

Contents

1

Philosophy of Science
An Introduction

Introduction

Dialectical materialism incorporates features of both absoluteness and relativity, of both an Aristotelian commitment to the immutable and independent and a Heraclitean belief in flux. To its defenders, this combination of opposite tendencies is a source of flexibility, strength, and truth; to its detractors, it is evidence of ambiguity, vagueness, and falseness.

Dialectical materialism has usually been discussed as if it were a uniquely Soviet creation, far from the traditions of Western philosophy. It is true that the term "dialectical materialism" comes from a Russian and not from Marx, Engels, or their west European followers. It is also true, of course, that Soviet dialectical materialism has acquired characteristics that are only explicable in terms of, first, its revolutionary, and later, its institutional setting. But the roots of dialectical materialism extend back to the beginning of the history of thought, at least to the Milesian philosophers and continue forward as subdued, changing, but reappearing strands in the history of philosophy. It is impossible to present here a discussion of the origins of dialectical materialism, which would constitute a large book in itself and would belong to the province of the professional philosopher; nonetheless, many similarities between dialectical materialism and previous currents of European philosophy will appear in the following pages.

The term "dialectical materialism" was first used, to the best of our knowledge, in 1891 by G. V. Plekhanov, a man frequently called the father of Russian Marxism. Marx and Engels utilized terms such as "modern materialism" or "the new materialism" to distinguish their philosophical orientation from that of classical materialists such as Democritus or thinkers of the French Enlightenment such as La Mettrie or Holbach. Engels did speak, however, of the dialectical nature of modern materialism. Lenin adopted the phrase used by Plekhanov, "dialectical materialism."

The basic writings of Marx, Engels, and Lenin on the philosophic and social aspects of science are Engels *Anti-Dühring*, printed first as a series of articles in 1877; his *Dialectics of Nature*, written in 1873–83 but not printed until 1925; his *Ludwig Feuerbach and the End of Classical German Philosophy*, published as a series of articles in 1886 and as a pamphlet in 1888; Marx's doctoral dissertation, written in 1839–41 and first published in 1902; pieces of the correspondence of Marx and Engels; a few sections of Marx *Capital*; Lenin *Materialism and Empirio-Criticism*, published in 1908; his Philosophical Notebooks, published in 1925–29 and later, in a complete form, in 1933; and fragments from his correspondence and speeches. Marx also left a number of unpublished manuscripts concerning science, technology, and mathematics, most of which are now in the Institute of Marxism-Leninism in Moscow. Some of these appeared in print only in the late 1960's. Together all these writings establish the basis of dialectical materialism as it is usually discussed in the Soviet Union, with the older writings obviously playing a more formative role than the newer ones.

In this rather large body of material written over a period of many decades by different authors for different purposes one can find a considerable diversity of viewpoints and even contradictions on fairly important questions. The dates of publication of the various works and the context in which each was composed are quite important in gaining an understanding of the evolution, modification, and structure of Soviet Marxist thought on the nature of science.

Although the primary interests of Marx and Engels were always in economics, politics, and history, they both devoted a surprisingly large segment of their time to the scrutiny of scientific theory, and cooperated in publishing their views on science. Engels described their background in the sciences:

> *Marx and I were pretty well the only people to rescue conscious dialectics from German idealist philosophy and apply it in the materialist conception of nature and history. But a knowledge of mathematics and natural science is essential to a conception of nature which is dialectical and at the same time materialist. Marx was well versed in mathematics, but we could only partially, intermittently and sporadically keep up with the natural sciences. For this reason, when I retired from business and transferred my home to London, thus enabling myself to give the necessary time to it, I went through as complete as possible a "moulting" as Liebig calls it, in mathematics and natural sciences, and spent the best part of eight years on it.*

Engels was much more important in elaborating the Marxist philosophy of nature than was Marx. This commitment to the study of the natural sciences as well as the social sciences was, in Engels's mind, a necessary consequence of the fact that man is, in the final analysis, a part of nature; the most general principles of nature must, therefore, be applicable to man. The search for these most general principles, based on knowledge of science itself, was a philosophic enterprise. Engels believed that by means of a knowledge of a philosophy that was materialistic, dialectical, and grounded in the sciences, both natural scientists and social scientists would be aided in their work. Those natural scientists who maintained that they worked without relying on philosophical principles were deluded; better to form consciously a philosophy of science, Engels thought, than to pretend to avoid one:

> *Natural scientists may adopt whatever attitude they please, they will still be under the domination of philosophy. It is only a question whether they want to be dominated by a bad, fashionable philosophy or by a form of theoretical thought which rests on acquaintance with the history of thought and its achievements.*

Engels's interest in the philosophy of science was so much more evident than Marx's that many scholars have maintained that it was Engels, not Marx, who was responsible for the concept of dialectical materialism; and that in bringing the natural sciences into the Marxist system, Engels violated original Marxism. Among the scholars holding this view are those who emphasize the young Marx as a theorist interested, not in universal systems, but specifically in man and his sufferings, a person whose first achievement was to present an explanation of the role of the proletariat in the modern world through the concept of alienated labor. Examples of exponents of this view are George Lichtheim, who wrote that dialectical materialism is a "concept not present in the original Marxian version, and indeed essentially foreign to it, since for the early Marx the only nature relevant to the understanding of history is human nature," and Z. A. Jordan, who maintained that dialectical materialism was a "conception essentially alien to the philosophy of Marx."

Scholars such as Lichtheim and Jordan have correctly emphasized the humanitarian ethic of the young Marx and the anthropological nature of his analysis, but they have erred in saying or implying that the idealistic young Marx was interested only in human nature, not physical nature. Marx's doctoral dissertation, written in 1839–41, several years before the now noted *Economic and Philosophical Manuscripts*, was suffused with the realization that an understanding of man must begin with an understanding of nature. Entitled *"The Difference between the Nature Philosophy of Democritus and the Nature Philosophy of Epicurus,"* the dissertation was a long discussion of the physics of the ancients, of the deviations from straight line descent in the atomic theory of Epicurus, of the nature of elementary substances and of elementary concepts such as time. Marx's attention to physical nature for an understanding of philosophy as a whole was entirely within the context of much of European thought; it was, further, an advantage rather than a disadvantage of his approach. Those recent writers who have tried to divest Marxism of all remnants of inquiry into physical nature have not only misrepresented Marx but have also deprived Marxism of one of its intellectual strengths. It is not necessary to restrict Marx's interests to ethics and economics to free him from vulgar materialism of the type of Vogt or Moleschott. Indeed, one of the points of Marx's dissertation was to show that Epicurus, although like Democritus a believer in atoms and the void, was not a strict determinist. The twenty-three-year-old Marx saw the atom as an abstract concept containing a Hegelian contradiction between essence and existence. Marx would later discard the philosophic idealism underlying this formulation, but there is no evidence that he ever abandoned his interest in physical nature itself.

As a young student of philosophy, Marx was affected by the metaphysical aspirations of almost all great philosophical systems prior to his time and accepted the necessity of making certain epistemological and ontological assumptions. In later years, he attempted to move away from metaphysics, a tendency of some significance since materialism (like all other philosophical systems, including pragmatism) is in the final analysis founded on metaphysical assumptions. At no known point, however, did he resist Engels's effort to bring nature explicitly into their intellectual system. Engels read the entire manuscript of his *Anti-Dühring* to Marx, who presented no objections and even contributed a chapter himself (not on natural philosophy, however) for inclusion in the book. At least as early as 1873, ten years before Marx's death, Engels began work on what became many years later *Dialectics of Nature*; their correspondence illustrated that the mature Marx shared Engels's interest in "modern materialism" in nature notwithstanding the fact that he usually yielded to Engels on issues concerning science. Another spot where Marx indicated his agreement with Engels's efforts occurred in *Capital*; Marx observed that the dialectical law of the transition of quantity into quality, applicable to economics, also applied to the molecular theory of modern chemistry. The thought behind Marx's somewhat casual observation on chemistry was his belief that in nature the whole is greater than the sum of its parts. It is a defensible and valuable principle defended by a large variety of thinkers, including contemporary dialectical materialists.

The point here is not that Marx's and Engels's views were identical, which has been maintained in the past both by Soviet scholars who wished to preserve the unity of dialectical materialism and by antiSoviet scholars who wished to condemn Marx with the albatross of Engels; rather, the main point is that to emphasize primarily the differences of two men whose views have a great many affinities and who did both consider themselves modern materialists is as much, if not more, of an inaccuracy as crudely to lump them together. It is one thing to say that Marx never committed himself to finding dialectical laws in nature to the extent to which Engels did; it is quite another to say that such an effort contradicted Marx's thought, particularly when Marx is known to have supported the effort on several occasions. Jordan called Marx a "naturalist" rather than a "materialist," meaning that Marx wished to avoid a metaphysical commitment to matter as the sole source of knowledge, but Jordan acknowledged elsewhere:

> *The materialist presuppositions which were shared by Marx might have included the principle of the sole reality of matter ("matter" being the term used to denote the totality of material objects, and not the substratum of all changes which occur in the world), the denial of the independent existence of mind without matter, the rule of the laws of nature, the independent existence of the external world, and other similar assumptions traditionally associated with materialism.*

Jordan pointed out that Marx did not regard knowledge as the mere passive reflection of external matter in the human brain; rather, Marx saw knowledge as a result of a complex interaction between man and the external world.

This epistemology does not deny materialism if one assumes that man is a part of the material world, but neither does it absolutely require it. There is a certain leeway in Marx's thought, permitting the supposition of naturalism instead of materialism, just as there is room in the Lenin of the *Philosophical Notebooks* (but not in the Lenin of *Materialism and Empirio-Criticism*) for several epistemologies. But despite these elements of latitude in Marx's thought, he never disclaimed "modern materialism," frequently accepted or used the term, supported Engels's elaboration of it, and in consequence, is, I believe, more accurately described as a materialist than as a naturalist.

The recent effort by many non-Soviet scholars to eliminate from Marxism an interest in physical nature can be explained, on the one hand, by their distaste for the ideological restrictions on science that were imposed in the Soviet Union, and on the other, by the general trend of philosophical thought in western Europe and North America. The interference of ideology with science in the Soviet Union, culminating in most people's minds in the Lysenko episode, led to a discrediting of the claims of Marxist philosophy in the natural sciences. Meanwhile, in the countries of western Europe and North America, metaphysical and ontological studies were out of fashion; dialectical materialism as an approach to nature was often seen as a vestige of archaic *Naturphilosophie*, an attempt to invade a realm that now belonged exclusively to the specific sciences. Many writers in Europe in the postwar generation were concentrating upon the individual man contained in existentialism rather than the soul-less determinism that most people assumed was contained in dialectical materialism. Even those deeply interested in Marxism, such as Sartre, usually paid little attention to detailed studies of physical nature.

Scholars still committed to Marxism often attempted to save it from *Naturphilosophie* by trying to separate the writings of Engels on science from those of Marx, an operation that is technically possible but that, as we have already seen, usually resulted in conclusions incorrectly restricting the breadth of Marx's interests. On the other hand, those scholars who were opposed to Marxism used the ideological incursions on science in the Soviet Union as important supports in their efforts to prove that Marxism was essentially a perversion of science, antirational and even anti-Western, ignoring the deeply Western origins of Marxism and the fact that the Lysenko affair had little to do with Marxism as a philosophy of science.

Frankly recognizing Marx's interest in science would not only be more accurate in a historical sense but would also eventually have a salutary effect upon Marxism as a political doctrine. The same man who wrote the *Economic and Philosophical Manuscripts* also inquired into atomism, the history of the concept of mathematical differentiation, criticized Newton's method of quadratures, and discussed the mathematics of Lagrange, Maclaurin, and Taylor. Science was not only an area of knowledge that Marx attempted to bring within his purview, but also a model to him as a methodology of investigation. As Engels said at Marx's graveside, Marx tried to do for human history what Darwin did for the organic world; indeed, Marx offered to dedicate *Capital* to Darwin as a sign of his esteem. Marx believed that his interpretation of history was scientific, not in a rigidly deterministic sense but in the morphological Darwinian sense. On the basis of his theory of evolution Darwin could not predict when a new species would arise, and on the basis of his theory of history Marx could not predict when a new form of society would arise. If Marxist materialism, either as an explanation of nature or of social history, is to be discussed in terms of science, the comparison must be with non-deterministic theories of low predictive power, such as the theory of evolution or probability theories. The analogy between Darwin's and Marx's theories could be easily exaggerated into a misleading form of "biological sociology," but there is reason to believe that Marx himself saw a similarity in terms of the predictive reliability of the two approaches.

One of the conclusions issuing from the observation that Marx considered his scheme scientific (in roughly the way Darwin's was) is that modification of the doctrine must therefore be a desired end, not a feared one. Nothing is more un-Marxist than a pathological fear of revision. Science, which must stand prepared to revise its conclusions in the light of new data, is in essence revisionist. A Marxism that took its goal of being scientific seriously instead of dogmatically would be a constantly evolving Marxism, an openly revised Marxism.

The current turn to humanism within Marxism is a development that all people aware of the importance of ethics and aesthetics topics not given their due in the Marxism of the past can only greet with enthusiasm. Marxism has been one of the great influences of the past century and will no doubt continue to be significant in the future; its contribution will be greater if its humanization is thorough. However, just as the humanism of the Renaissance is seen by historians of science as a development with contradictory effects on the development of science, so also might humanism in Marxism eventually have detrimental effects if it pushed science outside that realm of activities to which Marxism gives attention. Indeed, science is an intensely human activity; a sophisticated humanistic Marxism would be eager to bring it within its purview, not for the purpose of imposing a point of view, but for the derivation of additional knowledge of reality and for the appreciation of the crucial importance of science to the ethical and aesthetic system that it is attempting to create. Any conceptual scheme that hopes to be relevant to the modern world cannot afford to be uninterested in science. Science not only has philosophical content, but also requires careful evaluation not blind criticism from an ethical standpoint. Marxism has in the past possessed the advantage of being, as Sartre said, a totalizing "philosophy of our time," the single contemporary philosophical system that takes man in "the materiality of his condition." Whether Marxism as a philosophic system can lose its dogmatism without simultaneously losing its comprehensiveness is an unanswered question.

In the Soviet Union philosophers have not attempted to divest Marx of his interest in all of reality, including physical as well as human nature; they have not followed the trend elsewhere in abandoning the effort to construct comprehensive explanations of reality based on studies of nature itself. They have recognized that one of the most intellectually attractive aspects of Marxism is its explanation of the organic unity of reality; according to Marxism, man and nature are not two, but one. Any attempt to explain either will inevitably have implications for the other. But Soviet philosophers have frequently squandered this intellectual advantage by supporting a dogmatic philosophy, by raising it to a status of a political ideology used for the rationalization of the existing governmental bureaucracy. They have failed to recognize the essential intellectual revisionism contained within Marxism's claim to a scientific approach. As a result they have not been adept in connecting dialectical materialism with new interests arising in non-Soviet philosophy with which it is potentially compatible, such as process philosophy.

Engels and Lenin on Science

Although both Marx and Engels were interested in science from early ages, it is nonetheless true that Engels turned most seriously to science only after the Marxist philosophy of history had been fully developed. By 1848 their political and economic views were well formed, but Engels did not begin systematic study of the sciences, nor did Marx initiate his most detailed studies of mathematics, until some time later. Engels remarked that he took up the study of science "to convince myself in detail of what in general I was not in doubt that amid the welter of innumerable changes taking place in Nature, the same dialectical laws are in motion as those which in history govern the apparent fortuitousness of events.'

Just what the term "law" (*Gesetz*) meant to Engels is not altogether clear. He did not attempt a philosophical analysis of the many different meanings that have been given to such terms as "law of nature," "natural law," or "causal law," and he did not clearly indicate what he meant by "dialectical law." Engels's dialectical laws were considerably different from those laws of physics that, within the limits of measurement, permit empirical verification. Engels saw, for example, the dialectical law of the transition of quantity into quality in the observed phenomenon that water, after absorbing quantities of heat, experiences a qualitative change when it comes to a boil at 100 degrees centigrade. Such a change can, indeed, be empirically verified by heating many samples of water to 100 degrees. But Engels believed (and Marx agreed in *Capital*) that the same law describes the fact that "not every sum of money, or of value, is at pleasure transformable into capital. To effect this transformation, in fact, a certain minimum of money or of exchangevalue must be presupposed in the hands of the individual possessor of money or commodities." The latter case of the dialectical law of the transition of quantity into quality is rather different from the former, even though both are described as instances of the same law. The boiling of water seems to be simply a phenomenon adequately described by a normal physical law; the quantitative-qualitative dialectical relationship is best used in describing the much more abstract phenomenon that a large aggregate of an entity, taken as a whole, displays characteristics not displayed by its parts. In the case of economics, there is no way in which the law can be verified in every instance; if a certain accumulation of money occurred without its conversion into capital, one could merely say that the correct point had not yet been reached. In the case of the water, one not only possesses the description of *what* change is to occur, but information about *when* it is to occur.

Engels believed that nothing existed but matter and that all matter obeys the dialectical laws. But since there is no way of deciding, at any point in time, that this statement is true, the laws that he presupposed are not the same as usual scientific laws. It should be admitted that even in the case of "usual" laws in natural science the stated relationship, as a universal statement, is not subject to absolute proof. One cannot say, for example, that there will *never* be a case in which a standard sample of water heated to 100 degrees centigrade fails to boil. But when the violation of such laws does occur, it is, within the limits of measurement, apparent that something remarkable has happened.

The definition of "law" is a very controversial and difficult topic within the philosophy of science, and I shall not pursue it beyond noting that Engels's concept of dialectical laws was quite broad, embracing very different kinds of explanations. Indeed, he referred to the dialectical relationships not only as "laws," but also as "tendencies," "forms of motion," "regularities," and "principles."

Engels is known for two major works on the philosophy of science, *Anti-Dühring* and *Dialectics of Nature.* Since only the first of these was a finished book and appeared almost fifty years earlier than the second, it exercised the greatest influence on the formation of the Marxist view of nature. In *Anti-Dühring* Engels criticized the philosophic system advanced by Eugen Karl Dühring in his *Course in Philosophy.* Duhring was a radical lecturer on philosophy and political science at the University of Berlin, a critic of capitalism who was gaining influence among German social democrats. Engels disagreed with Dühring's claim to "a final and ultimate truth" based on what Dühring called "the principles of all knowledge and volition." The object of Engels's criticism was not Dühring's goal of a universal philosophic system, but the method by which he derived it and his claims for its perfection. Dühring's "principles" were to Engels a product of idealistic philosophy: "What he is dealing with are principles, formal tenets derived from *thought* and not from the external world, which are to be applied to nature and the realm of man and to which therefore nature and man have to conform." Engels believed, contrary to Dühring, that a truly materialistic philosophy is based on principles derived from matter itself, not thought. The principles of materialism, said Engels, are not the starting-point of the investigation, but its final result; they are not applied to nature and human history, but abstracted from them; it is not nature and the realm of humanity which conform to these principles, but the principles are valid in so far as they are in conformity with nature and history. That is the only materialistic conception of matter, and Herr Dühring's contrary conception is idealistic, makes things stand completely on their heads, and fashions the real world out of ideas, out of schemata, schemes or categories existing somewhere before the world, from eternity–just like *a Hegel.*

A number of writers have commented that this desire to counteract Dühring's idealistic philosophy pushed Engels's first philosophical work toward the positivistic position of maintaining that all knowledge must be composed of verifiable data derived from nature? They have frequently cited *Dialectics of Nature*, the later work, as containing an opposite metaphysical tendency, and have observed that the tension between these two strains in Marxist thought positivistic materialism and metaphysical dialectics has been present throughout its history. A tension between materialism and the dialectic has indeed existed within Marxism, and will be commented upon later, but the extent to which *Anti-Dühring* is positivistic and *Dialectics of Nature* metaphysical has been overdrawn. True, Engels in *Anti-Dühring* directed his chief criticism against a philosopher for not being materialist, while in *Dialectics of Nature* he, more in passing, admonished scientists (such as Karl Vogt, Ludwig Büchner, and Jacob Moleschott) for not being dialectical. But in both works Engels attempted to locate a balance point between reliance on the empirical findings of science, on the one hand, and the dialectical structure inherited from Hegel on the other. In *Anti-Dühring*, the reputedly positivistic work, Engels also presented some of his best-known discussions of the dialectic in nature, while in *Dialectics of Nature*, the work supposedly heavily Hegelian in inspiration, he stoutly defended the concept of the materiality of the universe.

If one turns from Engels's works on the philosophy of science to a consideration of his knowledge of science itself, one is likely to conclude that although essentially a dilettante, he was a dilettante in the best sense. For a person of his background he possessed a remarkable knowledge of the natural sciences. Engels's formal education never went beyond the *gymnasium*, but he immersed himself in the study of science at certain periods of his life; he was able, for example, to write a long chapter on the electrolysis of chemical solutions, including computations of energy transformations. He was familiar with the research of Darwin, Haeckel, Liebig, Lyell, Helmholtz, and many other prominent nineteenth century scientists. In retrospect his errors do not draw so much attention as his unlimited energy and audacity in approaching any subject and the high degree of understanding that he usually achieved. Even if one is not willing to accept J. B. S. Haldane's observation that Engels was "probably the most widely educated man of his day," he was, indeed, a man of impressive knowledge. Elements of naïveté and literalness are easily found, but they are less significant than his conviction that an approach to all of knowledge, and not just one portion, was necessary for a new understanding of man.

Indeed, a re-evaluation of Engels by historians of science is overdue. Engels's "errors" in science, as seen from the vantage of today his quaint descriptions of electricity, his discussions of cosmogony, his descriptions of the structure of the earth, and his assertion that mental habits can be inherited were usually the "errors" of the science of Engels's time. Engels was a materialist, and suffered from the tendency toward simplification that has plagued many materialists, but he was far more sophisticated than the popularizers of materialism of his day, who were usually scientists, such as Büchner and Moleschott. Those recent writers who have dismissed Engels's writings on science have usually forgotten the context of nineteenth century materialism in which they were written. Against the background of this materialism Engels appears as a thinker with a genuine appreciation of complexity and an awareness of the dangers of enthusiastic reductionism. He was, for example, convinced that life arose from inorganic matter, but he ridiculed the simple approach of the supporters of spontaneous generation who had in the 1860's suffered a defeat at the hands of Pasteur. Engels's attitude toward the origin of life has been praised by biologists even recently.

Lenin's writings on science are similar to Engels's, not only in terms of philosophical commitment, but also in several other secondary respects: He came to science after formulating his political and economic views; he first entered the field of philosophy of science for polemical reasons; he was responsible for two major works with somewhat different emphases; and his later, more sophisticated period is much less well-known than his earlier, relatively untutored phase.

The particular viewpoints of Lenin on the philosophy of science, as expressed in *Materialism and Empirio-Criticism* and in the *Philosophical Notebooks* will be discussed in the following sections on epistemology and dialectics, but at this point it is necessary to mention the fact that most non-Soviet discussions of Lenin's philosophic views are based on *Materialism and Empirio-Criticism*. The *Philosophical Notebooks*, which consist of abstracts, fragments, and marginal notes, were not published until the end of the twenties, and did not appear in English until 1961. Consequently, they have been neglected by Anglo-American students of Leninism. Yet to the extent that Lenin achieved sophistication in philosophy, that stage is revealed in the *Philosophical Notebooks*, where we have his comments on Hegel, Aristotle, Feuerbach, and other writers. As two recent editors of Marxist philosophy commented:

We are seeing a non-professional philosopher, but a professional social revolutionary, studying some of the most technically difficult and theoretically advanced philosophical works of all time. He seeks better to be able to handle complex concepts in the analysis of everchanging and moving social forces. In these notes we are able to see Lenin's mind at work, trying out every abstract idea he comes across in the effort to test it in the crucible of his own rich experience. His main concern was to reconstruct the Hegelian dialectics on a thoroughly materialist foundation.While Lenin was always the enemy of idealism, he opposed the off-hand dismissal of this type of philosophy. As against vulgar materialism, he insisted that philosophical idealism has its sources in the very process of cognition itself. His conclusion was that "intelligent idealism is closer to intelligent materialism than stupid materialism." Thus, these *Philosophical Notebooks* are an indispensable supplement to Lenin's previous philosophical works and observations. Indeed, they constitute a plea for a richer and fuller development of dialectical materialism.

The interpretation of the *Philosophical Notebooks* and their integration into Lenin's thought present particular problems for the historian. Lenin composed these fragments for himself alone, jotting down what first came to mind, and did not rewrite or rethink them. Obviously such materials must be treated more carefully than his published *Materialism and Empirio-Criticism.* Yet to rely upon the published work entirely would mean under estimating the full development of Lenin's thought. Lenin was quite aware of his shortcomings in philosophy in the earlier years; his efforts to overcome these deficiencies and his subsequent viewpoints emerge impressively in the Philosophical Notebooks.

The *Philosophical Notebooks* have exercised increasing influence in Soviet dialectical materialism since their publication, although they are still considered secondary to *Materialism and Empirio-Criticism.* As we shall see, this influence was usually in the direction of a greater appreciation of the subtleties of epistemology and of the dangers of reductionism. When the *Philosophical Notebooks* first appeared in the Soviet Union, they became elements in the debates between the dialecticians and the mechanists. In later years, they were frequently considered to be particularly suited for advanced students of dialectical materialism, partly because of their fragmentary and unsystematized nature, but even more, no doubt, because of the greater awareness that Lenin displayed there of the alternatives of epistemology.

Materialism and Epistemology

In the Marxist philosophy of science as presented by Engels, there is nothing in the objective world other than matter and its emergent qualities. This matter has extension and exists in time; as Engels remarked, level of being, the carrier of that motion is matter: "One day we shall certainly 'reduce' thought experimentally to molecular and chemical motions in the brain; but does that exhaust the essence of thought?"

According to Engels, man's knowledge flows from nature, the objective, material world. He saw two different epistemological schools in the history of philosophy: the materialists, who believe knowledge to derive from objective nature, and the idealists, who attribute primacy in cognition to the mind itself. As Engels said, "The great basic question of all philosophy, especially of more recent philosophy, is that of the relation of thinking and being." At this point in *Ludwig Feuerbach* Engels proceeded to link the epistemological problem of knowledge to the ontological one of the existence of God:

> *The question of the relation of thinking to being, the relation of spirit to nature the paramount question of the whole of philosophy has, no less than all religion, its roots in the narrow-minded and ignorant notions of savagery. The question: which is primary, spirit or nature that question, in relation to the Church was sharpened into this: Did god create the world or has the world been in existence eternally?*

A number of Engels's critics have pointed to what they call a "fatal flaw" in his reasoning at this point: his confusion of epistemology and ontology. There is no reason, they have said, for identifying idealism with a belief in God, or realism with atheism. A person could believe in objective reality and withhold judgment on the question of God or even consider God to be "objectively real." Within the framework of the problem of epistemology alone the critics are correct; there are more than two camps on the issue of cognition. In describing how man comes to know, one can emphasize the role of objective reality (realism); the role of matter (materialism); the role of the mind (idealism); or one can maintain that it is impossible to know how man comes to know (agnosticism). Furthermore, one's religious views are not determined by one's epistemology. But for Engels the ontological principle that all that exists is matter came before all others. Therefore, for him a God who could be objectively real to a person in terms of his epistemology but non-material in terms of ontology was nonsense.

The key to the Marxist philosophy of science is not its position on cognition, which contains considerable flexibility, as evidenced not only by Lenin's writings in the *Philosophical Notebooks* but even more so by subsequent developments (particularly in countries such as contemporary Yugoslavia), but its position on matter itself. What justification do we have for assuming that an ill-defined "matter" (later "matter" was equated with "energy") alone exists? The more thoughtful Russian Marxists such as Plekhanov (and perhaps Lenin at moments) have veered toward the position that the principle of the sole existence of matter is a simplifying assumption necessary for subsequent scientific analysis. Other Marxists, such as Engels, the Lenin of *Materialism and Empirio-Criticism*, and most Soviet philosophers, have maintained that the principle of materialism is a fact presented by scientific investigation. But as a result of the sensitivity of the subject, the issue of the justification for the belief in materialism has not been thoroughly investigated by philosophers in the Soviet Union.

To return to Engels's treatment of the opposition of idealism and materialism, we can see his merging of the problems of existence and cognition in the following quotation:

> *Contrary to idealism, which asserts that only our mind really exists, and that the material world, being, Nature, exists only in our mind, in our sensations, ideas and perceptions, the Marxist materialist philosophy holds that matter, Nature, being, is an objective reality existing outside and independent of our mind; that matter is primary, since it is the source of sensations, ideas, mind, and that mind is secondary, derivative, since it is a reflection of matter, a reflection of being.*

Engels's last phrase, "mind is a reflection of matter," strikes to the heart of the mind-matter relationship. In Russian Marxist philosophy the description of this relationship has been a major issue. Engels's term "reflection" was followed by Plekhanov's "hieroglyphs," Bogdanov's "socially-organized experience," and Lenin's "copy-theory." The copy-theory of Lenin, to be subsequently discussed, became the most influential model for Soviet philosophy.

Connected with Engels's view of the nature of the material world was his opinion on the attainability of truth about that world. Parallel to the existence of matter apart from mind was the existence, potentially, of truth about that matter. Scientists strive toward complete explanations of matter even though these explanations are never reached. The relationship between man's knowledge and truth, according to Engels, is asymptotic. It is not correct to say that Engels believed in the attainability of absolute truth. Only at the unattainable point of infinity in the relationship between man's knowledge and truth does an intersection obtain. Nonetheless, Engels believed in a cumulative, almost linear, relationship of knowledge to truth. Lenin, on the contrary, saw many more temporary aberrations in the upward march, and used the image of a "spiral movement" to describe the process.

Reinterpretations of Russian Marxist Views on Materialism and Epistemology

Among Russian Marxists the problems of epistemology and the philosophy of nature attracted considerably more attention than among west European Marxists. G. V. Plekhanov, Lenin's tutor in Marxism and later an opponent of the Bolsheviks, developed his "hieroglyphic" theory of knowledge in 1892 in his notes to his translation of Engels's *Ludwig Feuerbach.* Plekhanov wrote:

> *Our sensations are sorts of hieroglyphs informing us what is happening in reality. These hieroglyphs are not similar to those events conveyed by them. But they can completely truthfully convey both the events themselves and this is important also those relationships existing between them.*

The analysis presented by Plekhanov was an attempt to go beyond the common-sense realism implied by Engels's writings to a recognition of the difference between objects-in-themselves and our sensations of them. In Plekhanov's view there was a distinct difference so much so that he felt that these sensations "are not similar to those events conveyed by them." Nonetheless, he said, there is a correspondence between these events and our sensations. Thus Plekhanov went from a "presentational" theory of perception to a "representational" one. His epistemology was dualistic in a Lockean sense, but without the introduction of subjective qualities. It was still materialistic in that it assumed the existence of material objects outside the mind that reveal themselves in an indirect but trustworthy fashion by means of man's sensations.

It was important to Plekhanov that to each of man's sensations in the process of perceiving an object there be a materialistic correlate, and to each of the changes in a material object there be a sensational correlate. He said that one should imagine a situation in which a cube is casting a shadow on the surface of a cylinder:

> *This shadow is not at all similar to the cube: The straight lines of the cube are broken; its flat surfaces are bulged. Nevertheless for each change of the cube there will be a corresponding change of the shadow. We may assume that something similar occurs in the process of the formation of ideas.*

Plekhanov was aware that his epistemology was not scientifically provable, as the above words "we may assume" indicate. He discussed respectfully Hume's view that there was no way of proving that physical objects are anything more than mental images. Plekhanov's writings implied that by assuming the primacy of matter in cognition, he considered himself to be making a plausible and useful philosophic choice rather than coming to a scientific conclusion.

In the early twentieth century a controversy arose among Russian Marxists that ultimately led to the entry of Lenin into the field of epistemology. In the resulting *Materialism and Empirio-Criticism* Lenin criticized not only his immediate disputants, the "Russian Machists," but also Plekhanov. In order to introduce the controversy, some mention must be made of Ernst Mach.

The late nineteenth century's most formidable criticism of the philosophic belief in a material world independent of man's mind was contained in Mach's sensationalism. Mach was an Austrian physicist and philosopher who provided much of the impetus to the development of logical positivism and who prepared the way for the acceptance of relativity and quantum theory. His antimetaphysical views were equalled by those of his contemporary, the German philosopher Richard Avenarius, the proponent of the theory of knowledge known as empirio-criticism. Mach and Avenarius occupy a special place in Soviet Marxist philosophy, since they are the objects of copious criticism in Lenin *Materialism and Empirio-Criticism.*

In *Analysis of Sensations* Mach defended the view, already ancient among philosophers but now made particularly relevant to modern science, that the "world consists only of our sensations." According to Mach, space and time were as much sensations as color or sounds. A physical object was merely a constant sensation (or "perception," taken as a group of sensations). Mach followed Berkeley, then, in denying the dualism of sense perceptions and physical objects. But while Berkeley was a realist in the sense of assuming the reality of mental images and of an external God, Mach endeavored to introduce no elements into his system that were not scientifically verifiable. Therefore, he made no pronouncements about ultimate reality. According to his "principle of economy," scientists should select the simplest means of arriving at results and should exclude all elements except empirical data. Mach's approach employed on the practical scientific level, where he intended it to be utilized, would mean that a scientist would cease worrying about the "real" or "actual" nature of matter and would merely accept his sense-perceptions, working as carefully and thoroughly as he could. A theory that found a pattern in the data would be judged entirely on the basis of its usefulness rather than its plausibility in terms of other existing considerations. There might even be more than one "correct" way of describing matter (a concept to have influence later in discussions of quantum mechanics). Two explanations, working from opposite directions, could both be useful and could supplement each other, even if there seemed to be a contradiction between the two approaches.

Mach had shifted the emphasis from matter *reflecting* in the mind to the mind *organizing* the perceptions of matter. A group of Marxist philosophers soon followed Mach's lead. This school of Russian empiriocritics included A. Bogdanov (pseudonym of A. A. Malinovskii); A. V. Lunacharskii, the future commissar of education; V. Bazarov (V. A. Rudnev); and N. Valentinov (N. V. Vol'skii). Bogdanov, a medical doctor, was swayed by the lucidity and scientific nature of Mach's arguments, but dissatisfied with what he saw as their inconsistencies. If, as Mach maintained, sensations and objects are the same, why do two different realms of experience the subjective and the objective–continue to exist? Why are there two different sets of principles or "regularities" (*zakonomernosti*) in these different realms? Thus, in the objective world, there are such sensations as sight, sound, and smell. In the subjective realm are emotions and impulses: anger, desire, and so forth. Bogdanov defined objective sensations as those that are universally perceived, and subjective sensations as those that may be apparent to only one person or a small group of persons. Bogdanov then attempted to find the roots of this dualistic system and thereby unite them in a philosophical system called empiriomonism. The key to this development is the concept of the "organization of experience." To Bogdanov, the physical world equals "socially organized experience," while the mental world is "individually organized experience." Therefore, "if in the single stream of human experience we find two principally different conformities of law (*zakonomemosti*), then nevertheless, both of them arise in equal measure from our own organization: They convey two biological-organizational tendencies." Parenthetically, it is worthwhile to note that Bogdanov's emphasis upon organizational structure and the means of transmitting information would cause a new surge of interest in his work in the Soviet Union many years later when cybernetics and information theory were applied to psychology and epistemology.

Lenin's original entry into the field of philosophy was the result of his being disturbed by the views of Russian Marxist writers such as Bogdanov and Plekhanov. His first motivation was a tactical one; he wished to protect the Bolsheviks' claim to a materialistic view of nature and history. Only many years later did he become genuinely interested intellectually in problems of philosophy.

At first Lenin hesitated to commit himself in the philosophical fray; in order to avoid commenting on these difficult problems, he attempted to set a policy of philosophical neutralism within the Bolshevik faction of the Russian Social Democratic Party. As late as February 1908 the Bolsheviks continued to give equal treatment in their legal press to "Machism" and orthodox dialectical materialism. This position became intolerable for Lenin, however, as Bogdanov, Bazarov, Lunacharskii, and their friends published works increasingly sympathetic to the views of Mach; Lenin found the Bolsheviks being labeled "revisionists" while the rival Marxist faction, the Mensheviks, were winning the designation "orthodox."

In 1908 Lenin set himself to the task of writing a major work on philosophy in order, as he put it, "to find out what was the stumbling block to these people who under the guise of Marxism are offering something incredibly muddled, confused, and reactionary." The stumbling block, he found, was the influence of the latest developments of science upon philosophers, including Marxist philosophers such as Bogdanov.

By the early twentieth century many people believed that the foundations of materialism were being undermined by scientists themselves. The relative confidence of scientists of Marx's and Engels's time in their knowledge of nature had been replaced by perplexity. The investigation of the radiations of radium and uranium, resulting in the identification of alpha rays (helium nuclei) and beta rays (high speed electrons), had discredited the concept of non-divisible atoms. Such scientists as L. Houllevigue remarked, "The atom dematerialises, matter disappears." Henri Poincare observed that physics was faced with "a debacle of principles."

The rise of philosophical schools such as empirio-criticism on the continent and phenomenalism in England was largely a response to these and other developments in science. In Lenin's opinion, the philosophers following these trends were subordinating the search for truth about matter to attempts to provide convenient explanations of isolated perceptions. Idealism was again a threat, and Bishop Berkeley's theories were reborn, in the name of science rather than God.

In countering these new movements, Lenin stressed two tenets of his interpretation of dialectical materialism: the copy-theory of the mindmatter relationship and the principle that nature is infinite. It seems clear that Lenin regarded these principles as minimum requirements in order for dialectical materialism to have philosophical consistency or significance. He was not attempting to impose philosophy upon science, but to locate the bedrock of the materialist philosophy of science; he believed it impossible for science to contradict these principles.

By the "copy-theory" of matter Lenin meant that materialism is based on recognition of "objects-in-themselves" or "without the mind." According to him, "ideas and sensations are copies or images of these objects." Just *how* similar these ideas are to the objects themselves was left unsaid. There is some reason to believe that at the time of the writing of *Materialism and Empirio-Criticism* Lenin considered man's mental images to be quite similar to the corresponding objects. His epistemology was at this time close to that of common-sense realism; he criticized the "vagueness" of Plekhanov's hieroglyphic epistemology. Yet even in some of his remarks of *Materialism and Empirio-Criticism* Lenin indicated that the essential aspect of dialectical materialism was the principle of objectively existing matter, not the degree of correspondence between man's images and the objects of the material world. Indeed, he approached reducing the fundamentals of materialist epistemology to one principle: "Only one thing is from

Engels's viewpoint immutable the reflection by the human mind (when the human mind exists) of a world *existing and developing independently of the mind.*" Lenin added that this independent objective world can be known by man; "To be a materialist is to acknowledge objective truth, which is revealed to us by our sense organs."

It should be noticed that if a person accepted literally the last sentence in the above paragraph as Lenin's definition of materialism, he would be fully justified in saying that Lenin confused realism ("all that exists is real") with materialism ("all that exists is material") and that Lenin was, in fact, not a materialist, but a realist. For one could take Lenin's statement "To be a materialist is to acknowledge objective truth, which is revealed to us by our sense organs" and with complete justification change it to read "To be a realist is to acknowledge objective truth, which is revealed to us by our sense organs." Was Lenin, then, a realist rather than a materialist? An accurate answer to this question, one which took into consideration all of Lenin's writings, would have to be negative. Lenin always spoke of materialism, not realism, and he saw the difference, particularly in his later works; he supplemented his realist epistemology with an *assumption* of ontological materialism resulting from his belief in the conceptual value of such an assumption. The fact that Lenin's materialism was founded on an assumption has not been openly discussed in the Soviet Union, where dialectical materialism is usually portrayed as a provable doctrine, even an inevitable conclusion of modern science. Yet the best argument for materialism starts out with a recognition of its assumptive or judgmental character and a defense of such a minimal assumption or judgment as being consonant with all available evidence and persuasive to many scientists. This argument must, of course, leave room for the person who wishes to make a different initial assumption. Individual scientists are likely to have preferences for one or the other. The noted American biologist Hermann J. Muller recognized and approved the assumptive origin of Lenin's materialism in an article that he wrote in 1934 and that is included as an appendix in this book. To Muller, Lenin's assumption could be defended, further, on the basis of inductive judgment:

> *To those scientists who would protest that we should not make such prejudgements regarding scientific possibilities, on the basis of a prior "philosophical" assumption of materialism, but should rather follow in any direction in which the empirical facts of the case seem to be leading, we may retort, with Lenin, that all the facts of daily life, as well as those of science, together form an overwhelming body of evidence for the materialistic point of view and that therefore we are justified, in our further scientific work, in taking this principle as our foundation for our higher constructions. It too is ultimately empirical, in the better sense of the word, and it has the overwhelming advantage of being founded upon the evidence of the whole, rather than upon just some restricted portion of the latter.*

But Lenin in 1908 was not yet able to recognize the judgmental or preferential bases of materialism, although he would approach them later in his *Philosophical Notebooks*. In 1908 he was, instead, concentrating on a criticism of the Russian followers of Mach, and he naturally was more interested in revealing the vulnerable points in their analysis than in his own. He asserted that Bogdanov's idealistic philosophy actually concealed a belief in God, in spite of his repudiation of all religion. If, said Lenin, the physical world equals merely "socially organized experience," then the door is opened to God as an aspect of experience. "Idealism says that physical nature is a product of this experience of living beings," added Lenin, and therefore "idealism is equating (if not subordinating) nature to God. For God is undoubtedly a product of the socially organized experience of living beings."

In Lenin's opinion, the epistemological problem was not separated from the question of the nature of matter itself. The mental realm of experience is not distinct from the material realm, but is a result of it on a higher level. Matter itself is not at all threatened by Rutherford's dismantling of the atom because "the electron is as *inexhaustible* as the atom, nature is infinite, but it infinitely exists." Lenin believed that the expression "matter disappears" was an indication of philosophical immaturity by scientists and philosophers who did not understand that science will constantly discover new forms of matter and new principles of motion. The forms and principles of matter are taken as immutable only by naïve materialists who have fallen prey to metaphysics, Lenin asserted:

> *The recognition of immutable elements, "of the immutable substance of things," and so forth, is not materialism, but* metaphysical, *i.e., anti-dialectical, materialism. That is why the "subject-matter of science is endless," that not only the infinite, but the "smallest atom" is immeasurable, unknowable to the end,* inexhaustible. *In order to present the dialectical materialist standpoint, we must ask: Do electrons, ether* and so on *exist as objective realities outside the human mind or not? The scientists will also have to answer this question unhesitatingly; and they do invariably answer it in the* affirmative, *just as they unhesitatingly recognize that nature existed prior to man and prior to organic matter. Thus the question is decided in favor of materialism.*

Lenin believed that philosophies opposing science were based either on idealism or simple materialism, not dialectical materialism. He attempted to make dialectical materialism less vulnerable to criticism and less likely to retard science by drawing a line between it and simple materialism. Yet, if one judges by *Materialism and Empirio-Criticism*, one must conclude that this line was not drawn with any degree of clarity. Lenin did not even discuss in this work the laws of the dialectic, the principles that distinguish dialectical materialism from simple materialism. He merely maintained that dialectical materialism, a philosophical viewpoint, cannot be affected by the vacillations of scientific theory. Lenin labored to reforge the bond between the theory of dialectical materialism and the practice of science. While insisting on the materialist copy-theory, he also affirmed that nature is infinite. The division of particles into smaller particles could go on forever, he believed, but matter would never disappear.

In the notations that Lenin made six or seven years later during his further study of philosophy, he revealed a greater appreciation of the alternatives of epistemology. Although he did not repudiate his earlier copy-theory, he now saw the link between the material objects of the world and man's images as much more indirect. Indeed, he seemed to believe that in the highest forms of their development, materialist and idealist theories of cognition were linked in a unity of contradiction; they tended to pass into one another. Thus Hegel, whom he believed to be the greatest idealist in philosophy, arrived unwittingly at the threshold of dialectical materialism. Lenin's evolution was from another direction, from the side of materialism, but he approached the same spot as Hegel, the moment of unity between two philosophies. As Lenin observed, "the difference of the ideal from the material is not unconditional." The area where the distinction between idealism and materialism became nearly imperceptible was that of mental abstraction; in order to understand nature it is necessary for man not only to perceive matter but to construct a series of concepts that "embrace conditionally" eternally moving and developing nature. And these abstractions may include elements of fantasy:

> *The approach of the (human) mind to a particular thing, the taking of a copy (= a concept) of it is* not *a simple, immediate act, a dead mirroring, but one which is complex, split in two, zigzag like, which* includes in it *the possibility of the flight of fantasy from life; more than that: The possibility of the* transformation *(moreover, an unnoticeable transformation, of which man is unaware) of the abstract concept, idea, into a* fantasy *(in the final analysis = God). For even in the simplest generalization, in the most elementary general idea ("table" in general),* there is *a certain bit of* fantasy. *(Vice versa, it would be stupid to deny the role of fantasy, even in the strictest science; cf. Pisarev on useful dreaming, as an impulse* to *work, and on empty daydreaming.)*

The Lenin who is revealed in the above passage is not the one who is known to most students of Leninism; this Lenin recognizes the painful, halting, indirect path of knowledge, a path that includes clear reversals. He grants the useful role of fantasy "even in the strictest science." He sees this fantasy as an *inherent possibility* in scientific thought and is aware that in the final analysis it can lead to a belief in God. In his statement that the possibility of fantasy is included in the approach of the human mind to nature, he, like Plekhanov before, seemed to recognize that the rejection of idealism is not a matter of scientific proof but philosophic choice. And Lenin continued, of course, to choose materialism. The possibility of fantasy in science was, he believed, a result of the indirectness of man's knowledge. The fantasy in a given scientific theory was progressively reduced in scope, and the useful role it might play was temporary. But Lenin recognized that with the advent of new theories the possibilities of new fantasies, including useful ones, would arise.

This more flexible view of materialism, which sees it as a result of choice and not of proof, opened up room for its potential accommodation with some other philosophic currents, a development that, however, still has not occurred. If one considers, for example, some of the recent writings of W. V. O. Quine, it becomes apparent that there are similarities of argument. Quine wrote in his "Two Dogmas of Empiricism":

> *As an empiricist I continue to think of the conceptual scheme of science as a tool, ultimately, for predicting future experience in the light of past experience. Physical objects are conceptually imported into the situation as convenient intermediaries not by definition in terms of experience, but simply as irreducible posits comparable, epistemologically, to the gods of Homer. For my part I do, qua lay physicist, believe in physical objects and not in Homer's gods; and I consider it a scientific error to believe otherwise. But in point of epistemological footing the physical objects and the gods differ only in degree and not in kind. Both sorts of entities enter our conception only as cultural posits. The myth of physical objects is epistemologically superior to most in that it has proved more efficacious than other myths as a device for working a manageable structure into the flux of experience.*

The possibility for the convergence of the epistemological view represented by the quotation from Quine above and the epistemology of dialectical materialism is considerable. Holders of both views prefer a concept of "physical objects," and both find justification for this concept in pragmatic success. To be sure, what Quine calls "superior myth," the dialectical materialist has called "truth." But does the word "myth" here mean "that which is false" or "that which can not be proved"? And does the dialectical materialist really believe that the "truth" of his position can be illustrated, or is his position one that he assumes to be true for reasons similar to those for which Quine assumes his myth to be "superior"? And if the dialectical materialist may have some difficulty in defining "matter-energy," falling back eventually to Engels's reference to the "totality of things," Quine says with equal indefiniteness that "physical objects are postulated entities which round out and simplify our account of the flux of experience." In the final analysis there are differences between the positions, to be sure, but hardly of the type that would place dialectical materialism outside the realm of philosophy while leaving Quine's epistemology within.

Laws of the Dialectic

The discussion of dialectical materialism has so far centered on the last half of the term: materialism. The other half, the dialectic, concerns the characteristics of the development and movement of matter.

There are two rather different views of the dialectic that Soviet thinkers have, at different moments, taken; one is the belief not only that matter-energy obeys laws of a very general type, but that these laws have been identified in the three laws of the dialectic to be discussed below. This has been the official, ideological view expressed in the Soviet textbooks on dialectical materialism. The other view is that matterenergy does indeed obey general laws, but that the three laws of the dialectic are provisional statements to be modified or replaced, if necessary, as science provides more evidence. This has been an unofficial, philosophical view, appearing in the Soviet Union from time to time, particularly among younger scientists and philosophers.

The dialectic as applied by Engels to the natural sciences was based on his interpretation of Hegelian philosophy. This interpretation involved not only the well-known conversion of Hegelian philosophy from idealism to materialism, but also the reduction of Hegel's thought to a simple scheme of dialectical laws and triads.

In his *Science of Logic* Hegel spoke of "dialectic" as "one of those ancient sciences which have become most misjudged in modern metaphysics and in popular philosophy of ancients and moderns alike." Hegel believed that the way in which dialectic had previously been used had involved only two terms (dualisms, antinomies, opposites); he referred to Kant discussion of *Transcendental Dialectic* in his *Critique of Pure Reason*, in which Kant advanced the view that human reason is essentially dialectical in that every metaphysical argument can be opposed by an equally persuasive counter argument. Hegel saw a means of transcending this contradiction in a third position, "the second negative," which is "the innermost and most objective moment of Life and Spirit, by virtue of which a subject is personal and free."

Contrary to much opinion, Hegel never used the terms "thesisantithesis-synthesis" in the neat fashion so often attributed to him; he recognized, however, the importance of the thesis-antithesis contradiction in the writings of Kant, Fichte, and Jacobi, and he did sparingly use the term "synthesis" to indicate the moment of transcendence of such a polarity. But Hegel opposed reducing his analysis to a triadic formula, and warned that such a scheme was "a mere pedagogical device," a "formula for memory and reason." "The distinctions of the Concept of philosophical cognition can be realized," he said, "only by going through the whole process of self-knowledge."

Hegel did not provide a straightforward method of analysis that merely had to be "turned on its head" in order to become dialectical materialism. Engels's use of Hegel involved not only inversion, but also codification, a dubious reduction of rather obscure complexity. Nonetheless, many of the elements of Engels's dialectical materialism were indeed present in Hegel's thought. The very fact that Engels sought to simplify Hegel is not surprising many men, including Goethe, have condemned the great Prussian philosopher for being unnecessarily complex–but Engels's centering of attention upon the laws of the dialectic had the unfortunate effect of tying Marxism to three codified laws of nature rather than simply to the principle that nature does conform to laws more general than those of any one science, laws that may, with varying degrees of success, be identified.

To Engels the material world was an interconnected whole governed by certain general principles. The great march of science in the last several centuries had, as a regrettable by-product, so compartmentalized knowledge that the important general principles were being overlooked. As he observed, the scientific "method of work has also left us as legacy the habit of observing natural objects and processes in isolation, apart from their connection with the vast whole; of observing them in repose, not in motion; as constants, not as essentially variables; in their death, not in their life."

By "dialectics" Engels said that he meant the laws of all motion, in nature, history, and thought. He named three such laws: the Law of the Transformation of Quantity into Quality, the Law of the Mutual Interpenetration of Opposites, and the Law of the Negation of the Negation. These dialectical principles or laws were supposed to represent the most general patterns of matter in motion. Like Heraclitus, dialectical materialists believe that nothing in nature is totally static; the dialectical laws are efforts to describe the most general uniformities in the processes of change that occur in nature. The concept of the evolution or development of nature is, therefore, basic to dialectical materialism. The dialectical laws are the principles by which complex substances and concepts evolve from simple ones.

According to Engels, the laws were equally valid in science and human history. This universal applicability of the laws has served both as a source of strength and of weakness for dialectical materialism. On the one hand, the possession of the dialectic has given Marxists a conceptual tool of considerable power; many thinkers have been attracted by the Hegelian framework of dialectical materialism. The urge to possess a key to knowledge has been perhaps the strongest motivation in the history of philosophy. To oppose dialectical materialism to "Western thought," as many commentators have done in recent years, is to ignore the metaphysical foundation of most of European philosophy during the past several thousand years.

On the other hand, the universality of dialectical materialism has been frequently a disadvantage for its adherents. Many non-Soviet philosophers have turned away from it in the belief that it contains precisely those elements of Western philosophy that should have been abandoned before they were; dialectical materialism is a vestige, they say, of scholasticism. Rather than describing *how* matter moves, in the post-Newtonian sense, it attempts, in the Aristorelian sense, to explain *why* matter moves. Furthermore, the generality of the dialectic is achieved at the price of such diffuseness that to many critics its usefulness seems negligible. As one critic of the dialectic, H. B. Acton, remarked, the Law of the Negation of the Negation is "already general almost to the point of evanescence" when it is applied to such very different things as mathematics and the growing of barley; when the law is then extended to the transition from capitalist to communist society, "the only point of likeness appears to be the words employed." To this criticism dialectical materialists would reply that if one accepts the existence of one real world of which all aspects of man's knowledge are derivative parts, one should expect there to be at least a few general principles that are applicable to all those parts. Some of the more sophisticated dialectical materialists of the post-Stalin period would add that they are in principle prepared to reject the three laws of the dialectic enunciated by Engels if superior substitutes can be found, and there have been a few attempts to achieve this through the application of information theory.

The principle of the Transformation of Quantity into Quality derived from Hegel's view that "quality is implicitly quantity, and conversely quantity is implicitly quality. In the process of measure, therefore, these two pass into each other: each of them becomes what it already was implicitly."

Engels pointed to what he considered numerous examples of the operation of this law in nature. These were the cases when quantitative succession in a natural phenomenon is suddenly interrupted by a marked qualitative change. One example given by Engels was the homologous series of carbon compounds. The formulas for these compounds (CH_4, C_2H_6, C_3H_8, and so on) follow the progression C_nH_{2n+2}. The only difference among members of the progression, Engels observed, is the *quantity* of carbon and hydrogen. Nevertheless, the compounds have greatly differing properties, some being gases and others solids in their natural states. In these diverse properties Engels saw the Law of the Transformation of Quantity into Quality at work.

Perhaps the most unusual case that Engels cited as an example of the transformation of quantity into quality concerned Napoleon's cavalry during the Egyptian campaign. During the conflicts between French and Mameluke horsemen a curious relationship became apparent. Whenever a small group of Mamelukes would come upon a small group of Frenchmen in the desert, the Mamelukes would always win, even if somewhat outnumbered. On the other hand, whenever a large group of Mamelukes would come upon a large group of Frenchmen, the Frenchmen would always win, even if somewhat outnumbered. Engels's description can be represented in the following table:

Number of Mamelukes	**Number of Frenchmen**	**Victors**
2	3	Mamelukes
100	100	Even match
1500	1000	Frenchmen

The reason for the results was that the Frenchmen were highly disciplined and trained for large-scale maneuvers, but were not veteran horsemen. The Mamelukes had been on horses from the earliest age, but knew very little about discipline and tactics. Hence, a qualitative-quantitative relationship existed that yielded differing results at different quantitative levels.

To Marx and Engels, Darwin's theory of evolution was an important illustration of the principle of the transition of quantity into quality. This tenet as a part of the Hegelian dialectic preceded Darwin, of course, but Marx and Engels considered Darwinism a vindication of the dialectical process. In the course of natural selection, different species developed from common ancestors; this transition could be considered an example of accumulated quantitative changes resulting in a qualitative change, the latter change being marked by the moment when the diverging groups could no longer interbreed.

In the interpretation of science, the principle of the transition of quantity into quality has been important in the Soviet Union as a warning against reductionism. Reductionism here means the belief that all complex phenomena can be explained in terms of combinations of simple or elemental ones. A reductionist would maintain that if a scientist wishes to understand a complex process (growth of crystals, stellar evolution, life, thought), he must build up from the most elemental level. Reductionism tends to emphasize physics at the expense of all other sciences. It is a view that was often supported by nineteenth century materialists and even today continues to have considerable strength. Reductionism is highly criticized by Soviet dialectical materialists, who carefully distinguish themselves from earlier materialists. In the biological sciences in particular, the quantity-quality relationship has been interpreted in the Soviet Union as foreclosing the possibility of explaining life processes most of all, thought in elementary physicochemical terms. The development of matter during the history of the earth from the simplest non-living forms up through life and eventually to man and his social organizations is regarded as a series of quantitative transitions involving correlative qualitative changes. Thus, there are "dialectical levels" of natural laws. Social laws cannot be reduced to biological laws, and biological laws cannot be reduced to physicochemical laws. In dialectical materialism the whole is greater than the sum of its parts. This principle has been a valuable guard against simplistic explanations in materialistic terms, but it has also, on occasion, skirted the opposite danger of organicism or even vitalism.

The principle of the transition of quantity into quality, therefore, distinguishes dialectical materialism from mechanistic materialism. A materialist, similar to Democritus, would say, for example, that the human brain is essentially the same as an animal's brain but is organized in a more efficient manner. According to this line of thinking, then, the difference is merely quantitative. The Marxist materialist, however, would say that the human brain is distinctly different from that of an animal and that this difference is a qualitative change resulting from accumulated quantitative changes during the course of the evolution of man from lower primates. Therefore, human mental processes cannot be reduced to those of other animals. Indeed, life processes in general cannot be totally reduced to physical and chemical processes if the latter are defined in contemporary terms. This emphasis on the qualitative distinctiveness of complex entities from simple ones has led dialectical materialists in recent years to become interested although cautiously in such concepts as "integrative levels" and "organismic biology," approaches widely discussed in Europe and America in the thirties and forties and displaying new vigor with the advent of cybernetics.

The attitudes of Soviet philosophers toward explanation of organic processes illustrate the complex and perhaps even contradictory nature of dialectical materialism. The dialectical materialist says, in effect, "There is nothing but matter, but all matter is not the same." Some critics have seen this expression as a paradox existing at the very root of dialectical materialism. Berdyaev wrote, for example,

"Dialectic, which stands for complexity, and materialism, which results in a narrow onesidedness of view, are as mutually repellent as water and oil." Of course, one can easily note that almost every philosophical or ethical system contains tensions: The strain existing between the ideals of individual freedom and social good has existed in much of Western thought without destroying the value of that body of thought. Similarly, the tension between complexity and simplicity in dialectical materialism is by itself of small consequence in judging the adequacy of the system for the problems to which it is put. For the practicing scientist this tension has the merit of providing him with the confidence that nature can be fruitfully approached, while warning him against assuming that success in one area or at one level will answer ultimate questions.

Thus, the tension between complexity and simplicity that inheres in the principle of the transition of quantity into quality is simply a permanent feature of dialectical materialism that, operating at different times in different ways, both strengthens and weakens it. This tension was an important source of the disputes that arose in Soviet philosophy in the 1920's.

A partial rationalization of this dichotomy is offered by the second principle of the dialectic, the Law of the Mutual Interpenetration of Opposites, sometimes called the Law of the Unity and Struggle of Opposites. Hegel gave his views on this principle in terms of "positive" and "negative":

Positive and negative are supposed to express an absolute difference. The two however are at bottom the same: the name of either might be transferred to the other. Thus, for example, debts and assets are not two particular, self-subsisting species of property. What is negative to the debtor, is positive to the creditor. A way to the east is also a way to the west. Positive and negative are therefore intrinsically conditioned by one another, and are only in relation to each other. The north pole of the magnet cannot be without the south pole, and *vice versa*. If we cut a magnet in two, we have not a north pole in one piece and a south pole in the other. Similarly, in electricity the positive and the negative are not two diverse and independent fluids. In opposition, the different is not confronted by any other, but by its other.

By the principle of the unity of opposites Engels meant that harmony and order are found in the resulting synthesis of two opposing forces. Engels saw the operation of this law in the rotation of the earth around the sun, which resulted from the opposing influences of gravitational and centrifugal forces. The same law governed the formation of a salt by the chemical interaction of an acid and a base. Other examples of the unity of opposites cited by Engels were the atom as a unity of positive and negative particles, life as a process of birth and death, and magnetic attraction and repulsion.

The Law of the Unity and Struggle of Opposites is an explanation within dialectical materialism for the energy inherent in nature. To the question, How did the matter in the world acquire its motion? dialectical materialism answers that matter possesses the property of self-movement as a result of the contradictions or opposites present in it. Thus, it is not necessary for dialectical materialists to postulate the existence of a First Mover who set the planets, molecules, and all other material objects in motion. This concept of self-movement derived from internal contradictions was also present in the thought of Hegel, who commented in his *Science of Logic*, "contradiction is the root of all movement and life, and it is only in so far as it contains a contradiction that anything moves and has impulse and activity."

The Law of the Negation of the Negation is closely connected with the second law, since negation is supposedly the process by which synthesis occurs. Negation, according to Hegel, is a positive concept, an affirmation. The constant struggle between the old and the new leads to superior syntheses. In its most general sense, the principle of the negation of the negation is merely a formal statement of the belief that nothing in nature remains constant. Everything changes; each entity is eventually negated by another.

Engels commented, "Negation in dialectics does not mean simply saying no, or declaring that something does not exist, or destroying it any way one likes, and long ago, Spinoza said:

> Omnis determinatioest negatio *every limitation or determination is at the same time a negation." Engels considered the principle of the negation of the negation extremely important to dialectical and historical materialism; he wrote that it is a law of development of "Nature, history and thought," and a law that "holds good in the animal and plant kingdoms, in geology, in mathematics, in history and in philosophy." He gave a number of examples of the law: the negation of capitalism (which was a negation of feudalism) by socialism; the negation of plants such as orchids by artificially altering them through cultivation, yielding better seeds and more beautiful blooms; the negation of butterfly eggs by the birth of the butterflies, which then lay more eggs; the negation of barley seed by the growth of the plant, which then yields more barley grains; differentiation and integration in calculus; and the mathematical process of squaring negative numbers.*

It is clear that "negation" meant a number of different things to Engels: replacement, succession, modification, and so forth. The last example above deserves more comment. Engels proposed taking the algebraic symbol a and negating this quantity by making it a, then "negating the negation" by multiplying by a, obtaining a^2. He said that a^2 represented a "synthesis of a higher level" since it was a positive number of the second power. One may ask, Why did Engels multiply in order to negate the negation, instead of adding, subtracting, or dividing? Why did he multiply by a instead of some other figure? The obvious reply is that Engels picked a particular example that suited his purpose from the myriad of examples available. One of his discussions of the (evidently to prove the Law of the Mutual Interpenetration of Opposites) caused one anguished mathematician to write to Marx complaining about Engels's "wanton attack on the honor of -1."

The dialectical laws in Marxist philosophy remained virtually as Engels elaborated them for many years. In the period immediately after the Russian Revolution, Soviet philosophers neglected the dialectical laws. At that time neither Engels *Dialectics of Nature* nor Lenin *Philosophical Notebooks* were known to them. The latter work, published separately in 1933, introduced one change in the Soviet treatment of the dialectical laws: Lenin considered the Law of the Unity of Opposites the most important of the three. Engels had listed the laws in the following order: (1) the Law of the Transformation of Quantity into Quality, (2) the Law of Mutual Interpenetration of Opposites, and (3) the Law of the Negation of the Negation. Lenin chose the order (2), (3), (1). Lenin even hinted that the transformation of quantity into quality was really only another description of the unity of opposites; if the two were truly synonymous, only two of the three primary laws would remain.

Before this discussion leaves the subject of the dialectic, a few words must be said about the "categories." In dialectical materialism the term "categories" is employed to refer to those basic concepts that are necessary in order to express the forms of interconnection of nature. In other words, while the laws of the dialectic just discussed are attempts to identify the most general uniformities of nature, the categories are those concepts that must be employed in expressing these uniformities. Examples of categories given in the past in Soviet discussions of the dialectic have been concepts such as matter, motion, space, time, quantity, and quality.

Nowhere does dialectical materialism reveal its affinities with traditional philosophy more clearly than in its emphasis on categories, although dialectical materialists have given the classical categories a new formulation and meaning. The word "category" was first used as a part of a philosophical system by Aristotle. In his treatise *Categories* Aristotle divided all entities into the following ten classes: substance, quantity, quality, relation, place, time, posture, state, action, and passion. Objects or phenomena that belonged to different categories were considered to have nothing in common and, therefore, could not be compared. In his writings Aristotle frequently listed only some of the above ten categories, with no indication that others had been omitted. Aristotle apparently considered the exact number of categories and the terminology best suited for describing them open questions. Following Aristotle, many thinkers relied on a system of a priori categories of varying number and nature as a base for their philosophical systems. Medieval philosophers usually considered the original ten categories of Aristotle complete, ignoring Aristotle's latitude on the issue.

The two greatest modifiers of the Aristotelian concept of categories were Kant and Hegel. Kant based his categories not on particular subjects or entities, but on different types of judgments or propositions. To him, categories applied to logical forms, and not to things in themselves. "Quality" to Kant meant not "bitter" or "red," as it did to Aristotle, but logical relationships such as "negative" or "affirmative." "Quantity" meant to him not "five inches long" but "universal," "particular," and "singular." The Kantian radical reform of the Aristotelian categories is a complex subject of continuing discussion to philosophers.

Soviet philosophers have borrowed heavily from the approaches of Aristotle and Kant, adding to them Hegel's belief that the categories are not absolute. They consider Aristotle's writings amenable to the interpretation that the categories are reflections of the general properties of objectively existing objects and phenomena, although, as one Soviet text comments, "he did not always hold to this view, and, moreover, he did not succeed in revealing the inner dialectical connection of the categories." Kant's major contribution to an understanding of the categories, in the opinion of Soviet philosophers, was his research into the logical functions of the categories and into the role of thought in refining sense perceptions. But Kant, they continue, made the great error of eliminating all connection between the categories and the objective world, looking upon them as the product of reason. Hegel's achievement in understanding the categories, according to Soviet dialectical materialists, was his realization that the categories are not static, but are in the process of development and are connected with each other. Thus, for example, the category "quantity" can grow into "quality." The major distinction of the dialectical materialist approach to the categories is its heavy emphasis on natural science. Since, according to Marxism, being determines consciousness and not consciousness being, the material world, as reflected in man's consciousness, determines the very concepts by which he thinks that is, the categories. Thus, "In order for the materialist dialectic to be a method of scientific cognition, to direct human thought in search of new results, its categories must always be located at a level with modern science, with its sum total of achievements and needs."

Since man's knowledge of the material world changes with time, so then will his definition of the categories. The *Short Philosophical Dictionary*, published in Moscow in 1966, defined the categories as "the most general concepts, reflecting the basic properties and regularities of the phenomena of objective reality and defining the character of the scientific-theoretical thought of the epoch." The same source listed as examples of the categories matter, motion, consciousness, quality and quantity, cause and effect, and so on.

The inclusion of the "and so on" at the end of the list of categories is an important indication of the flexibility of the categories within dialectical materialism. As with Aristotle, the list is purposely kept open and subject to revision. Lenin remarked that "if *everything* develops, then doesn't this refer to the most general *concepts* and *categories* of thought as well? If it doesn't, then that means thinking is not connected with being. If it does, then it means that there is a dialectic of concepts and a dialectic of knowledge having an objective significance." The same approach is reflected in the following description in the *Short Philosophical Dictionary*: "The categories are regarded as flexible and changing because the very properties of objective phenomena are also mobile and changing. The categories do not appear at once in a completed form. They are formed during the long historical process of the development of knowledge." Thus, the categories develop with science itself.

The avowed elasticity of the categories gives, indirectly, room for interpretation of the dialectical laws themselves, since the categories are the terms in which the laws are expressed. In this study the possibility of revising the categories will be particularly relevant in the discussion of cosmology, where certain authors reinterpreted the term "infinity" after 1956 by examining the categories. "Time" and "space" had been listed as categories in the texts of the fifties, and these concepts were re-examined. Another area where the categories were scrutinized was quantum mechanics: Here the concept of causality, or the category of "cause and effect," was actually modified.

Unity of Theory and Practice

Another aspect of dialectical materialism that has relevance for science is not so much an integral part of the intellectual structure of the system as it is a methodological principle; this aspect is the unity of theory and practice. During a considerable portion of Soviet history the unity of theory and practice meant for scientists that they should give their research a clear social purpose by tying it to the needs of Soviet society. The strength of this recommendation has varied greatly in different fields and at different times. The unity of theory and practice can be traced back to Marx's opposition to speculative philosophy; he hoped to transcend philosophy by "actualizing" it. One of his best known sentences referred to this effort to build a conceptual theory that would result in concrete achievements: "The philosophers have only *interpreted* the world in various ways; the point however is to *change* it."

Engels believed that the unity of theory and practice was connected with the problem of cognition. The most telling evidence, he believed, against idealistic epistemologies was that man's knowledge of nature resulted in practical benefits; man's theories of matter "worked" in the sense that they yielded products for his use. As Engels commented, "If we are able to prove the correctness of our conception of a natural process by making it ourselves, bringing it into being out of its conditions and using it for our own purposes into the bargain, then there is an end of the Kantian incomprehensible 'thing-in-itself.'" Thus, practice becomes the criterion of truth. Of course, Engels admitted that many theories or explanations "worked" while being incomplete or based on false assumptions. The Babylonians were able to predict certain celestial phenomena through the use of tables, with almost no knowledge of the location and movement of the bodies themselves. Every scientific theory at any point in time contains false assumptions and lacks important evidence; many useful theories, such as Ptolemaic astronomy, are "overthrown." But Engels maintained that the successful application of a theory about nature indicated that it contained somewhere within it a kernel of truth.

During the Soviet period the principle of the unity of theory and practice acquired an influential moral significance. When the Soviet Union was attempting to industrialize rapidly, scientists were under considerable pressure to pursue research that would help in the effort. For geologists to study the strata of the earth merely to speculate about its age was considered immoral at a moment when they could have been identifying valuable mineral deposits. For mathematicians to develop theories that seemed to have no conceivable application was believed to be a squandering of the investment that society made in their educations.

This emphasis on practical return, highly typical in a different form among businessmen in other countries, would have significance in the controversy over genetics. Yet its operational meaning is unclear. Engels's belief that the approach to truth is evidenced by the benefit from man's theories did not provide a method for choosing among new hypotheses. It is clear that a time element is involved in the unity of theory and practice. Almost no scientific theory gives immediate practical benefit. Time is required not only for the development of industrial processes that would utilize the theory, but for the exploration of the theory's implications that have practical consequences. When a theory is first developed, it may have no apparent applications at all but may turn out much later to be an invaluable approach to a practical problem. In the years after Stalin's death an appreciation of the indirectness of the link between theory and practice grew in the Soviet Union.

It should be recognized that every country, and particularly those such as the Soviet Union that feel the existence of external threats, will sacrifice certain luxuries in order to meet the most urgent practical needs. The emphasis that the Soviet Union placed in science upon the unity of theory and practice in the thirties and forties was more a result of these considerations than the opposition to speculative philosophy contained in Marxism.

Looking back over the system of Soviet dialectical materialism, we see, on the most general level, that it represents a natural philosophy based on the following quite reasonable principles and opinions:

- The world is material, and is made up of what current science would describe as matter-energy.
- The material world forms an interconnected whole.
- Man's knowledge is derived from objectively existing matter.
- The world is constantly changing, and, indeed, there are no truly static entities in the world.
- The changes in matter occur in accordance with certain over-all regularities or laws.
- The laws of the development of matter exist on different levels corresponding to the different subject matters of the sciences, and therefore one should not expect in every case to be able to explain such complex entities as biological organisms in terms of the most elementary physicochemical laws.
- Matter is infinite in its properties, and therefore man's knowledge will never be complete.
- The motion present in the world is explained by internal factors, and therefore no external mover is needed.

- Man's knowledge grows with time, as is illustrated by his increasing success in applying it to practice, but this growth occurs through the accumulation of relative not absolute truths.

The history of thought clearly shows that no one of the above principles or opinions is original to dialectical materialism, although the total is. Many of the above opinions date from the classical period and have been held by various thinkers over a period of more than two thousand years. Today many working scientists operate, implicitly or explicitly, on the basis of assumptions similar to the above principles (hence the Soviet view that outstanding non-Marxist scientists are often at least implicit dialectical materialists). Yet the common currency of many of the most general principles of dialectical materialism does not devaluate it. First, these principles have been more fully developed and more closely linked to science in the writings of dialectical materialists than in any other corpus of literature. Furthermore, unexceptional as some of their opinions may at first glance seem, dialectical materialists have their opponents even on these broadest principles. Their commitment to the primacy of matter is rejected by many, probably most, philosophers. Materialism has never been a philosophy of the majority of philosophers at any point in the history of Western philosophy; its most ardent advocates have not usually been professional philosophers. In addition, dialectical materialists disagree not only with their most obvious opponents theists and idealists but also with materialists of the old type, the thoroughgoing reductionists who believe that all science will eventually be absorbed by physics. In sum, dialectical materialism despite what some observers regard as the unobjectionable character of its most general principles is still a controversial world view, one that enjoys the explicit support of only a small minority of philosophers and scientists in the world. When one adds to these intellectual obstacles the political liability deriving from dialectical materialism's support by the bureaucracy of a non-democratic state, it is not surprising that dialectical materialism has won relatively few supporters outside the Soviet Union. Yet it should be noticed that in intellectual terms dialectical materialism is a legitimate and valuable point of view, far more interesting than non-Soviet scientists and philosophers have usually assumed.

Soviet dialectical materialism as a philosophy of science draws upon both Russian sources and traditional European philosophy. The Soviet contribution has been primarily one of emphasis on the natural sciences as determining elements of philosophy. In the opinion of Soviet philosophers, dialectical materialism both helps scientists in their research and, in turn, is ultimately affected by the results of that research. Their critics have occasionally maintained that such a description of the relation of science and philosophy is no description at all. Exactly what meaning is carried in the statement "Philosophy influences science and is, in turn, influenced by it"?

An answer that accurately weighs the mutual influence of philosophy and science is difficult, but it is clearly true that such a mutual influence exists. Furthermore, that interaction is an important element in the genesis and elaboration of scientific schemes. Questions may revolve around the degree of influence of philosophy upon science, or the mechanisms by which such influence is transmitted, but the existence of the interaction can not be questioned. Throughout the history of science, philosophy has significantly affected the development of scientific explanations of nature, and, in turn, science has influenced philosophy. Scientists inevitably go beyond empirical data and proceed, implicitly or explicitly, on the basis of one or another philosophy. Philosophers, on the other hand, have been forced by the evolution of science to revise basic concepts underlying their philosophic systems, such as the concepts of matter, space, time, and causality.

Moments when philosophy has importantly influenced science can be found from the earliest points in the history of science, and similar influences continue today in all countries. The early teachings of the Ionian natural philosophers, based on a naturalistic or non-religious approach to nature, were heavily modified by the post-Socratic Greeks in the name of philosophic viewpoints that assumed the necessity of a divine being for an understanding of the cosmos. Benjamin Farrington commented that astronomy was "Pythagoreanized and Platonized within a few generations of the Ionian dawn." He also observed that "astronomy did not really make its way with the Greek public until it had been rescued from atheism." Here was an example in which a philosophic world view influenced a scheme of scientific explanation.

Many others have followed. The historian of science Alexandre Koyré maintained that Galileo was a Platonist in his understanding of nature, and that this view had important influences on his scientific development. Koyré's view has been criticized, but not from the stand-point of a denial of philosophic influence on Galileo. Newton's explanation of nature was presented within a religious framework that made it more acceptable to the society of his time and yet also revealed something important about Newton's internal convictions and presuppositions. Descartes actually postponed publication of his *Principia Philosophiae* in an effort to fit an orthodox and religious interpretation to his view of nature; this orthodoxy accorded with his views, but the fitting process was not obvious. The impact of German nature philosophy upon European scientists in the early decades of the nineteenth century is well-known, and such historians of science as L. Pearce Williams would even see nature philosophy as an important ingredient in the origin of field theory, maintaining that convertibility of forces "was an idea that was derived from nature philosophy and one to which the Newtonian system of physics was, if not hostile, at least indifferent." In each of these cases interaction between science and philosophy is one of the prime topics of study for the historian of science.

The impact of philosophy upon science has continued through the present day; it should not be regarded as a vestige of the past that will hopefully soon be overcome if it has not already been vanquished. Einstein even wrote, "In our time physicists are forced to concern themselves with philosophical questions to a greater degree than physicists of previous generations." Einstein frequently acknowledged his personal debt to philosophical criticisms of science; out of such criticisms arose a revolution in twentieth century science.

We stand so close to the development of contemporary science that we may not at first discern the interaction of philosophy and science, but it is certainly there. As an example, the new concepts in quantum mechanics and relativity physics of this century not only had a philosophic background, but in turn exerted a considerable influence on subsequent philosophy in Western countries in the first half of this century. These were countries in which many different philosophic viewpoints were expressed but the most popular ones favored religion over atheism and idealism over materialism. Consequently, it is not surprising that a number of influential scientists and philosophers in these countries seized upon the new physics and attempted to build philosophical systems justifying their religious and epistemological points of view. The uncertainty principle was to some of them an opportunity for the defense of freedom of will, while the rise of relativity physics signaled to many of them the end of materialism. In the Soviet Union the defenders of materialism answered back in full measure indeed, in more than full measure criticizing idealistic and religious points of view. Each side went far beyond conclusions that were intellectually justified, treating the opposition as if its position were groundless. The outcome of this debate was to illustrate that neither side possessed the clear superiority of argument that it claimed. This gradually became apparent to many writers, and the quality of their arguments began to improve. The Soviet authors, the main concern of this study, were able to develop a dialectical materialist interpretation of the universe based on the very principles of contemporary science that their opponents attempted to use against them.

The fact that emerges from these considerations is that science and philosophy have interacted at all times and places, not merely in the ancient past or in the contemporary Soviet Union. Soviet science is a part of world science, and the type of interaction of philosophy and science that can be found in Soviet scholarly writings (those of intellectuals, not of Party activists) is not essentially different from the interaction of science and philosophy elsewhere. But since the philosophical tradition in the Soviet Union is different from the tradition in western Europe and the United States, the results of that interaction have not been identical with the results of similar interactions in other geographical areas.

Thus, the significance of dialectical materialism is not so much its insistence on this mutual interaction of philosophy and science many of its critics would readily grant such a relation but the way in which this interaction has actually occurred in the Soviet setting. The dialectical materialism in the Soviet Union today is not the same as the dialectical materialism there thirty years ago, and the impact of developments in science is one of the reasons for its change. But then neither is science itself in the Soviet Union the same as it was thirty years ago, and one of the many influences upon it has been dialectical materialism. Although the Communist Party has attempted at times to control this interaction much more so twenty years ago than now it quite obviously was not able to do so.

2

Philosophy of Mathematics in the Twentieth Century

Introduction

To understand twentieth century philosophy of mathematics, it is therefore necessary first to have some knowledge of Kant's ideas and of the ideas that were at the heart of the nineteenth century reactions to his views. We shall therefore devote the remainder of this introduction to surveying these ideas.

We begin with Kant and the *Problematik* he established for mathematical epistemology. This *Problematik* was focused on the reconciliation of two apparently incompatible features of mathematical thought: namely, its rich *substantiality* as a science, which gives it the appearance of something that arises from sources external to the human intellect, and its apparent certainty or *necessity,* which gives it the appearance of something that is independent of the one external source which is best founded and understood–namely, sensory experience.

To resolve this difficulty, Kant formulated a theory of knowledge which imported much of what had traditionally been thought of as information arising from external sources (specifically, the basic spatial characteristics of sensory thought, and the temporal characteristics of both sensory and non-sensory thinking) into the human mind itself, representing it as the product of certain deep, standing traits of human cognition. At the centre of this theory was a certain conception of judgement which represented the intersection of two different schemes for classifying propositions. On the first of these, propositions were sorted according to the type of knowledge of which they admitted; those which required sensory experience were called *a posteriori*, those which did not were called *a priori*. On the second, they were sorted according to whether or not their predicate terms were contained in their subject terms (in the sense that one *thinking* the subject term would, as a part of that very act itself, also *think* the predicate term). Those judgements in which the subject term contained the predicate term in this sense were to be called *analytic*. Those in which no such containment obtained were either falsehoods, because there was no connection between the subject and predicate concepts at all, or they were *synthetic* truths. In true synthetic judgements, the subject and predicate concepts were joined not by a relation of containment, but rather by a relation of *association*. The association of a predicate with a subject provided for their being thought together in tandem, though it did not, like containment, require that a thinking of the predicate term of a judgement be a constituent part of any thinking of its subject term.

Kant erected his mathematical epistemology upon these distinctions between a priori and a posteriori and analytic and synthetic judgements. He attempted to explain what he referred to as the 'certainty' or 'necessity' of mathematical judgements by showing that our knowledge of them is a priori. Such knowledge, he argued, derives from two standing capacities of the human mind. One of these, which Kant referred to as our a priori intuition of space, was taken to function as a formal constraint on sensory experience by forcing it to be represented in a Euclidean space of three dimensions. The other, the so-called a priori intuition of time, served formally to constrain both sensory and non-sensory experience by representing it as temporally ordered. Both the a priori intuition of space and the a priori intuition of time therefore functioned to control the senses rather than the other way round. It was because of this that Kant believed judgements arising from them (i.e. the judgements of geometry and arithmetic) to be impervious to falsification by sensory experience.

This, in brief, was Kant's proposal for accounting for the necessity of mathematics. He proposed to account for its substantiality by establishing that its judgements are synthetic rather than analytic in character. If mathematical judgement is synthetic in character, then it cannot be seen as consisting in the mere apprehension of a *containment* relation between its subject and predicate concepts. Rather, it must be seen as the fusion in thought of two analytically unrelated concepts through one of two means: either the conjunction of concepts provided for by repeated sensory experience, or the invariant and inevitable association provided for by an a priori structuring of our minds in such a way as to bring the two together *in thought*. Such a joining of analytically unrelated concepts, in which the thinking of the predicate concept is not, logically speaking, strictly required for the thinking of the subject concept, was, in Kant's view, the essential ingredient of substantiality in judgement. This notion of analytically unrelated concepts necessarily joined in thought enabled Kant to frame an account of the substantiality of mathematical judgements which would allow mathematical judgement to be necessary but, at the same time, not limit the degree and kind of the informativeness of mathematical judgements to the degree and kind of complexity that logical containment relations are capable of displaying. In Kant's estimation, this latter was a limitation that it was important to avoid.

Kant adopted a similarly synthetic view of the nature of mathematical *reasoning*. He maintained that mathematical (as opposed to logical or analytical) inference possesses the same rich substantiality of character that distinguishes mathematical from logical or analytical judgement. He also argued that the connection between the premiss(es) and conclusion of a mathematical inference calls for synthetic rather than analytic means of bonding.

To illustrate the point, he elaborated upon an elementary case of geometrical inference; namely, that inference in ordinary Euclidean geometry which takes one from a premiss to the effect that a given figure is a triangle to the conclusion that the sum of its interior angles is equal to that of two right angles. He maintained that no amount of analysis of the concept of a triangle could ever reveal that its interior angles sum to two right angles. Rather, he said, in order to arrive at such a conclusion (i.e. a conclusion that extends our knowledge of triangles beyond what is given in the definition of the concept itself) we must rely primarily not on the definition of the concept, but on the means by which triangles are presented to us in intuition. In other words, we must construct a triangle in intuition, and then extract the conclusion not from the mere concept of a triangle but rather from *the universal conditions governing the construction of triangles* in our intuition. 'In this fashion', said Kant, the mathematical reasoner arrives at his conclusion 'through a chain of inferences guided throughout by intuition'.

These, in brief, are Kant's proposals for the resolution of what he took to be the central problems facing the philosophy of mathematics. But though the problems themselves have remained a staple of twentieth century thinking on the subject, Kant's particular proposals for their resolution have not. What caused this decline in the popularity of Kant's ideas was, primarily, the emergence of challenges to his conception of geometry and his view of the relation between geometry and arithmetic that arose in the nineteenth century. It is to these ideas that we now turn, beginning with geometry.

On Kant's conception, geometry was the product of an a priori intuition of space which specified the space in which human spatial experience was 'set', so to speak. The character of this a priori visual space was that spelled out in the Euclidean axioms for three-dimensional space. In calling Euclidean three-dimensional space the space of human visual experience, one does not mean, of course, that it is the only space that is intelligible or logically coherent to the human mind. Visualizability is one thing, intelligibility or logical coherence another. Kant's position was that Euclidean three-dimensional space is the only space that is *visualizable* by humans.

Not long after Kant elaborated his views in the *Critique of Pure Reason,* mathematicians expressed doubts about them. Gauss, for example, clearly stated his doubts concerning the a priori character of geometry in a letter written in 1817 to Olbers (cf. [2.60], 651-2). He restated the same view in an 1829 letter to Bessel (cf. [2.59], VIII:200) and added that it had been his view for nearly 40 years. In (my translation of) his words:

My innermost conviction is that geometry has a completely different position in our a priori knowledge than arithmetic we must humbly admit that, though number is purely a product of our intellect, space also has a reality external to our intellect which prohibits us from being able to give a complete specification of its laws a priori.

Later, in a letter written in 1832 to Bolyai's father (cf. [2.59], VIII: 220-21), he reiterated this view, saying that Bolyai's results provided a proof of the incorrectness of Kant's views.

It is precisely in the impossibility of deciding a priori between [Euclidean geometry] and S [the younger Bolyai's non-Euclidean geometry] that we have the clearest proof that Kant was wrong to claim that space is only the form of our intuition.

There thus arose among nineteenth century thinkers the belief (given special impetus by the work of Bolyai and Lobatchevsky) that there are fundamental epistemological differences between geometry and arithmetic. Put briefly, the difference is that arithmetic is more and geometry less central to human thought and reason. Arithmetic, on this view, was taken to be wholly a product or creation of the human intellect; geometry, on the other hand, was taken to be determined at least in part by forces external to the human intellect. The difference was implied by a broad epistemological principle (which we might refer to as the *creation principle*) to the effect that what the mind creates or produces of itself is better known to it than that which comes from without.

Belief in the epistemological asymmetry of arithmetic and geometry (though not necessarily Gauss's particular conception of its character) thus became a central tenet of nineteenth century thinking concerning the nature of mathematical knowledge. It also became a prime force shaping the major movements of twentieth century philosophy of mathematics. In the main, two basic kinds of reactions emerged, corresponding to the two basic ways of accommodating this asymmetry. One was to retain a Kantian conception of arithmetic (as based on an a priori intuition of time) and adopt a non-Kantian conception of geometry (as based on an a priori intuition of space). The other was to take a non-Kantian view of arithmetic while retaining a Kantian conception of geometry. The former of these two tactics was essentially that which was adopted by the intuitionists Brouwer and Weyl, while the latter became the central idea motivating the logicism of Frege and Dedekind. Hilbert's finitist programme, the third main movement of twentieth century philosophy of mathematics, in a way adopted and in a way rejected both. It maintained both the epistemological symmetry of arithmetic and geometry and their fundamentally a priori character. It rejected, however, Kant's proposed a priori intuitions of space and time as their bases.

The powerful confirmation of belief in the epistemological asymmetry of arithmetic and geometry that was provided by the nineteenth century discovery of non-Euclidean geometries was therefore a major factor contributing to the decline of Kant's positive views in the philosophy of mathematics and the emergence of major alternatives in the twentieth century. The second major factor contributing to the weakening of Kant's influence in twentieth century philosophy of mathematics was the dramatic development of logic during the latter part of the nineteenth and the early part of the twentieth centuries. This included the introduction of algebraic methods by Boole and DeMorgan, the improved treatment of relations by Peirce, Schröder and Peano, the replacement of the Aristotelian analysis of form based on the subject-predicate relation with the more fecund analysis of form based on Frege's general notion of a logical function, and the advances in formalization brought about by the introduction (by Frege, Russell and Whitehead, and Peano) of precisely defined and managed symbolic languages and systems.

These developments took logic to a point well beyond what it was in the time of Kant, and this caused some to judge that it was the relatively under-developed state of logic in Kant's time that was primarily responsible for his belief in the need for a *synthetic* basis for mathematical judgment and inference. Russell, for one, took such a position, arguing that though Kant's views may have seemed reasonable given the sorry state of logic in his day, they would never have been given a serious hearing had our knowledge of logic been then what it is now. (N.B. But though Russell saw the enrichment of the analysis of logical form brought about by the modern logic of relations and the functional conception of the proposition as being of particular importance to the correction of Kant's deficiencies, he also believed that certain developments in mathematics proper were of great importance. Chief among these were (i) the arithmetization of analysis by Weierstrass, Dedekind and others; and (ii) the discovery by Peano of an axiomatization of arithmetic. These led to what Russell regarded as a codification of pure mathematics within a certain axiomatic system of arithmetic (viz. second-order Peano arithmetic), and so provided for its 'logicization'. Russell reckoned the significance of these developments for Kant's philosophy of mathematics to be as great as that of the discovery of non-Euclidean geometries.

For the most part, Russell's views on these matters were taken over by the logical empiricists, who, like Russell, were much impressed with the new logic, and who were also attracted to a logicism like Russell's, because it allowed them to resolve the difficulties that mathematics had traditionally posed for empiricist epistemologies. The new logic, then, in being seen as the basis for the working out of Russell's sweeping form of logicism, eventually led to the resurgence of empiricist epistemologies for mathematics, and these, quite clearly, represented a significant departure from Kantian mathematical epistemology. In addition, it posed what has proven to be an enduring challenge to the Kantian view that mathematical *reasoning* is essentially distinct from logical reasoning.

This completes our synopsis of the major influences shaping twentieth century philosophy of mathematics. The longer story, which we shall now tell in greater detail, is, for the most part, the story of the ebb and flow of Kant's ideas as they met and interacted with new developments in geometry, logic, science and philosophy.

Early Period and the Development of The Three 'ISMS'

We begin our discussion with the first three decades (what we are calling the 'early period'), which, if it was not *the* most active productive period, was certainly one of the most such in the entire history of the subject. The major developments of this period were the three great 'isms' of recent philosophy of mathematics: logicism, intuitionism and (Hilbert's) formalism. All of these, we shall argue, were profoundly influenced by Kantian ideas. In the case of logicism, however, one must take care to distinguish Frege's from Russell's version. Frege's had much closer ties to Kantian epistemology than did Russell's. Indeed, it attempted to retain many of Kant's most important ideas, including, as we shall see, certain of his ideas regarding the nature of reason.

Logicism

Frege was moved by the discovery of non-Euclidean geometries, and dedicated to the task of explaining what he took to be the main upshot of this discovery-namely, the asymmetry between arithmetic and geometry as regards their basicness to human thought. Geometrical thinking, though widely applied in human thought, was not so widely applied as to suggest that it is not based on a Kantian type of a priori intuition. Thus, Frege supported Kant's geometrical epistemology. Arithmetic, on the other hand, was too pervasively applicable in human thought to be ascribed plausibly to the working of a similar faculty of intuition. No, its epistemological source must be sought elsewhere-ultimately, as Frege saw it, in a reconceived faculty of reason.

The basics of this viewpoint were evident in Frege's writings from the very start. Thus, already in his 1873 doctoral dissertation, he emphasized that 'the whole of geometry rests, in the final analysis, on principles that derive their validity from the character of our intuition'. And, in his 1874 *Habilitationsschrift* [2.47], he expanded this observation to include his view of the relation between geometry and arithmetic *vis à vis* their dependency on intuition.

It is quite clear that there can be no intuition of so pervasive and abstract a concept as that of magnitude (Größe). There is therefore a noteworthy *(bemerkenswerter)* difference between geometry and arithmetic concerning the way in which their basic laws are grounded. The elements of all geometrical constructions are intuitions, and geometry refers to intuition as the source of its axioms. Because the object of arithmetic is not intuitable, it follows that its basic laws cannot be based on intuition.

The same basic point concerning the 'unintuitedness' of the objects of arithmetic is made in the *Grundlagen,* where Frege remarks that:

> *In arithmetic we are not concerned with objects which we come to know as something alien from without through the medium of the senses, but with objects given directly to our reason and, as its nearest kin, utterly transparent to it.*

There Frege broached the question of the relative places occupied in our thinking by empirical, geometrical, and arithmetical laws. His conclusion is that arithmetic laws are deeper than geometrical laws, and geometrical laws deeper than empirical laws. He arrives at this conclusion by conducting a thought experiment in which he considers the cognitive damage that one might expect to be done by denying each of the various kinds of laws. Denying a geometrical law, he concludes, stands to do more extensive damage to a person's cognitive orientation than denying a physical law. For it would lead to a conflict between what people can conceive and what they can spatially intuit. It would bring severe disorientation to a person's cognition. It would force them, for example, to deduce things that formerly they had been able just to 'see'. And it would even make the deductions strange and unfamiliar. It would not, however, result in a global breakdown of their rational thinking. Such global breakdown in one's rational functioning is rather that which would follow from a denial of arithmetical law. Denying an arithmetical law would not only keep one from seeing what he had formerly been able to see, it would, according to Frege, prohibit his engaging in deduction or reasoning of any sort. In his words, it would bring about 'complete confusion', so that 'even to think at all would seem no longer possible'.

Frege sought to explain this projected global breakdown in rational thought by arguing that the scope of arithmetical law, unlike that of physical and geometrical law, is universal. It governs not only that which is physically actual and that which is spatially intuitable, but, indeed, all that which is *numerable*-and that, according to Frege, is the widest range possible, extending to all that which is in any coherent way thinkable or conceivable. The laws of arithmetic must, therefore, he concluded, 'be connected very intimately with the laws of thought' that is, the laws of logic.

This alleged difference in the pervasiveness to thought of arithmetic and geometry thus became, in Frege's thinking, the (or at least a) fundamental datum for the philosophy of mathematics. He also believed it to be a datum Kant had overlooked. For, had he been aware of it, Frege felt sure, Kant would never have tried, as he did, to stretch essentially the same epistemology to cover both arithmetic and geometry. Rather, he would have tried to do justice to the 'observable' differences in depth-to-rational-thought of arithmetic and geometry.

Kant's ultimate shortcoming, Frege believed, was that he had acknowledged only two basic sources of knowledge–sensation and understanding. This allowed room only for a distinction between sensory and a priori knowledge. It did not allow for a distinction–at least not a distinction of kind–between different subspecies of a priori knowledge. Frege, on the other hand, distinguished between sensory experience, the source of our knowledge of natural science, intuition, the source of our geometrical knowledge, and reason, which Frege described as the source of our arithmetical knowledge. This modification of Kant's general epistemology was, Frege believed, necessary if one was to account for the perceivable differences in the relative pervasiveness of arithmetic and geometry.

Actually, it is not clear that Kant's epistemology did not enable him to do something of the same sort. Certainly it *did* distinguish two types of experience, 'inner' and 'outer', and noted that the one (viz., inner) made use of intuitional resources (viz., the a priori intuition of time) that are more pervasive than those (viz., the a priori intuition of space) upon which the other is based. Adding to this the fact that Kant maintained that arithmetical thinking is based on the more pervasive intuition of time and geometrical thinking on the less pervasive intuition of space, it would seem that the distinction between inner and outer experience in Kant is capable of effecting something of at least the same general kind of asymmetry between arithmetic and geometry that Frege believed to be so important to mathematical epistemology. Frege seems never to have considered this point.

We add this remark, however, mainly as an aside. For, clearly, there are important differences between Frege and Kant concerning the pervasiveness of arithmetic. Kant, for example, despite acknowledging arithmetic to be widely applicable, nonetheless held it to be limited to that which is *experienceable*. He did not take it to apply to the whole of what is (rationally) imaginable or *conceivable*. Consequently, though he judged arithmetical law to be a priori in character, he also judged it to be synthetic. Frege, on the other hand, believed arithmetic to apply to all that is conceivable, and it was precisely in this departure from Kant that he was led to regard it as analytic rather than synthetic in character.)

Frege thus disagreed with Kant concerning the pervasiveness to rational human thought of arithmetical thinking. This disagreement cannot, however, be taken at face value in explaining why Kant held a synthetic and Frege an analytic conception of arithmetic. For the two employed different conceptions of the notions of analyticity and syntheticity. Thus, to comprehend better the true differences separating Kant and Frege, we must look more carefully at the definitions each used in formulating the key notions of his position.

Kant defined an analytic truth as a truth in which the predicate 'belongs to' the subject as something 'covertly contained' in it, and a synthetic truth as one that is not analytic. He did not characterize the analytic/synthetic distinction, as he did the a priori/a posteriori distinction, in terms of the characters of the possible justifications of a judgement. Frege, on the other hand, did exactly that. In his scheme, both the analytic/synthetic and the a priori/a posteriori distinctions are parts of a classificatory system regarding the different kinds of *justifications* a given judgement might have.

Each truth, Frege believed, possesses a kind of *canonical proof* or justification. This is a proof which, in its ultimate premisses, goes all the way back to the 'primitive truths' of the subject to which the theorem belongs. It gives 'the ultimate ground upon which rests the justification for holding [the theorem proven] to be true' (Ibid., section 3). It thus presupposes an ordering of truths, and its objective is precisely to retrieve that segment of the given ordering which links the proposition to be proven to those ur-truths of its subject which are responsible for its truth. It aims, in other words, at revealing what might be called the *grounds* of the proven proposition's truth–its Leibnizian *Sufficient Reason,* as it were.

A proposition or judgement is said to be analytical, in this scheme, if its canonical proof contains only 'general logical laws' and 'definitions'. It is said to be synthetic if its canonical proof contains at least one premiss belonging to 'some special science'. It is considered a posteriori if its canonical proof includes an 'appeal to facts, i.e. to truths which cannot be proved and are not general'. And, finally, it is considered to be a priori if its canonical proof uses exclusively 'general laws, which themselves neither need nor admit of proof' (Ibid.) (in other words, if knowledge of it can arise from the Fregean faculty of reason alone).

Frege believed that finding the canonical proofs of arithmetical truths would reveal an intimate connection between them and the basic laws of thought (i.e. the 'general logical laws'). At the same time, however, he was keenly aware of the Kantian objection to such a proposal; namely, that it makes it difficult to account for the epistemic productivity or substantiality of arithmetic. Indeed, immediately after having broached the view that arithmetic is analytic in section 15 of the *Grundlagen* [2.49], Frege went on in section 16 to note that the chief difficulty facing such a view is to explain how 'the great tree of the science of number as we know it, towering, spreading, and still continually growing' can 'have its roots in bare identities'. He thus clearly saw his main task as that of explaining how analytic judgement and analytic inference can yield an epistemic product having the robustness that arithmetic appears to have.

His response can be seen as divided into two parts. The first of these consists in the giving of an account of the 'objectivity' of analytic judgements that does not appeal to sensation or intuition. The second concerns the more general problem of explaining how one might get a conclusion that extends the knowledge represented by the premisses of an inference out of premisses that can be inferentially manipulated only by purely logical means.

Regarding the former, Frege's idea was to ascribe special properties to concepts, or to the objectively existing *thoughts* which, via the context principle are prior to them. Numbers were then to be defined in terms of concept extensions, and concept extensions to be treated as 'logical objects' (which we somehow grasp by grasping the concepts of which they are extensions). The epistemologically salient features of this arrangement were summed up in the following remark from the *Grundlagen.*

In arithmetic we are not concerned with objects which we come to know as something alien from without through the medium of the senses, but with objects given directly to reason and, as its nearest kin, utterly transparent to it.

And yet, for that very reason, these objects are not subjective fantasies. There is nothing more objective than the laws of arithmetic. In later writings, Frege elaborated a bit–but only a bit–on his notion of concept extensions as *logical objects*.

It is futile to take the extension of a concept as a class, and make it rest, not on the concept, but on single things...the extension of a concept does not consist of objects falling under the concept, in the way, e.g., that a wood consists of trees it attaches to the concept and to the concept alone the concept takes logical precedence to its extension.

Thus, what made a class into a logical object, in Frege's view, was only its relation to the concept of which it formed the extension. He had ultimately, however, to establish the sense in which logical objects exist, since, in his view, they were not actual (i.e. not spatial or 'handleable'). Here he had only analogies to offer, citing such examples as the axis of the earth and the centre of mass of the solar system. These illustrated his generally negative characterization of logical objects as objects 'independent of our sensation, intuition and imagination, and of all construction of mental pictures, memories and earlier sensations, but not independent of reason'.

Frege had also to establish that logical objects deserve to be called 'logical'. This he did not do in the *Grundlagen,* being at that time unsure whether he needed concept extensions or just concepts. He pursued the matter to a (for him) satisfactory end only in the 1891 lecture 'Function und Begriff, where he argued (i) that the notion of concept-extension can be reduced to that of the range-of-values of a function; and (ii) that this latter notion is clearly a logical notion.

Among the more important things that Frege's belief in the logical precedence of concepts to their extensions allowed him to do was to reduce knowledge of infinities to logical knowledge. Along with this he had also to accept a restriction on how we come to *acquire* concepts; namely, that we do so by means other than abstraction from the particulars falling under them. Such a view of concept acquisition was of the utmost importance to his logicism. For, were concepts to be obtainable only via such a process of abstraction, knowledge of the number concept would likewise be obtainable only through prior knowledge of the particulars falling under it. If that were so, however, one would have to give a prior account of how it is that we come by knowledge of the particulars from which knowledge of the abstracted concept is derived. And in order to keep this account from destroying the 'logical' character of numbers, one would have to make sure that it made no appeal to the likes of sensation or Kantian intuition. Moreover, even if it were successful in avoiding appeals to Kantian intuition, such an abstractive account of concept acquisition would cause severe problems for the knowledge of infinite sets. For it is hardly plausible to believe that we could either obtain separate

The discovery by Russell that this way of thinking of the relationship between concepts and (logical) objects is subject to paradox therefore threw Frege's entire 'improvement' of Kant's arithmetical epistemology into crisis. For without a principle that makes concepts prior to their extensions, a Fregean philosophy of arithmetic would have great difficulty in developing an appropriately non-intuitional model of cognition for our knowledge of concept extensions. And without a non-intuitional model of our knowledge of concept extensions, the major novelty (i.e. the major non-Kantian element) would be missing from Frege's proposed explanation of the epistemic robustness or substantiality of arithmetic.

Russell's discovery thus raised the problem of how we might come to apprehend logical objects (and thus numbers), even if it is assumed that they exist. Without a Frege-type scheme of comprehension, which sees apprehension of numbers as derived from apprehension of concepts, and which allows concepts to be apprehended without any prior non-conceptual apprehension of particulars, one is hard-pressed to avoid at least some minimal appeal to non-conceptually based knowledge of sets or extensions-a knowledge which is hard to account for without making some appeal to sensation or intuition. Russell's paradox therefore raised grave problems for the epistemology of Frege's logical objects.

But even supposing these problems to have been solved, there were, by Frege's own admission (indeed, his own insistence!), serious difficulties still to be overcome in explaining the epistemic productivity of mathematical *inference*. To manage these, Frege appealed to the general phenomenon of *Sinn* and to the possibility of rearranging (or 'recarving') the contents of a proposition in such a way as to expose contents that had been hitherto undetected.

Concepts featured prominently in this explanation too. In particular, it included an appeal to an assumed relationship between concepts and propositions which allows a proposition to be both understood and known even though not all the concepts contained in it are apprehended. This crucially important feature of the relationship between propositions and their constituent concepts was taken to be based on the principles that (i) in order to apprehend a proposition, one need only know a *definition* of its immediate constituent concepts; and that (ii) knowing a definition of a concept does not require that all content tacit in it be apprehended. It is the recovery of tacitly contained content (i.e. discovery of its presence and character) that thus allows the conclusion of an analytic inference to represent something more, by way of cognitive accomplishment, than is represented by the apprehension and knowledge of its premisses. As Frege himself put it, the identification and utilization of such content amounts to something more than merely 'taking out of the box again what we have just put into it'. For what we put into an inferential 'box' is the knowledge of concepts we use in order to arrive at understanding and knowledge of its premises. What, on Frege's account, we are capable of extracting from such a box is not only judgements formed from those concepts, but also judgements formed from concepts identified, indeed *formed,* through the 'carving up' or conceptual rearrangement of the premises.

However, if Frege required analytic inference to be epistemically productive, he also required (at least some of) it to be rigorous. Therefore, in some sense, he demanded that analytical judgement be *incapable* of concealing content. Indeed, he himself insisted that his logicism required the giving of utterly rigorous proofs for the laws of arithmetic. It is 'only if every gap in the chain of deductions is eliminated with the greatest care', he said, that we can 'say with certainty upon what primitive truths' they rest. And it is only through seeing with certainty the truths upon which the truths of arithmetic rest that we would be in a position to judge whether or not those grounds are logical in character.

What Frege seems not to have seen so clearly, however, is that we can be certain that a canonically proven proposition is analytic only to the extent that we can be certain that its premisses do not tacitly contain any synthetic content. At any rate, how certainty on this score is to be attained is something about which he seems to have said little. He firmly believed that there are propositions-the so-called 'basic laws' of arithmetic–that are at once so rich as to be capable of delivering the whole of arithmetic and also so clearly analytic as to self-evidently conceal no synthetic content. What he took to be the justification of this confidence is less clear. For, in the first place, he believed that all propositions are analytic. Second, the propositions he saw as making up the 'basic laws'-namely, the so-called logical identities of the form 'A is A'-

were transparently analytic. This is so because, for Leibniz, analyticity consisted in containment of the predicate of a proposition in its subject, and propositions of the form 'A is A' satisfy this containment requirement in a manner in which none clearer or more certain can be conceived. Leibniz therefore had a natural stopping point for his reduction to analytic truth. Frege, on the other hand, having adopted a more complex and sophisticated definition of analyticity, seems to have lost the ability to identify a class of truths which were as clearly and certainly analytic as Leibniz's 'identities'. As a result, he lacked as clear a point at which to terminate the reduction of arithmetic laws to analytic truths.)

Frege's conception of mathematical inference was thus faced with two apparently competing demands: on the one hand, the need to endow analytic judgments with tacit content so as to enable analytic inference to be epistemically productive; and, on the other, the need to restrict the mechanisms producing tacit content in such a way as to guarantee that synthetic content can never be tacitly contained in what passes for analytic content. In the end, I believe, he failed to meet these two demands adequately. He did not succeed in providing a set of basic laws and a criterion of tacit content the pair of which were guaranteed to permit only the production of analytic truths as tacit contents of the basic laws. Nor did he manage to ensure that the epistemic productivity sustainable by means of his mechanisms of tacit content production are capable of matching those which may be observed to hold in arithmetic.

The first failure was clearly illustrated by Russell's paradox, which shows that the latent content concealed by Frege's axioms (in particular, his axiom of comprehension) could include not only synthetic truths, but even analytical falsehoods! The second failure became the pivotal feature of the intuitionists' critique of logicism.

Russell's logicism was quite different from Frege's. In the first place, it was not motivated primarily by the discovery of non-Euclidean geometries with its attendant belief in the epistemological asymmetry between geometry and arithmetic. Nor was it based on belief in such things as logical objects, and the associated division of cognition into faculties of sense, intuition and reason. Nor, finally, did it restrict its logicism to arithmetic, but, rather, extended it to the whole of mathematics, and even to certain areas outside traditional mathematics. Rather, it took as its starting points (i) a certain general definition of mathematics; (ii) a methodological principle to pursue ever further generalization in science; and (iii) a belief that pursuing this principle in mathematics would eventually lead one to a most general science of all, namely, logic. It was fuelled in these pursuits by the then rapid and impressive advances in symbolic logic.

In the opening paragraph of *The Principles of Mathematics*, Russell offered the following definition of pure mathematics: 'Pure mathematics is the class of all propositions of the form "*p* implies *q*", and neither *p* nor *q* contains any constants except logical constants'. He then went on to describe his logicist project as that of showing 'that whatever has, in the past, been regarded as pure mathematics, is included in our definition, and that whatever else is included possesses those marks by which mathematics is commonly though vaguely distinguished from other studies'. Russell also maintained that, in addition to asserting implications, propositions of pure mathematics are characterized by the fact that they contain variables, indeed, variables of wholly unrestricted range.

Russell planned to defend this last claim, which he acknowledged as being highly counterintuitive, by showing that even such apparently variable–free statements as '1+1=2' can be seen to contain variables once their true meaning and form is revealed. The discovery (or, better, *recovery*) of the true meaning and form of such statements was made possible by the vast enrichment of the basic stock of logical forms made available through the work of Peirce, Schröder, Peano and Frege. Using this work, Russell produced analyses of the deep forms of ordinary mathematical statements. '1+1=2', for example, was analysed as 'If *x* is one and *y* is one, and *x* differs from *y*, then *x* and *y* are two.' Analysed in this way, Russell maintained, the supposedly non-implicational, variable free '1+1=2' is seen both to contain completely general variables and to express an implication, just as his logicist theory predicted would be the case.

Of course, 'if x is one and y is one, and x and y are different, then x and y are two' does not express a genuine proposition at all since it contains free variables. It expresses instead what might be called a proposition form or a proposition schema. Russell called it a '*type* of proposition', and went on to say that 'mathematics is interested exclusively in *types* of propositions' rather than in individual propositions *per se*. On this view, the business of mathematics is to determine which propositions can be generalized (i.e. which constants can be turned into variables), and then to carry this process of generalization out to its maximum possible extent. This maximum will have been reached when we have penetrated to a level of propositions whose only constants are *logical* constants and whose only undemonstrated propositions are the most basic truths whose only constants are logical constants. The logical constants themselves, as a class, were to be characterized only by enumeration. Indeed, by their very nature they admitted only of this kind of characterization, since any other kind of characterization would be forced to make use of some element of the class to be defined.

At bottom, then, Russell's logicism was motivated by a view of mathematics that saw it as the science of the most general formal truths; a science whose only indefinables are those constants of rational thought (the so-called *logical* constants) that have the widest and most ubiquitous usage and whose only indemonstrables are those propositions which set out the most basic properties of those indefinable terms. In his view, this provided the only precise description of what philosophers have had in mind in describing mathematics as an a priori science. Mathematics is thus in the business of generalization. Its aim is to identify those truths that remain true when their non-logical constants are replaced by variables. This process of generalization may require some analysis in order to find the genuine form of the sentence to be generalized. But once that form is found, the generalization process should ultimately lead to the realization that the mathematical truth in question expresses a formal truth whose variables are completely general and whose only constants are logical constants.

Ideally, proper method in mathematics requires pursuit of this process of generalization to the ultimate degree. At that point, Russell believed, we will find formal truths of maximum generality–truths of a generality so great as to render them incapable of further generalization–truths, that is, that are so general that they would become non-truths were any of their constants to be replaced, even through conceptual analysis, by variables. This, in Russell's opinion, was the only point at which the method of mathematics (i.e. the pursuit of maximal formal generalization) can properly and naturally be brought to a close. He also believed that it is in this domain of formal truths of the utmost generality, and in this domain alone, that we can rightly expect to meet what are properly regarded as *laws of logic.*

According to Russell, these laws are justified inductively from their consequences.

> The propositions from which a given proposition is deduced generally give the reason why we believe the given proposition. But in dealing with the principles of mathematics, this relation is reversed. Our propositions are too simple to be easy, and thus their consequences are generally easier than they are. Hence we tend to believe the premises because we can see that their consequences are true, instead of believing the consequences because we know the premises to be true...thus the method in investigating the principles of mathematics is really an inductive method, and is substantially the same as the method of discovering general laws in any other science.

Thus, contrary to what Kant had maintained, the pursuit of greater generality (or what Kant referred to as 'unification') has a natural and fairly inevitable stopping point; specifically, that level of judgements having broad scope, entirely general variables and utterly ubiquitous constants.

Frege and Russell, then, though they agreed in their rejection of Kantian intuition as the basis of mathematical knowledge, none the less differed with regard to their estimates of the proper scope of logicism and the nature and origins of its basic laws. They also differed on the important question of our knowledge of the infinities with which mathematics deals, and of how, exactly, that knowledge is related to our knowledge of concepts.

Unlike Frege, Russell did not believe that concepts alone can give rise to extensions or sets. Indeed, he responded to his own antinomy by offering a conception of set which assumed a class of individuals as given *prior to* the generation of sets by concepts. In his view, before there can be a rich universe of sets, there must first be a totality of individuals that is given by means other than grasp of a concept. Using this domain of individuals as a base, comprehension by concepts (or what Russell referred to as 'prepositional functions') was then supposed to operate according to predicative principles of collection. There was thus an order of 'priority' of *types* or levels induced among entities, with the domain of individuals constituting the lowest level and the upper levels being formed by application of comprehension operations to the entities lying at prior levels.

This way of thinking of set comprehension differs radically from the way Frege thought of it. Fregean comprehension did not presume an ordering of entities according to some 'priority' ranking, and it was not restricted to collection of entities formed at prior levels. Perhaps even more importantly, it did not posit a '0th-level' domain of entities as somehow given prior to all comprehension by concepts. Indeed, in Frege's scheme, the whole idea was to avoid the need of having a 'starting' collection–particularly an infinite one–to serve as the raw material from which comprehension by concepts is to get off the ground. For, in Frege's view, having a non-conceptually comprehended domain of individuals required something like Kantian intuition, and this is exactly what he hoped to avoid (since he did not see how knowledge of such a domain could rightly be seen as *logical* knowledge).

Russell, though he showed some sensitivity to this difficulty, seems never to have settled on a means of resolving it. In his earlier writings he sometimes spoke as if any statement positing a domain of existents, and therefore any axiom positing a domain of individuals, is not to be regarded as a truth of pure mathematics *per se,* but rather as an hypothesis whose consequences are to be investigated.

Later, however, he stated both that he believed such a view to be mistaken and that he himself had never held such a view. There are also systematic elements of his thought that would (or at least should) have led him to reject such a view. Chief in this regard was his belief in the need for the 'regressive' method in the foundations of mathematics.

Use of the regressive method allows the truth of a principle to be inferred from its usefulness in deductively unifying a recognized body of truths. Hence, to the extent that the postulation of a domain of individuals (e.g. an axiom of infinity) has utility as a means of deductively organizing the recognized truths of mathematics, it, too, inherits a certain plausibility and so deserves to be 'detached' and asserted as a truth on its own right. Russell's 'regressive' method thus elevated axioms of existence to the status of justified assertions, and so made them more than mere 'hypotheses' to be used as antecedents of conditionals whose consequents are propositions whose proof requires their use. There would therefore seem to be a tension between Russell's adoption of the 'regressive' method in mathematics and that part of his logicism which saw axioms of existence (and, in particular, his axiom of infinity) as mere hypotheses to be put into the antecedents of conditionals.

But if there were differences between Russell and Frege regarding the nature and justification of the basic laws of mathematical thought, there was substantial agreement between them as regards the nature of mathematical inference. In particular, there was agreement on the points that the inferences of mathematics are all to be strictly logical in character, and that this is necessary in order to satisfy the demands of rigour.

It is not always easy to see the similarities between their views, however, because they used different definitions of analyticity and syntheticity. For Frege, a synthetic inference was one in which the conclusion could not be extracted from the premisses by any re-carving of their contents, but rather required something like an infusion of intuition in order to connect the conclusion with the premiss(es). For Russell, on the other hand, an inference was synthetic, and, so, epistemically productive (at least in a minimal way), if its conclusion constituted a *different proposition from* its premisses. Thus, many inferences that Frege would have classified as 'analytic', would have been classified by Russell as 'synthetic'.

The standard of syntheticity in inference set forth by Russell was weaker than that set forth by Frege. Consequently, many inferences satisfying Russell's condition would not satisfy Frege's. Indeed, judged by the lights of Russell's definition, even the elementary inferences of syllogistic reasoning, which Frege classified as analytic, would have been counted as synthetic by Russell. This was all to the good so far as Russell was concerned. For it enabled him to meet Kant's challenge to explain the epistemic substantiality of mathematical reasoning while at the same time maintaining, in opposition to Kant, that the inferences involved in such reasoning are of a purely formal, logical nature and make no appeal to intuition. If growth of knowledge through inference is essentially a matter of thereby obtaining a justified judgement whose propositional content is simply *distinct* from those of one's previously justified judgements, then even very elementary logical inferences can be epistemically productive.

The 'logicization' of mathematical inference was thus, in Russell's opinion, nothing to be balked at epistemically. What had kept previous generations of thinkers, and, in particular, Kant, from embracing it was simply the relatively impoverished state of logic prior to the late nineteenth century. The old logic with its meagre stock of subject-predicate forms may have been inadequate to the riches of mathematical reasoning, but the new logic with its robust functional conception of form had changed all this. With its help mathematical reasoning could finally be logicized in its entirety, and 'a final and irrevocable refutation' of the Kantian doctrine that mathematical inference makes use of intuition be given.

Effectively countering Kant's intuitional conception of mathematical inference was thus an important element of both Frege's and Russell's logicist programmes. They seemed to believe that this could be accomplished simply by deriving large bodies of mathematical theorems from specified axioms by purely logical means. On reflection, however, this seems to be mistaken. Kant may well have under estimated the power of logical inference. His main point, however, was not that there are mathematical proofs that have no logical counterparts whatsoever. Rather, it was that such counterparts, even if they were to exist, would not preserve the epistemologically essential features of the mathematical proofs of which they are the 'logicized' counterparts.

Against this essentially epistemological point, detailed *tours de force* (e.g. Frege's *Grundgesetze* and Russell's *Principles of Mathematics* and *Principia Mathematica*) which locate logical counterparts for mathematical proofs on even a grand scale can have but little effect. For the claim to be met is not that there are no counterparts, but rather that they are epistomologically inadequate substitutes for the mathematical proofs they are to replace.

In summary, let us return to our original claim that logicism in this century arose from two very different sources, to wit, the discovery of non-Euclidean geometries and the development of symbolic logic. Frege's logicism owed the bulk of both its motivation and character to the former, while that of Russell was due primarily to the latter. This is the basic reason why Frege's logicism, unlike Russell's, was able to preserve a remarkable degree of fidelity to the precepts of Kantian epistemology.

Frege did not, however, agree with Kant's idealist conception of the faculty of reason. To get a realist account, though, he had to get the right sorts of objects into the picture. They had to be independent of the human mind in order to insure the objectivity of arithmetic; but they also had to be intimately related to the basic operation of the human mind in order to avoid an appeal to intuition and thus to account for the greater pervasiveness of arithmetic as over against geometry. His solution was the logical object, the ur-form of which was the class-as-concept-extension. Through its essential relation with concepts, it could be brought close to reason. But through the objectivity of concepts it could also be made objective.

Frege's idea of giving a realist rather than idealist treatment of Kant's faculty of reason foundered on Russell's paradox. Russell's reaction to his paradox was rather different. Far from causing him to give up logicism, it led him instead to seek another basis for it–a methodological basis whose chief principle was one enjoining pursuit of maximal generality in one's theorizing, including one's mathematical theorizing. In the presence of his belief that mathematical claims express generalizations, this principle led him in a natural way to a logicist conception of mathematics. In the end, however, Russell's paradox proved to be nearly as great an impediment to Russell's logicism as it was to Frege's. For just as Frege was unable to find a way to fit classes that do not descend from concepts into his realist logicism of logical objects given directly to reason, so, too, was Russell unable to find a satisfactory way of justifying laws asserting the existence of such classes as genuinely logical laws.

Intuitionism

Like Frege's logicism, the intuitionism of the early part of this century was also dominated by (i) the idea that what the mind brings forth purely of itself cannot be hidden from it; and (ii) the belief that the existence of non-Euclidean geometries reveals important epistemological differences between geometry and arithmetic. The direct predecessors of the intuitionists appear to have been Gauss and Kronecker, who interpreted the discovery of non-Euclidean geometries differently than Frege. For whereas Frege proposed a realist modification of the creation principle in order to account for the apparent differences between arithmetic and geometry brought to light by the discovery of non-Euclidean geometry, Gauss and Kronecker, and the intuitionists after them, interpreted the difference between arithmetic and geometry in the light of the creation principle (i.e. principle (i) above), of which they adopted an idealist reading.

Thus, instead of maintaining Kant's synthetic a priori conception of geometry and trying to account for the difference between geometry and arithmetic by establishing arithmetic as analytic, the early intuitionist rejected Kant's synthetic a priori conception of geometry and proposed to account for the differences between arithmetic and geometry by seeing the former as a priori and the latter as a posteriori. As Gauss and Kronecker emphasized, arithmetic is purely a product of the human intellect, whereas geometry is determined by things outside the human intellect. Years later, Wey would reiterate the same theme, remarking that 'the numbers are to a far greater measure than the objects and relations of space a free product of the human mind and therefore transparent to the mind'.

Brouwer, too, expressed similar ideas, identifying as the primary cause of the demise of intuitionism since the time of Kant the refutation of Kant's belief in an a priori intuition of space by the discovery of non-Euclidean geometries. At the same time, however, he advocated resolute adherence to an a priori intuition of time, and even argued that from this intuition one could recoup a system of geometric judgements via Descartes' 'arithmetization' of geometry. He considered the 'primordial intuition of time'–which he described as the falling apart of a life–moment into a part that is passing away and a part that is becoming as the 'fundamental phenomenon of the human intellect'. From this intuition one can pass, via a process of abstraction, to the notion of 'bare two-oneness', which Brouwer regarded as the basal concept of all of mathematics. The further recognition by the intellect of the possibility of indefinitely continuing this process then leads it through the finite ordinals, to the smallest transfinite ordinal, and finally to the intuition of the linear continuum (i.e. to that unified plurality of elements which cannot be thought of as a mere collection of units since the relations of interposition which join them is not exhausted by mere interposition of new units). In this way, Brouwer believed, first arithmetic and then geometry (albeit only analytic geometry), via reduction of the former to the latter through Descartes' calculus of coordinates, come to be qualified as synthetic a priori.

The early intuitionists thus retained a semblance of adherence to Kant's belief in the synthetic a priority of arithmetical knowledge while denying his belief in the priority of our knowledge of the base characteristics of visual space. They were also staunchly Kantian in their conception of mathematical inference. Poincare and Brouwer, in particular, devoted considerable attention to this point. Indeed, Poincaré, who carried on a well-known debate with Russell in the early years of this century, made the role of logical inference in mathematical proof the centrepiece of his critique of logicism. So, too, in effect, did Brouwer, though his critique was aimed at the use of logical reasoning in classical mathematics generally, and not just at the logicists' programmatic demand regarding the logicization of proof.

At the heart of the view of proof that both criticized is a conception of evidence–the *classical* conception evidence–which sees it as essentially a means of determining the (classical) truth value of a proposition. On this view, evidence is a relatively 'malleable' commodity. Its effects extend to a variety of propositions other than that which forms its direct content. This comes about as a result of subjecting the content of a piece of evidence to logical analysis, which is used to extract 'new' contents from the original content. By this means, the justificatory power which the evidence provided for its content can be transferred to the analytically extracted content. Hence, one and the same piece of evidence can be used to identify the truth value of a variety of different propositions. This holds, moreover, despite the fact that there is no parallel analysis directed at the evidence itself whose purpose is to reveal a separable part of the evidence whose content is precisely the new content brought forward by means of the analysis of its content. On the classical view, then, the prepositional content of a piece of evidence can be 'detached' from that evidence itself. Applying logical analysis to that 'detached' content, one can then transfer the warrant attaching to it to any of the new propositions extracted by means of that analysis.

Both Brouwer and Poincare reacted sharply to this view of inference. Brouwer's reaction was based on the view that mathematical knowledge is essentially a product of introspective experience. The extension or development of such knowledge can therefore not proceed via logical extrapolation of its content, since such extrapolation does not guarantee any similar extension of the experience having the extrapolated content as its content. Extension of genuinely mathematical knowledge thus requires the extension of the mathematical experience serving as the evidence for a given content into a mathematical experience of another content. In other words, inference is not to be seen as a matter of logically extracting new contents from old and thence transferring warrant from old to new. Rather, it is to be seen as a process of *experientally transforming* an introspective construction having one content into an introspective construction having another.

Brouwer thus held that one can never 'deduce a *mathematical* state of things' by means of *logical inference.* He memorialized this view in his so-called *First Act of Intuitionism,* in which he declared that mathematics should be completely separated from 'mathematical language and hence from the phenomena of language described by theoretical logic, recognizing that intuitionist mathematics is an essentially languageless activity of the mind having its origin in the perception of a move of time'.

Brouwer thus adhered to a basically Kantian conception of mathematical reasoning according to which extension of mathematical knowledge via inference requires development of a new intuition underlying that inference. Poincaré, too, adopted such a conception of inference, though his view differed in certain respects from Brouwer's. Mathematical reasoning, as he put it, has a 'kind of creative virtue' by which its conclusions go beyond its premisses in a way that the conclusions of logical inferences do not go beyond their premisses. Logical inference from a mathematically known proposition, therefore, though it may yield *some* kind of extension of that knowledge, will none the less typically not yield an extension of the genuinely mathematical knowledge thereby represented. In short, in order for mathematical knowledge that p to be extended to mathematical knowledge that q, it is not enough that p be seen logically to imply q. Rather, p must be seen both to be mathematically different from q and to *mathematically* imply q. In other words, the 'movement' from premiss to conclusion in a mathematical inference is a case of joint comprehension of the premisses and conclusion by a common mathematical 'universal' which is seen to persist in the 'differences' through which it 'moves'.

For Poincaré, then, mathematical reasoning consisted in the synthesis of different propositions by a single, distinctively mathematical structure or architecture. Thus, as with Brouwer, so, too, with Poincaré, we find a view of mathematical reasoning which contrasts sharply with the logicists' conception of mathematical reasoning.

The views of mathematical reasoning or inference of Brouwer and Poincaré are thus Kantian in the sense that they reject the idea that genuine mathematical inference can be logical. They also represent a modification of Kant's views, however. For Kant suggested that by means of genuinely mathematical reasoning from a given set of premisses, one can obtain conclusions that are actually *unattainable* by means of purely logical (i.e. purely analytical or discursive) reasoning from those same premisses. Such an idea, however, seems not to have figured at all in the arguments of Poincare and Brouwer. What they were stressing was a difference in *epistemic quality* between logical and mathematical reasoning–a difference which, in their view, would persist even if the two types of reasoning might prove to be result-wise equivalent. This emphasis on epistemic quality was based on their belief in a difference between the epistemic condition of one whose reasoning is founded on topic-neutral steps of logical inference and one whose inference rests on topic-specific insights into the given mathematical subject at hand. Reasoning of the latter sort presupposes a knowledge of the local 'architecture' of a subject. Reasoning of the former sort does not. To use Poincaré's own figures of speech, the difference is (i) like that between a writer who has only a knowledge of grammar versus one who also has an idea for a story; or (ii) like that between a chess player who has knowledge only of the permissible moves of the several players versus one who has a tactical understanding of the game as well.

The intuitionists were thus at odds with the logicists over the question of the nature of mathematical reasoning. The heart of their disagreement, moreover, was not a dispute concerning which logic is the right logic, but rather a deeper difference regarding the role that *any* logical inference–classical or non-classical–has to play in mathematical reasoning. They were, in other words, divided over the Kantian question of whether intuition has an indispensable role to play in mathematical inference. The intuitionists sided with Kant in holding that it does. The logicists took the contrary view.

In the intuitionism of Brouwer, Poincaré and Weyl we thus find an attempt to work out a modified form of Kant's specifically mathematical epistemology. So far, the modifications noted include (i) the jettisoning of Kant's use of spatial intuition as a fundament for mathematical knowledge; and (ii) the extension and elaboration of his use of temporal intuition as a basis for arithmetic and, relatedly, the reduction of geometry to arithmetic via appeal to Descartes' 'arithmetization' of geometry.

There is, however, one final modification to be noted, and that concerns the intuitionists' (in particular, Brouwer's and Weyl's) conception of existence claims. It is perhaps the most significant of all the modifications made and consists in a shift from Kant's conception of existence claims and our knowledge of them to a conception of existence claims that is more like that found in such post-Kantian romantic idealists as Fichte, Schelling and Goethe. The basic non-Kantian element of this view was the introduction of a non-sensory, purely intellectual form of intuition. This was conceived as a form of self-knowledge whose key epistemic feature was its immediacy–an immediacy expressing the romantic idealists' concern with the epistemic effects of representation. They saw representation as the basic source of error and uncertainty in cognition and therefore advocated its avoidance.

Their reasoning was basically Kantian. That is, they began with the Kantian premiss that no idea or concept (more generally, no representation) contains the being or existence of that which it represents and concluded from this that no concept or idea (more generally, no representation) could, in and of itself, give the existence of anything falling under it. Indeed, representations only tend to increase the epistemic distance between the knower and the object to be known since they leave the being of the object still to be given while adding grasp of the representation to those things that must be accomplished before the object can be known.

What was wanted, therefore, was some kind of representationless knowledge of being. For the paradigm case of such knowledge, the romantic idealists turned to our knowledge of our willing and acting selves. Their model for knowledge of existence thus became one of self-knowledge; in order to know that something exists, the knower must *live* it or *be* it. In other words, she must incorporate it into herself so that her knowledge of its existence becomes that of her knowledge of her own existence. As Schelling said, 'the proposition that there are things external to us will only be certain to the extent that it is identical with the proposition I exist, and its certainty can only match that of the proposition from which it derives.'

Brouwer, it seems, adopted this romantic idealist conception of knowledge of existence. His so-called *First Act of Intuitionism* can indeed be seen as issuing a call for the mathematical knower to turn into himself and to shun the epistemic indirection of the classical view of mathematics with its involvement in the representation of mathematical thought–that is, mathematical *language*. Thus he reminded us:

> *You know that very meaningful phrase 'turn into yourself'. There seems to be a kind of attention which centres round yourself and which to some extent is within your power. What this Self is we cannot further say; we cannot even reason about it, since-as we know-all speaking and reasoning is an attention at a great distance from the Self; we cannot even get near it by reasoning or words, but only by 'turning into the Self' as it is given to us. Now you will recognize your Free Will, in so far [as] it is free to withdraw from the world of causality and then to remain free only then obtaining a definite Direction which it will follow freely, reversibly.*

Here we clearly see the romanticist idea that representation impedes knowledge–an idea that was expressed in strikingly similar terms by Fichte, who said:

> *Look into yourself. Turn away from everything that surrounds you and towards your inner life. This is the first demand that philosophy makes on its followers. What matters is not what is outside you, but only what comes from within yourself.*

Though we lack the space adequately to argue for it here, we believe that Brouwer adhered to this romantic idealist conception of knowledge in his mathematical epistemology. He believed that mathematical existence consisted in construction, that construction was a kind of autonomous 'interior' activity and that mathematical knowledge was therefore ultimately a form of self-knowledge. The key point was summed up well in Weyl's remark (quoted earlier) that arithmetic is a free creation of the human mind and therefore especially transparent to it.

For Brouwer, then, existence claims were to be backed by exhibitions of objects (of the type claimed to exist), where these exhibitions were, at bottom, acts of *creation* by the mathematical subject. He thus departed from Kant's receptive conception of our knowledge of existence claims whose main idea was that judgements of existence must be forced upon a passive cognitive agent and not be the product of its own creative or inventive activity.

Hilbert's position

In the third major 'ism' of the early period, Hilbert's so-called formalism, we find another form of Kantianism, and one which contrasts with the intuitionist position in at least three important respects. The first concerns the conception of our knowledge of existence claims that was adopted. The second concerns the epistemic importance placed on spatial or quasi-spatial intuition in the foundations of mathematics. The third concerns the distinction between genuine judgements and regulative ideals that figured so prominently in Kant's general epistemology.

As was noted above, Brouwer and Weyl conceived of the act of exhibition required for knowledge of an existence claim as, ultimately, an act of creation by the exhibiting subject. The epistemic significance of this act was taken to be based on the special access that a creating subject is supposed to have to his creations. This reduced the epistemic distance between the exhibitor and the exhibited object to that between the willing, acting subject and himself–a distance which, according to romantic idealist lights, is the desirable, optimal or perhaps only tolerable distance to have separating the mathematical knower from the objects of her existential judgements. It also, however, created an irreducible asymmetry between the exhibiting agent and all other agents as regards their knowledge of the exhibited object. Indeed, that was an essential part of the intuitionists' point–namely, that mathematical knowledge is ultimately a form of self-knowledge, and that it is indeed only self-knowledge that possesses the epistemic qualities that we want mathematical knowledge to have.

Hilbert consciously adopted a conception of mathematical knowledge that was more in keeping with what he thought of as the ideal of *objectivity*. He rejected the intuitionists' focus on the inner life and self-knowledge as too subjective a basis on which to found mathematical knowledge. In opposition to the epistemic *individualism* of the intuitionists, Hilbert opted for a more *communitarian* conception of knowledge. Indeed, he believed that it was the very 'task of science to liberate us from arbitrariness, sentiment and habit and to protect us from the subjectivism that already made itself felt in Kronecker's views and find its culmination in intuitionism'.

In Hilbert's view, therefore, there was to be a public domain of objects to which all members of the human (or at least the human-scientific) epistemic community were to have equal access. Hilbert thus stressed the fact that the objects of finitary intuitions were to be *recognizable.* This meant that those intuitions could be re-enacted and confirmed by other intuitions, including other intuitions of the exhibitor's as well as intuitions of non-exhibitors. Consequently, the exhibitor of a finitary object would have no essential epistemic advantage over the non-exhibitor as regards knowledge of the object exhibited.

In Hilbert's finitism, therefore, the 'constructivist' demand that objects claimed to exist be exhibited was to serve the role of taking the object of exhibition out of the exhibitor's head and putting it in the public domain where both exhibitor and non-exhibitor, alike and equally, would be able (and, indeed, required) to judge the object by its intersubjectively confirmable effects. Intuitionistic and finitistic exhibition were thus two very different things. For while the whole intent of the former was to tap the epistemic power of the supposedly special relation of intimacy that a creative subject was believed to have with respect to his own creative acts and intentions, the latter was intended to function as part of a more communitarian scheme of knowledge–a scheme in which the exhibitor had no epistemic advantage over the non-exhibitor. This parity between exhibitor and non-exhibitor is the kind of thing that is necessary if there is to be meaningful epistemic co-operation (e.g. division of epistemic labour) between them and if there is to be a way of monitoring the quality of each contributor's contribution. Epistemic co-operation, in turn, is desirable because through it the total amount of knowledge at the disposal of the individual community member can be expected to exceed that obtainable by that member himself, acting exclusively on his own.

There are, then, we believe, large and important differences between the finitist and intuitionist conceptions of what is to be accomplished through exhibition. So much so, indeed, that we doubt there is much to be accomplished by describing them both as having adhered to a 'constructivist' conception of existence claims.

The second point of contrast between Hilbert and the early constructivists (which, like the preceding one, we can only mention and not develop here) concerns the very different roles they accorded to spatial intuition. Contrary to both the early constructivists (in particular, Kronecker, Brouwer and Weyl) and Kant, all of whom limited spatial intuition to geometry, Hilbert identified a type of spatial intuition which he took to be the basis of *arithmetical* knowledge. This was the position of his so-called 'finitary standpoint' according to which the basis of our arithmetical (and perhaps also our geometrical) knowledge is a kind of a priori intuition in which the shapes or forms of concrete signs are 'intuitively present as immediate experience prior to all thought' and 'immediately given intuitively, together with the objects, as something that neither can be reduced to anything else nor requires reduction'.

Hilbert thus proposed replacing Kant's a priori intuitions of space and time, which he viewed as so much 'anthropological garbage', with a single intuition which was taken to provide a framework of shapes or forms in which our experience of concrete signs was embedded. This intuition, being 'prior to' all thought as its 'irremissible pre-condition', was the source of all our a priori knowledge.

The third main point at which Hilbert's Kantianism contrasted with that of the early constructivists was in its use of certain key elements of Kant's general (as opposed to his specifically mathematical) epistemology. Of particular importance here is Kant's distinction between genuine judgements and regulative ideals. Hilbert took this distinction as the basic model for his division of classical mathematics into a *real* and an *ideal* part. The real propositions and proofs were taken to be the genuine judgements and evidence of which our knowledge is constituted. Ideal propositions, on the other hand, though they served to stimulate and guide the growth of our knowledge, were none the less not considered to be a part of it. They did not describe things that are 'present in the world'. Nor were they 'admissible as a foundation of that part of our thought having to do with the understanding. They corresponded instead to *ideas* 'if, following Kant's terminology, one understands as an idea a concept of reason which transcends all experience and by means of which the concrete is to be completed into a totality'.

Hilbert's ideal sentences are therefore not to be likened to the indirectly verifiable 'theoretical sentences' of a realistically interpreted scientific theory familiar to us from logical empiricist epistemology. Rather, they are to be interpreted instrumentalistically, as having the same general regulative function as Kantian ideas of reason. The objects and states of affairs described in the 'theoretical sentences' of a realistically interpreted science clearly do not 'transcend all experience'. Kant's ideas of reason, on the other hand, do.

Hilbert's ideal propositions thus function as regulative devices. They do not 'prescribe any law for objects, and [do] not contain any general ground of the possibility of knowing or of determining objects as such'. Rather, they are 'merely subjective law(s) for the orderly management of the possessions of our understanding, that by comparison of its concepts it may reduce them to the smallest number'.

Hilbert also followed Kant in maintaining that the use of ideal methods should be *epistemically conservative*. They should, that is, be only more efficient means of producing real judgements which could, nonetheless, in principle (though less efficiently) be developed through the exclusive use of real methods. As Kant put it:

> *Although we must say of the transcendental concepts of reason that they are only ideas, this is not by any means to be taken as signifying that they are superfluous or void. For even if they cannot determine any object, they may yet, in a fundamental and unobserved fashion, be of service to the understanding as a canon for its extended and consistent employment. The understanding does not thereby obtain more knowledge of any object than it would have by its own concepts, but for the acquiring of such knowledge it receives better and more extensive guidance.*

Similarly in Hilbert. Ideal methods, he said, play an 'indispensable' and 'well-justified' role 'in our *thinking*'. They should not, however, be permitted to generate any real result that does not agree with the dictates of real evidence itself. Their role is rather that of enabling us to retain in our reasoning those patterns of inference in terms of which we most readily and efficiently conduct our inferential affairs.

These patterns are the patterns of classical logic. Thus, Hilbert's introduction of the so-called ideal elements was ultimately for the sake of preserving classical logic as the logic of our mathematical reasoning. Introduction of ideal methods was made necessary by the fact that there exist certain real propositions (referred to by Hilbert as *problematic* real propositions) that do not abide by the principles of classical logic. By this it is meant that when these propositions are manipulated by the principles of classical logic, they produce conclusions that are not real propositions. In order to obtain, then, a system that both contains the real truths and also has classical logic as its logic, Hilbert believed it necessary to add the *ideal* propositions. He also believed this to be the minimal modification of real mathematics necessary to restore it to its epistemically optimal classical logical state.

However, in thus restoring mathematical reasoning to its classical logical state, Hilbert observed that the logical operators were no longer being conceived of and employed in a semantical or contentual way as expressions for operations on meaningful propositions. Rather, they were being used in a purely syntactical way as part of a larger computationo-algebraic device for manipulating formulas. As he put it:

> *We have introduced the ideal propositions to ensure that the customary laws of logic again hold one and all. But since the ideal propositions, namely, the formulas, insofar as they do not express finitary assertions, do not mean anything in themselves, the logical operations cannot be applied to them in a contentual way, as they are to the finitary propositions. Hence, it is necessary to formalize the logical operations and also the mathematical proofs themselves; this requires a transcription of the logical relations into formulas, so that to the mathematical signs we must still adjoin some logical signs, say*
>
> & → ~

and or implies not

We thus find here a final step of abstraction from meaning in Hilbert's ideal mathematics–namely, abstraction from the meanings of the logical constants. It was made necessary by the decision to preserve the psychologically natural laws of classical logic as the laws of mathematical reasoning; a decision which, in turn, was the result of trying to preserve the most effective 'canon' available to us for the development of our real mathematical judgements. Ultimately, then, this 'formalism' of Hilbert's, with its radical abstraction from meaning, derived from his Kantian conception of the distinction between the real and ideal propositions according to which he saw the cognitive or epistemic value of the ideal elements as residing in their utility as instruments for extending our real judgements.

At the same time, however, it is also almost certainly this same radical abstraction from meaning that has tempted so many to misdescribe Hilbert's position as a *formalist* position in the sense of one which sees mathematics as a 'game' played with symbols. The idea behind this 'game' imagery, presumably, is that when every trace of meaning is obliterated, as in Hilbert's view of ideal mathematics, mathematics becomes ultimately a symbol-manipulation activity conducted according to certain rules; rules which, moreover, answer not to anything as serious as a concern for objective truth, but rather only to such less weighty concerns as a subjective or psychological urge for logical unity in our thinking. Even such a well-positioned and astute interpreter of Hilbert as Weyl eventually succumbed to the temptations of this description of Hilbert's views. In our opinion, however, such an interpretation fails to take account both of Hilbert's overall Kantian epistemology and of certain quite specific remarks he himself made regarding the syntactical character of ideal reasoning. Hence, while we see no particular reason to deny the title of formalism to Hilbert's position, we would nonetheless insist that it is formalism of a quite different kind than the 'game-played-with-symbols' formalism. Hilbert stated his view forcefully in the following remark.

The formula game that Brouwer so deprecates has, besides its mathematical value, an important general philosophical significance. For this formula game is carried out according to certain definite rules, in which the *technique of our thinking* is expressed. These rules form a closed system that can be discovered and definitively stated. The fundamental idea of my proof theory is none other than to describe the activity of our understanding, to make a protocol of the rules according to which our thinking actually proceeds. Thinking, it so happens, parallels speaking and writing: we form statements and place them one behind another. If any totality of observations and phenomena deserve to be made the object of a serious and thorough investigation, it is this one.

This suggests that the rules of the so-called 'game' of ideal reasoning are nothing other than the basic laws of human thought. The heart of Hilbert's proof theory, and the heart of the 'formalism' of his later thought, was thus the belief that much of human mathematical thought is, at bottom, formal-algebraic or syntactical in character. Indeed, as he remarked elsewhere, the custom in mathematical thought, and in scientific thought generally, is to make 'application of formal thought-processes *(formaler Denkprozesse)* and abstract methods'. In fact, he noted, Even in everyday life one uses methods and conceptual constructions which require a high degree of abstraction and which only become intelligible by means of an unconscious application of the axiomatic method. Examples are the general process of negation and the concept of infinity.

What emerges from all this is an idealistically oriented formalism whose goal is to locate and defend (as a sound regulative device) the basic 'forms' of human thought. These forms of thought, which might be thought of as *theory-forms,* represent high-level commonalities of form that our thinking about a wide variety of subjects share. It is less clear whether, in speaking of 'the techniques of our thinking' as being expressible in a 'closed system' of rules that can 'be discovered and definitively stated', Hilbert meant a single system of rules which gives a general algebra of thought, or whether he was thinking of a plurality of different theory-forms, the repository of which is classical mathematics. In either case, however, we obtain a formalism whose forms are fundamentally forms of *thought*-forms of thought, moreover, which, despite their syntactical character, are none the less deep expressions of the nature of human reasoning and therefore much more than a mere 'playing' of a 'game' with symbols.

For Hilbert, then, the ideal methods of thinking constituted a logical mould to whose contours our minds are shaped in their inferential dealings. This makes their use inviting, if not unavoidable. But inviting or not, the legitimacy of ideal reasoning still depends on the satisfaction of a certain condition–namely, its consistency, or, more specifically, its *finitarily demonstrable consistency with the finitarily provable propositions.* As is well-known, however, it is precisely the satisfaction of this requirement that was called into question by Godel's discovery of his celebrated incompleteness theorems in 1931.

The proofs of these theorems featured a technique (commonly referred to as the 'arithmetization' of metamathematics) for representing the concepts and statements of the metamathematics of a given formal system T in that portion of a formal theory of arithmetic that contains the elementary theory of recursive operations on the natural numbers. For present purposes, the important feature of this fragment of arithmetic is that it appears to be contained in what Hilbert regarded as the *finitary* part of number theory. For that reason, it also appears to be contained in those ideal theories of classical mathematics which it was Hilbert's concern to defend as legitimate.

What Gödel was able to show was first that for any formal system T containing the elementary fragment of arithmetic spoken of above, if T is consistent, then there is a sentence G of the language of T such that neither G nor $\neg G$ is a theorem of T. Using the proof of this *first incompleteness theorem,* Gödel was then able to prove a *second incompleteness theorem* by formulating in T a sentence, *Con*T, of which there is reason to say that it expresses the claim that T is consistent, and of which it can be proven that it is not provable in T if T is consistent. From this second theorem, and the assumption that T contains finitary arithmetic, it is then concluded that the consistency of T is not provable by finitary means. From this conclusion it is in turn inferred that no system I of ideal mathematics that contains T is such that its real-consistency can be proven by finitary means, and from this, finally, it is concluded that Hilbert's intended defence of the ideal reasoning of classical mathematics cannot be carried out.

In the beginning, Godel shied away from this conclusion, maintaining (with characteristic caution) that his second theorem did 'not contradict Hilbert's formalistic viewpoint' since 'it is conceivable that there exist finitary proofs that *cannot* be expressed' in the classical systems for which that theorem had been proved to hold. Eventually, however, he was persuaded by Bernays that these reservations were unwarranted, at which time he then accepted the view that his second theorem did indeed refute Hilbert's programme as it was originally conceived by Hilbert.

The Later Period

This completes our discussion of the developments of the early period. We turn now to the later period (i.e. the period after 1931), where we shall begin by considering the changes it brought to the 'isms' of the early period.

Hilbert's Formalism

The above-stated argument against Hilbert's programme using Gödel's theorems gained nearly universal acceptance in the later period and has indeed become a commonplace amongst twentieth century philosophers of mathematics. The few challenges there have been to it have been mainly of two types: (i) those that seek to revive Hilbert's programme by arguing for a less restrictive conception of fmitary evidence (and, hence, a more potent base from which to launch the search for a finitary proof of the real-consistency of ideal mathematics) than was originally intended by Hilbert; and (ii) those that seek a more restricted body of ideal methods whose real-consistency needs to be proven.

The means used in giving the proof of real-consistency required by Hilbert's programme ought to be extended to means reaching beyond that which is formalizable in what has commonly been recognized as the natural formalization of Hilbert's finitary standpoint (namely, the theory known as *Primitive Recursive Arithmetic*, or PRA). Among those, some (e.g. Gentzen [2.61], Ackermann [2.1], and, on one reading, Godel [2.64]) have questioned the correctness of identifying the finitary with what is formalizable in PRA, arguing that finitary reasoning extends well beyond that which is formalizable in PRA, and includes such things as forms of transfinite induction which go beyond even what is provable in ordinary first-order Peano arithmetic (PA).

The basic idea of this line of thought is that there are types of reasoning which (a) are not codifiable in PRA, but which nonetheless (b) share the same characteristics believed to give finitary evidence its distinctive epistemological credentials, and which (c) allow us to establish the consistency of much of the ideal reasoning of classical mathematics that cannot be secured by means of proofs formalizable in PRA. It is therefore argued that an extension of what is to be counted as admissible reasoning in constructing the required consistency proofs of the various ideal systems of classical mathematics is in order, and that a significant partial realization of Hilbert's original aims can thus be attained.

Others (e.g. Kreisel [2.90] and Feferman [2.39]) in the first camp have argued not so much for a reconsideration of what should be counted as finitary evidence, as for a liberalization and refinement of what are the epistemically gainful means of proving consistency, whether or not they are properly classifiable as finitary. The basic idea here is that the simple distinction between real and ideal methods does not begin to do justice to the rich scheme of gradations in epistemic quality that separate the various kinds of evidence available for use in metamathematical proofs. Hence, this simple distinction should be replaced by a more refined scheme which distinguishes not only the finitary and the non-finitary, but also the various 'grades' of both constructive and non-constructive methods (and the various reducibility relations which exist between the distinguished kinds of non-constructive methods and the distinguished kinds of constructive methods). When this is done, it is claimed, results amounting to a substantial partial realization of a generalized Hilbert's programme can be achieved.

The same basic conclusion is arrived at by a quite different line of reasoning in the so-called programme of 'reverse mathematics' of Friedman and Simpson. The strategic idea of this programme is basically the opposite of that of Kreisel and Feferman. It does not aim at *beefing up* the methods available for constructing the requisite consistency proofs, but rather at *cutting down* the systems of ideal reasoning whose consistency needs proving. This is to be done by giving a more exact characterization of the core of ideal reasoning that is truly indispensable to the reconstruction of the essential results of classical mathematics.

The reverse-mathematical revision of Hilbert's programme thus begins by isolating those results of classical mathematics that are believed to constitute its 'core'. It then seeks to find the weakest possible natural axiomatic theory capable of formalizing this core. The hope is that this minimal system will eliminate unnecessary strength present in the usual axiomatizations of the core (generally, some version of second-order arithmetic) and therefore that its real-consistency will prove to be more susceptible to finitary proof than that of the usual systems.

So far, significant partial progress along these lines has been achieved. In particular, it has been shown that (i) there is a certain sub system (known as WKL 0) of PA^2 (i.e. second-order Peano arithmetic) which embodies a sub stantial portion of classical mathematics; (ii) all the II1 theorems (i.e. theorems equivalent to some formula of the form

'$\forall x\ \phi x$'

where

'ϕx'

is a recursive formula) of WKL 0 are provable in PRA; and (iii) the proof of (ii) can itself be given in PRA. Assuming the codifiability of finitary reasoning in PRA and the importance of the III class of real truths, this amounts to a finitary proof of the real-consistency of an important part of classical ideal mathematics. This, in turn, would constitute a significant partial realization of Hilbert's programme.

In addition to these two alternatives, it is possible to describe, at least in philosophical outline, a third approach which seems in certain important respects to be closer to Hilbert's original ideas than either of them. The key element of this third alternative, which is absent from each of the other two approaches just described, takes its lead from the Kantian character of Hilbert's conception of ideal mathematics. In particular, it stresses the point that Hilbert's ideal methods, like Kant's ideas of pure reason, are recommended solely by the efficiency that their instrumental use is supposed to bring to the development of our real judgements.

This means, among other things, that ideal propositions and inferences that fail to bring with them discernible improvements in efficiency (when compared to their real counterparts proving the same results) do not belong to that part of ideal mathematics that need, in principle, be defended by Hilbert. In other words, ideal elements that fail in any significant way to increase the efficiency of the development of our real knowledge have, in principle, no claim to be included among those ideal elements whose real-consistency the Hilbertian must defend. Therefore, in identifying the elements (i.e. axioms and rules of inference) of an ideal system I for whose defence the Hilbertian is to be held accountable, it must be borne in mind that they must figure in some significant way in the production of efficiency; that is, each must be an essential ingredient in some ideal derivation ?1 of a real theorem ôR such that (i) ?1 (together with the necessary metamathematical proof of soundness for I) is more efficient than any real proof of ôR, and (ii) ?1 is the only derivation in I that significantly improves upon the efficiency of the real proofs of ôR. If an item (e.g. an axiom, rule of inference, etc.) of I possesses none of the virtues of efficiency for which ideal elements are in general prized, then, in principle, it can and should be eliminated from I. With all such eliminations made, one would expect the prospects for a finitary proof of It's consistency to have been improved. Therefore, the question of whether an ideal system is comprised wholly of elements that are essential in the above-indicated sense ought to be of prime importance in determining the make-up of those ideal theories for whose defence the Hilbertian is ultimately taken to be responsible.

Yet, despite its clear importance for the proper reckoning of the ultimate responsibilities and prospects of Hilbert's programme, this question has been either ignored or overlooked by those writing on the subject. Simpson, for example, readily admits that the proofs of standard theorems in WKL 0 and

$$WKL_0^+$$

are sometimes 'laborious' and 'much more complicated than the standard proof(s)'. Yet he takes no notice of the potential this feature of reverse mathematics has to undo its entire rationale for effecting a partial realization of Hilbert's programme. To the extent that the proofs in WKL 0 and

WKL_0^+

are *more* laborious than the 'standard' proofs of the same ideal

It would be unfair, however, to lay too much blame on the reverse mathematicians. For the questions they have neglected have been generally neglected by those writing on Hilbert's programme. This includes philosophers, too (indeed, perhaps primarily), and not just logicians. All have failed properly to emphasize two fundamental points: (i) that the prospects for Hilbert's programme can adequately be assessed only when a suitably accurate means of comparing the complexity of real and ideal proofs has been developed and those systems containing the gainful ideal proofs have been identified; and (ii) the complexity metric figuring in (i) is capable of measuring not only the kind of complexity (call it *verificational complexity*) that is encountered when one sets about the task of determining of a *given* item whether or not it is an ideal proof of a certain kind, but also, and, indeed, primarily, a kind of complexity (call it *inventional complexity*) which constitutes the complexity involved in *discovering* an ideal proof of the desired kind in the first place. Failure to appreciate points (i) and (ii) has, it seems to me, led to inadequate attention being paid to the development of *appropriate* metrics for measuring the complexity of ideal proofs and comparing the complexity thus measured to that for the corresponding real proofs. Without the development of such a theory of complexity, however, it does not seem to me possible to render a compelling final assessment of Hilbert's programme–and by that I mean the programme of Hilbert's *original* philosophical conception.

Logicism

Logicism re-emerged in the 1930s and 1940s as the favoured philosophy of mathematics of the logical empiricists. I say 're-emerged' because the positivists did not develop a logicism of their own in the way that Dedekind,

Frege and Russell did. Rather, they simply appropriated the technical work of Russell and Whitehead (*modulo* the usual reservations concerning the axioms of infinity and reducibility) and attempted to embed it in an overall empiricist epistemology.

This empiricist turn was a fairly novel development in the history of logicism, and it represented a radical departure from both the original logicism of Leibniz, which was part of a larger rationalist epistemology, and the more recent logicism of Frege, which was strongly critical of empiricist attempts to accommodate mathematics. It was, perhaps, less at odds with Russell's logicism with its imputation of a common methodology linking mathematics and the empirical sciences.

Like all empiricists, the logical empiricists, too, struggled with Kant's idea that mathematics is immune to empirical revision. More accurately, they struggled with the Kantian problematic concerning how to account for the apparent certainty and necessity of mathematics while at the same time being able to explain its seeming robust informativeness. Their choice of strategies for trying to accommodate these two data was to empty mathematics of all non-analytic content while, at the same time, arguing that analytic truth can be 'substantial' and non-self-evident.

The logical empiricists thus sacrificed the strict empiricist claim that all knowledge is evidensorily based on the senses. Their empiricism was, therefore, a liberal empiricism, an empiricism making use of a distinction like that of Hume's between 'relations of ideas' and 'matters of fact'. The exact distinction they utilized was one calling for the separation of those propositions whose truth or falsity is determined by the meanings of their constituent terms (and is therefore independent of contingent fact) and those propositions whose truth or falsity is dependent upon contingent, empirical matters of fact. They then appealed to this distinction in arguing that the truths of logic are analytic in character. And from this, and the technical work of Russell and Whitehead, they concluded that the truths of mathematics are analytic.

Thus, though they accepted the traditional Kantian idea that mathematical judgements are immune to empirical revision, and though they made central use of something like Kant's analytic/synthetic distinction in formulating their account of mathematics, the logical empiricists none the less rejected the distinctive thesis of Kantian mathematical epistemology; namely, that knowledge of mathematics is synthetic a priori in character. Indeed, the denial that *any* knowledge is synthetic a priori in character was a central ingredient of the logical empiricists' epistemological outlook.

Their mathematical epistemology came under heavy attack by Quine in the 1950s. Quine's attack was based on a criticism of the pivotal distinction of the empiricists between analytic and synthetic truths. According to Quine, the basic unit of knowledge or judgement–the basic item of our thinking that is tested against experience–is science as a whole. Since mathematical and logical statements are inextricably interwoven parts of the larger body of science, they therefore too, at least to some extent, must derive their confirmation or disconfirmation from empirical sources. Consequently, the statements of logic and mathematics cannot rightly be regarded as true by virtue of meanings alone, if by that there is intended some contrast with statements regarded as true in virtue of the facts. A distinction between truths of meaning (i.e. analytic truths) and truths of fact (i.e. synthetic truths) cannot therefore be maintained. Yet, without some such distinction, the logicism of the logical empiricists does not have a hope of succeeding.

Within a relatively brief period of time, this critique of Quine's became a major influence in the philosophy of mathematics and, under the weight of that influence, the logicism of the logical empiricists began to sink into oblivion. There have, however, been a few attempts to revive (or, perhaps better, to exhume) logicism along other lines. The most systematic and detailed (if, perhaps, not the most convincing) of these is that given in the two-volume work of David Bostock. This is not, however, so much a defence of logicism as it is an attempt to determine a best case for it, so that its plausibility as a philosophy of arithmetic might finally be judged. Indeed, he ends up concluding that logicism is of strictly limited viability as a philosophy of arithmetic. The main point of his argument is that there is no unique reduction of arithmetic to logic, and that this creates problems for any logicism (like, say, that of Frege or Russell) that wants to identify numbers with *objects*.

More nearly defences are the recent attempts by Hilary Putnam, Harold Hodes, Hartry Field and Steven Wagner to establish the plausibility of modified forms of logicism. Putnam and Hodes both offer defences of a type of logicist position also known as 'ifthenism' or 'deductivism'. The former argues that, though statements of existence in mathematics are generally to be seen as statements asserting the existence of structures, they are not to be taken as assertions of the actual existence of these structures, but only of their possible existence. Existence statements are therefore, at bottom, logical claims, and they are to be verified by generally logical means (and, in particular, by syntactical consistency proofs).

Hodes takes a somewhat different approach, arguing, in a manner reminiscent of Frege, that arithmetic claims can be translated into a second-order logic in which the second-order variables range over functions and concepts (as opposed to objects). In this way, commitment to sets and other specifically mathematical objects can be eliminated, and, this done, arithmetic may be considered a part of logic.

Field offers what might be regarded as an epistemological form of logicism. He is concerned to defend the claim that mathematical knowledge is (at least largely) logical knowledge. He also argues, contrary to Quine and Putnam (*see* below), that one can be a realist with respect to physical theory without being a realist with respect to mathematics.

Field begins his argument that mathematical knowledge is logical knowledge by defining mathematical knowledge as that knowledge which 'separates a person who knows a lot of mathematics from a person who knows only a little mathematics'. He then goes on to claim that what separates these two kinds of knowers is '*not* that the former knows many and the latter knows few' of such claims as those for which mathematicians commonly provide proofs of (i.e., of those claims, such that there are prime numbers greater than a million). Rather, he goes on to say, 'insofar as what separates them is knowledge at all, it is knowledge of various different sorts'.

Some of it is empirical knowledge (e.g. knowledge of what is commonly accepted by mathematicians and what they regard as the proper starting point for an inquiry). The bulk, however, is knowledge 'of a purely logical sort-even on the Kantian criterion of logic according to which logic can make no existential commitments'. In the end, Field concludes, mathematical knowledge for the most part comes down to knowledge that certain sentences do and certain other sentences do not follow from a given set of axioms.

Generally, such knowledge would either be understood in a semantical way (as knowledge about a class of models) or in a syntactical way (as knowledge about a class of formal proofs or derivations). Neither of these, however, qualifies as logical knowledge in the sense Field wants. For since models are just a particular kind of mathematical object, knowledge of them must be just a particular kind of mathematical knowledge. Similarly for syntactical derivations. For whether they are conceived of abstractly or concretely, they cannot be known to exist on purely logical grounds, since (so the reasoning goes) pure logic cannot assert the existence of things.

Field identifies the chief task facing this analysis as that of accounting for the applicability of mathematics to physics, and he divides this task into two sub-tasks, namely: (a) that of showing the mathematics applied to be 'mathematically good'; and (b) that of showing that the physical world is such as to make the mathematical theory particularly useful in describing it. Both tasks must be carried out without appeal to the truth of (any parts of) mathematics if Field's logicist ideal is to be met, and it is to the demonstration of this that Field's writings on the subject are primarily devoted.

For present purposes, however, we shall restrict ourselves to a question concerning the overall place of logical reasoning in mathematical proof. This, we noted earlier, was a point of central concern in Kant's mathematical epistemology, and was taken over as a principal motivating factor of the intuitionist epistemologies of Brouwer and Poincaré. Their belief, and the crux of their disagreement with logicism, was that mathematical and logical inference are fundamentally different–that the former is not only not reducible to the latter, but that it is indeed antithetical to it. As Poincaré put the point: mathematical reasoning has a kind of creative virtue that distinguishes it from the epistemically more 'sterile' reasoning of logic; the logical reasoner's grasp of mathematics is like the grasp of elephanthood that naturalists would have were their knowledge to be restricted entirely to what they had observed through microscopic examination of their tissue.

When the logician shall have broken up each demonstration into a multitude of elementary operations, all correct, he still will not possess the whole reality; this I know not what which makes the unity of the demonstration will completely escape him. In the edifices built up by our masters, of what use to admire the work of the mason if we cannot comprehend the plan of the architect? Now pure logic cannot give us this appreciation of the total effect; this we must ask of intuition.

Poincare therefore believed in a distinctively *mathematical* kind of reasoning. Without it one cannot account for the manifest differences in epistemic condition which separate the logical from the genuinely mathematical reasoner. He would, therefore, have denied Field's assumed starting point that mathematical knowledge is largely logical knowledge and also his claim that the modalities pertaining to mathematical reasoning are *logical* rather than distinctively mathematical in character. Both points, we believe, raise serious challenges for Field's position.

Steven Wagner offers both a different conception and a different defence of logicism. He seeks to root mathematics in what he takes to be the universal needs and urges of (ideally) rational agents. His argument begins with the presumption of an idealized rational enquirer who seeks not only bodily survival but also understanding. He then goes on to say that counting is indispensable to such a being both in the sense of being partially constitutive of that person's rationality and in the sense of being necessary to his or her assimilation and processing of the elementary data of experience and thought. Nobody, he maintains, can get by without answering a host of 'How many?' questions. Thus, the rational agent must develop a system of counting.

From a system of counting there will eventually emerge a system of calculation. This is so because (i) calculation is essentially a refinement of counting (i.e. it functions to advance those cognitive and bodily ends that are served by the capacity to count); and (ii) there is a rational urge to improve those capacities that one possesses, but possesses in what is perhaps only too rudimentary a form.

Thus, from counting, the ideally rational agent passes to calculation. And from calculation, the argument continues, he or she will pass to something like number theory. This is due to the facts that (i) part of the urge to understand consists of an urge to unify, simplify and generalize; and (ii) efforts to unify, simplify and generalize one's system of arithmetic calculation will inevitably lead one to something like the arithmetic of the natural numbers. Therefore elementary number theory may be described as a kind of rational necessity.

As a final stage of rational development, the continued need and capacity to simplify, unify and generalize will eventually lead the ideally rational enquirer to some form of analysis and set theory. Wagner's 'second generation' logicism thus states that (i) any ideally rational enquirer will be under pressure to develop forms of arithmetic and set theory; and that (ii) the theorems of such theories as are developed in response to this pressure will be analytic in the sense that *any rational being would have a reason to accept them.*

Wagner's position differs in significant ways from both the metaphysical logicism of Frege and the methodological logicism of Russell. It also differs from Field's logicism-particularly over the question of the in-principle dispensability of mathematics as a device for processing empirical information. Field believes that it is dispensable. Wagner, on the other hand, appeals to an alleged explanatory role that at least certain mathematical theories (e.g. elementary number theory, analysis and set theory) play, and he treats that explanatory role as part of the ultimate justification for those theories. He thus treats mathematics as ultimately indispensable for the full cognitive processing (i.e. the full explanation) of empirical information.

Intuitionism

Turning finally to post-Brouwerian intuitionism, there are two main developments to note: (i) the various attempts, initiated by Heyting's work, to formalize intuitionistic logic and mathematics and to develop it into something roughly comparable in power to classical mathematics; and (ii) the more recent attempts by Dummett and his followers to derive a philosophical justification for intuitionism from a kind of Wittgensteinian conception of the meanings of mathematical sentences.

Since (i) is only very loosely connected with philosophical concerns, we shall say little about it here. The reader desiring a relatively up-to-date presentation of technical findings may consult any of a number of recent surveys of the subject. We shall confine our discussion of (i) to the idea that seems to have been at the centre of Heyting's attempts to formalize intuitionistic reasoning; namely, the supposed distinction between 'a logic of existence' and 'a logic of knowledge'–a distinction Heyting regarded as fundamental to intuitionistic thought.

What Heyting meant by a 'logic of existence' is a logic of statements about objects whose existence is to be understood as being independent of human thought. Intuitionists, and other constructivists too, for the most part, reject the idea of a realm of thought–independent mathematical objects. Likewise, they reject the idea that mathematical propositions are true or false independently of human thought. They hold, instead, the view that mathematical propositions express the results of certain kinds of mental constructions. A logic of such propositions–a 'logic of knowledge', in Heyting's terminology–is therefore one whose theorems express relations between mental constructions; specifically, relations of latent containment among them (where one construction is taken latently to contain another just in case going through the process of effecting the one would automatically either effect the other or put one in a position to effect it).

Given this general outlook, Heyting argued, there is no essential difference between logical and mathematical theorems. Both serve essentially to affirm that one has succeeded in performing mental acts satisfying certain conditions. The former are distinguished only by their relatively greater generality.

Intuitionist logic was thus intended by Heyting to serve as an articulation of those most general patterns of latent containment relating our mental mathematical constructions. Since, however, our mental constructional activities constitute 'a phenomenon of life, a natural activity of man', containment relations between them are not conceptually based but rather based on the relations which bind our activities together *behaviourally,* as it were. Thus, a logical theorem to the effect that we know a proof of A only if we know a proof of B would express neither a conceptual connection between our knowledge of a proof of A and our knowledge of a proof of B, nor, except *per accidens,* a conceptual connection between A and B. Rather, it would express a natural fact concerning our (perhaps idealized) mental constructional life, namely, that it is characterized by a disposition to transform proofs of A into proofs of B. As Heyting himself put it, 'mathematics, from an intuitionistic point of view, is a study of certain functions of the human mind'. As such, it is akin to history and the social sciences.

Michael Dummett has offered both a different conception

and a different defence of intuitionism. He conceives it as a view concerning what is to be regarded as the proper *logic* of mathematics. He argues that an adequate account of the meanings of mathematical propositions reveals that it is intuitionist logic (i.e. the logic set out by Heyting) that is the proper logic of mathematics. This account bears certain affinities to the views on meaning set out by Wittgenstein in his *Philosophical Investigations*. In particular, like Wittgenstein's view, it equates the meaning of a sentence with its canonical use. In mathematics, Dummett believes, canonical use consists in the role that an assertion plays in the central activity of proof. Hence, to know the meaning of a mathematical sentence is ultimately to know the conditions under which it would be proved or refuted.

Dummett's intuitionism is thus far removed from that of Poincaré and Brouwer both in substance and motive. Its defence is also, we believe, open to certain doubts. Dummett argues that since (i) 'that knowledge which, in general, constitutes, the understanding of language must be implicit knowledge'; and (ii) 'implicit knowledge cannot meaningfully be ascribed to someone unless it is possible to say in what the manifestation of that knowledge consists-there must be an observable difference between the behaviour or capacities of someone who is said to have that knowledge and someone who is said to lack it' (Ibid.)–that it therefore follows that (iii) 'a grasp of the meaning of a statement must, in general, consist of a capacity to use that statement in a certain way, or to respond in a certain way to its use by others'.

Premiss (ii), we believe, is inappropriate as a premiss in an argument for (iii). For it seems to beg the crucial question, namely, that concerning whether implicit knowledge (and hence knowledge of meaning) is ultimately to be taken as consisting in a capacity for behaviour or in, say, something like one's being in a mental or psychological or neural state that *underlies* such behaviour. Everyone, it would seem, must agree that (ii') implicit knowledge cannot legitimately be ascribed to someone unless such ascription is warranted by the best total explanation of his or her observed behaviour. But (ii') carries with it no guarantee that, as (ii) requires, an ascription of implicit knowledge be traceable to some specific piece or pieces of speaker behaviour or, indeed, that it will even be traceable to the entire corpus of the speaker's behaviour. Speaker behaviour will, after all, typically under-determine its best total explanation. Nor is there any reason to believe in advance that the best total explanation of the speaker's behaviour will not differ from its rivals in its ascriptions of implicit knowledge to the speaker.

Thus, to avoid begging the question, premiss (ii) of Dummett's argument would seemingly have to be replaced by something like (ii'). Such replacement, however, would block valid passage to (iii), and (iii) is at the heart of Dummett's defence of intuitionism. To salvage his defence would therefore seemingly require finding something like our (ii') that would (in the company of premiss (i), of course) validly imply (iii). It is not at all clear that this can be done, however.

As noted above, Dummett's defence of intuitionism appeals to certain ideas of Wittgenstein's, or, at any rate, ideas that Dummett and others have attributed to Wittgenstein. It would, however, we believe, be a mistake to identify Wittgenstein's ideas on the philosophy of mathematics too closely with intuitionism. For though it does bear affinities to those forms of constructivism which stress the autonomy of mathematics as a human creation (where creation is understood to consist ultimately of acts of will or decision), it also bears certain affinities to a more conventionalist interpretation. Moreover, in the end, it resists categorization as *either* a constructivist or a conventionalist philosophy.

Wittgenstein's central idea seems to have been that there is a basic kind of construction in mathematics known as a *Beweissystem.* A *Beweissytem* is made up of proofs and theorems, but its theorems function as rules of logical syntax rather than descriptive truths concerning the terms of the system. These rules constitute autonomous acts of will by which are laid down the rules according to which we agree to play a certain 'language game' involving the terms introduced by the system. Mathematics as a whole, Wittgenstein maintained, is a 'motley' of such *local* activities or games.

For Wittgenstein, then, mathematical proofs (including even the simplest ones) do not play the role of *compelling* us to accept their conclusions. Nor are mathematical propositions in any other way 'forced' upon us as being true. 'Acceptance' of a mathematical statement does not therefore represent an acknowledgment of its truth, but rather represents a *decision* on our part to count something as a convention of a certain language game. Similarly, proof does not function to *remove* doubt through penetration of the truth, but rather to *exclude* doubt as a logical possibility through the establishment of a theorem as a *norm* governing a certain language game.

Thus, though Wittgenstein saw mathematics as essentially a human creation, he did not understand this in the *descriptive* way that constructivists traditionally have understood it. Nor did his normative conception of our mathematical creations lend itself to any of the usual conventionalist interpretations. For those all call for the reduction of mathematical truths to logical truths and this requires the 'co-operation' of the logical forms of mathematical truths, so to speak.

Logical form does not, however, appear to form the same kind of constraint on the institution of norms or rules. Moreover, even if it did, there is a big difference between a statement's being a logical truth and its being a norm of logical syntax. The former has to do with whether a statement passes a certain test of truth invariance (i.e. that it be stable under a certain variety of different semantical interpretations of its semantically variable parts), while the latter has to do with whether something is to be seen as one of the *rules* constituting a certain *rule-governed activity*. Sentences failing the requisite tests of the truth-invariance test could none the less, it seems, still at least in principle play the role of rules of a linguistic game. Consequently, it seems that Wittgenstein is not well classified as a traditional conventionalist, though there is a conventionalist element in his philosophy.

Later Developments

Among later developments not having primarily to do with the three 'isms', perhaps the major one is that stemming from Quine's criticism of the logical empiricists in the 1940s and 1950s mentioned above. This criticism resulted in a new empiricist philosophy of mathematics shorn of even those few Kantian ideas that were retained by the logical empiricists–and this new empiricism has been perhaps the dominant theme in the philosophy of mathematics ever since.

The logical empiricists, as was noted earlier, retained Kant's commitment to the necessity of mathematical truth (though they conceived of this necessity as essentially consisting in immunity from empirical revision). Indeed, it was precisely for the sake of accommodating this 'datum' of mathematical epistemology that the logical empiricists put so much effort into nurturing and preserving the Kantian distinction between analytic and synthetic judgements. In their view, only such a distinction could sustain the division of mathematical judgements into those that are and those that are not subject to empirical revision.

Quine and Putnam swept even this final Kantian datum aside, offering instead a general empiricist epistemology in which all judgements, those of mathematics and logic as well as those of the natural sciences, are seen as evidentially connected to sensory phenomena and therefore subject to empirical revision. Central to their argument was an observation borrowed from Duhem; namely, that in order to connect science with sensory evidence, logic and mathematics are inevitably required. They are therefore part of what gets confirmed when connection between a theory and a confirming phenomenon is made, and part of what gets disconfirmed when connection between a theory and a disconfirming phenomenon is made. They are, in sum, of an epistemological piece with natural science and, so, broadly speaking, empirical.

To accommodate the lingering conviction that there is at least some difference regarding sensitivity to empirical evidence between mathematics and logic, on the one hand, and natural science, on the other, Quine argued that while both are subject to empirical revision, the precise extent or degree to which this is true is different in the two cases. He backed this view with a pragmatic conception of rational belief revision, and a view of the totality of our cognitive holdings according to which it forms a 'web' whose various parts are interdependent and ordered with respect to their importance to the preservation of the general structure of the web. At the centre of the web lie logic and mathematics, and the beliefs of natural science and common sense for the most part fan out from this towards the periphery where the entire scheme encounters sensory experience.

According to Quine's pragmatic conception of belief revision, it is to be governed by a concern to at all times maximize our overall predictive and explanatory power. A scheme of beliefs which optimizes predictive and explanatory power also optimizes our ability cognitively to cope with our environment(s), and production of such a scheme is generally aided by policies of revision which minimize, both in scope and severity, the changes that are made to a previously successful conceptual scheme in response to recalcitrant experience. If the beliefs of mathematics and logic are therefore typically less subject to empirical revision than are the beliefs of natural science and common sense, this is because revising them generally (albeit not, in Quine's view, invariably) tends towards greater upheaval in the conceptual scheme than does revision of our common sense and natural scientific beliefs. Quine therefore accommodates the traditional Kantian datum that mathematics is necessary by treating logic and mathematics not as being altogether impervious to empirical revision, but only as being more so, on average, than either natural science or common sense. Quine thus merges the epistemologies of mathematics and natural science into a single, albeit quantitatively differentiated, empiricist whole.

In merging mathematics and science into a single explanatory system, Quine's empiricism also induces a realist or platonist view of mathematics. It sees the world as populated by those entities needed to staff the best theory of the totality of our experience. These include not only the medium-sized objects of ordinary experience and the theoretical entities of our best current physical science, but also mathematical entities, since, as was noted above, mathematical claims play an integral role in our best total theory of experience.

Quine's views have been challenged on various grounds. Field, for example, has challenged Quine's claim that the roles played by natural science and mathematics are essentially the same and cannot be differentiated. On his view there is an immense difference between the role played by mathematics and the role played by natural science in our overall conceptual scheme. Mathematics, he believes functions basically as a logic, and its theorems do not make claims that are substantive in the way that the laws of natural science are.

Another criticism is that Quine's epistemology leaves unaccounted for those parts of mathematics that are *not* somehow involved in the explanation of sensory experience. It is not clear, however, how serious an objection this is since it is not clear how much, if any, of even the least elementary, most abstract mathematics would play *no* contributory role in the simplification and explanation of sensory experience.

A final criticism is that by Parsons, who has argued that treating the *elementary* arithmetical parts of mathematics (e.g. the truth that 7 + 5 =12) as being on an epistemological par with the hypotheses of theoretical physics fails to capture an epistemologically important difference regarding the different kinds of evidentness or 'obviousness' displayed by the two. The kind of evidentness displayed by an elementary arithmetic proposition like '7+5=12' is not the same as that displayed by even such highly confirmed physical hypotheses as 'The earth moves about the sun', says Parsons. To put it roughly, the latter is more highly *derivative* than the former. Consequently, Parsons concludes, it is not plausible to regard the two claims as based on essentially the same kind of evidence and, so, the empiricist epistemology of Quine and Putnam must be regarded as being of questionable adequacy for at least the more elementary parts of mathematics.

In addition to Quine and Putnam, there have been others who have suggested different kinds of mergings of mathematics with the natural sciences. Of those, some have been empiricists, others not. Kitcher, for instance, has presented a generally empiricist epistemology for mathematics in which *history* and *community* are important epistemological forces. Gödel, too, offered an epistemology in which mathematical justification was conceived along lines structurally similar to those of justification in the natural sciences.

We do have something like a perception also of the objects of set theory, as is seen from the fact that the axioms force themselves upon us as being true. I don't see any reason why we should have less confidence in this kind of perception, i.e., in mathematical intuition, than in sense perception.

It should be noted that mathematical intuition need not be conceived of as a faculty giving an *immediate* knowledge of the objects concerned. Rather it seems that, as in the case of physical experience, we *form* our ideas also of those objects on the basis of something else which is immediately given. Only this something else is *not,* or not primarily, the sensations. That something besides the sensations actually is immediately given follows from the fact that even our ideas referring to physical objects contain constituents qualitatively different from sensations or mere combinations of sensations, e.g., the concept of object itself, whereas, on the other hand, by our thinking we cannot create any qualitatively new elements, but only reproduce and combine those that are given. Evidently, the 'given' in underlying mathematics is closely related to the abstract elements contained in our empirical ideas. It by no means follows, however, that the data of this second kind, because they cannot be associated with actions of certain things upon our sense organs, are something purely subjective, as Kant asserted. Rather they, too, may represent an aspect of objective reality, but, as opposed to the sensations, their presence in us may be due to another kind of relationship between ourselves and reality.

Godel then went on to claim that this use of and need for intuition obtains both in high-level abstract areas of mathematics such as set theory, and in the elementary areas of mathematics such as finitary number theory. He noted, moreover, that even without appeal to intuition, mathematics, like the natural sciences, makes use of what are essentially inductive means of justification.

It has no intrinsic necessity at all, a probable decision about its truth is possible also in another way, namely, inductively by studying its 'success'. Success here means fruitfulness in consequences, in particular in 'verifiable' consequences, i.c., consequences demonstrable without the new axiom, whose proofs with the help of the new axiom, however, are considerably simpler and easier to discover, and make it possible to contract into one proof many different proofs. The axioms for the system of real numbers, rejected by the intuitionists, have in this sense been verified to some extent, owing to the fact that analytical number theory frequently allows one to prove number-theoretical theorems which, in a more cumbersome way, can subsequently be verified by elementary methods. A much higher degree of verification than that, however, is conceivable. There might exist axioms so abundant in their verifiable consequences, shedding so much light upon a whole field, and yielding such powerful methods for solving problems that, no matter whether or not they are intrinsically necessary, they would have to be accepted at least in the same sense as any well-established physical theory.

Higher-level mathematical hypotheses are thus taken to be inductively justified by the simplifying, or, more generally, explanatory effects they have on lower-level mathematical truths. In a later remark, Gödel extended this characterization of inductive justification in mathematics to include not only the 'fruitful' (i.e. simplificatory and explanatory) organization of lower-level mathematical results, but also the 'fruitful' organization of principles and facts of physics.

This extension is significant because it makes a place in Gödel's mathematical epistemology for what is essentially empirical justification of mathematical truths. This does not, however, make him an empiricist like Quine. For Gödel allowed only that some of our knowledge of mathematical truths might arise from empirical sources. He remained staunchly opposed to any suggestion that all mathematical justification must or even can ultimately rest on sensory experience. Indeed, he described the empiricist view of mathematics as too 'absurd' to be seriously maintained. More positively, he maintained that there is a fundamental epistemic phenomenon (viz., that of certain axioms– including certain high-level set theoretical axioms– 'forcing' themselves upon us as 'being true') that cannot be accounted for by an empiricist epistemology for mathematics. This same phenomenon is apparently also that which caused Godel to reject idealism and accept a Platonist conception of mathematics. In our estimation, neither this phenomenon nor the related question of its status as a 'datum' for the philosophy of mathematics have received the attention they deserve.

Both the empiricist Platonism of Quine and the non-empiricist Platonism of Godel have been important influences on recent work in the field. So, too, has been an argument given by Benacerraf in the early 1970s. According to that argument, mathematical epistemology faces a general dilemma. It must, on the one hand, give a satisfactory account of the truth of mathematics and, on the other, offer a satisfactory account of how mathematics can be known. This constitutes a dilemma, in Benacerraf's opinion, because while to get a satisfactory account of the truth of mathematics seemingly requires that we bring abstract objects into the picture as referents of the singular terms used in mathematical discourse, to get a satisfactory account of mathematical knowledge seemingly requires that we avoid such reference. His argument makes use of the following key claims: (i) the semantics for mathematical language should be continuous with the semantics for non-mathematical language; (ii) the deep logical form of a mathematical expression should not be treated as too different from its surface grammatical form; (iii) the semantics for non-mathematical language is referential; and (iv) the best referential semantics for mathematical language uses abstract objects as referents.

As a result of the alleged need to construe the semantics of mathematical language as referential in character, we are obliged to see the *grounds of truth* of a mathematical sentence as residing in the properties of those abstract objects to which its referring terms make reference. Thus, for example, we are obliged to say that what makes '7 + 5 =12' *true* are the properties of the abstract objects 7, 5 and 12, the characteristics of the addition operation as an operation on abstract objects, and the features of the identity relation as a relation between abstract objects.

On the other hand, any epistemology for mathematics that is to avoid susceptibility to Gettier-type problems must link the *grounds of truth* of a mathematical sentence with the *grounds of belief* in it. In other words, if a given belief is to count as genuine knowledge, there must be a certain causal relationship between that which makes it true and our state of belief in it. This causes a dilemma because there is seemingly no way of securing the aforementioned causal connection while at the same time securing a reasonable referential semantics for mathematical discourse. There are mathematical epistemologies–in particular, various Platonist epistemologies–which allow for a plausible account of the truth of mathematical sentences. And there are mathematical epistemologies (e.g. various formalist epistemologies) which allow for a plausible account of how we might come to know mathematical sentences. There is, however, no known way of securing both a plausible account of mathematical truth and a plausible account of mathematical knowledge.

A great deal of recent work in the philosophy of mathematics has been devoted to trying to resolve this dilemma. So, for example, there are Field, both of which offer anti-Platonist resolutions of the dilemma, and Maddy, who attempts to work out an epistemology which is at once Platonistic and yet naturalistic. To date there is no general consensus on which approach is the more plausible.

An earlier argument of Benacerraf's on another topic has been similarly influential in shaping recent developments. Perhaps more than any other single source, this paper has served as the inspiration of that recent position known as 'structuralism'. Applied to mathematics generally, structuralism is the view that in mathematics we do not have objects with an 'internal' composition arranged in structures, we have only structures. The objects of mathematics, that is, the entities which our mathematical constants and quantifiers denote, are structureless points or positions in structures. As positions in structures, they have no identity or features outside of a structure.

Such a view, Resnik claims, is in keeping with the fact that 'no mathematical theory can do more than determine its objects up to isomorphism', a fact which seems to have led mathematicians increasingly to the view that (i) 'mathematics is concerned with structures involving objects and not with the "internal" nature of the objects themselves' (Ibid.), and that (ii) we are not 'given' mathematical objects in isolation, but rather only in structures.

Benacerraf himself applied the idea only to arithmetic, and, in particular, to the question, raised by Frege and others, of the 'deeper' ontological characteristics of the individual numbers. Frege, as is well-known, held the view that numbers are objects. Benacerraf opposed this, arguing that such questions as 'Are the individual numbers really sets?' are spurious. 'Questions of the identification of the referents of number words should be dismissed as misguided in just the way that a question about the referents of the parts of a ruler would be seen as misguided.' He then went on to complement this by noting that:

'Objects' do not do the job of numbers singly; the whole system performs the job or nothing does. I therefore argue that numbers could not be objects; for there is no more reason to identify any individual number with any one particular object than with any other.

...in giving the properties of numbers...you merely characterize an abstract structure *and the 'elements' of the structure have no properties other than those relating them to other 'elements' of the same structure.*

The primary motivation for such a view, apart from the desire for a more descriptively adequate account of mathematics, is apparently epistemological in nature. Knowledge of the characteristics of individual abstract objects would seem to require naturalistically inexplicable powers of cognition. Knowledge of at least some (e.g. finite) structures, on the other hand, could conceivably be explained as the result of applying such classical empiricist means of cognition as *abstraction* to observable physical complexes. Such abstracted structures would then become part of the general scientific framework and, as such, they could be extended and generalized in all manner of ways as the search for the simplest and most highly unified overall conceptual scheme is pursued.

Though Benacerraf and the other recent structuralists take no note of the fact, the basic idea behind their position enjoyed widespread popularity among philosophers of mathematics in the late nineteenth and early twentieth centuries. Indeed, Dedekind's essay *The Nature and Meaning of Number* contains an expression of the same basic idea on which the argument of Benacerraf is centred. There Dedekind writes:

Similarly, Weyl contains a striking expression–indeed, a strong generalization–of the structuralist idea that mathematical objects have no mathematically important features beyond those they possess as elements of structures. He writes: 'A science can only determine its domain of investigation up to an isomorphism mapping. In particular it remains quite indifferent as to the 'essence' of its objects. The idea of isomorphism demarcates the self-evident insurmountable boundary of cognition'.

Structuralism as a general philosophy of mathematics is criticized by Parsons. There it is argued that there must be some mathematical objects for which structuralism is 'not the whole truth'. The objects of which Parsons speaks are those he refers to as 'quasiconcrete'. These are so-called because they are directly 'instantiated' or 'represented' by concrete objects. Examples are geometrical figures, symbols (construed as types) whose instances are written marks or uttered sounds, and the so-called 'stroke numerals' and the like of Hilbert's finitary arithmetic. Such quasi-concrete entities are among the most elementary mathematical objects there are and they are therefore of considerable importance to the foundations of mathematics. Yet they cannot be treated in a purely structuralist way, because their 'representational' function cannot be reduced to the purely intrastructural relations that they bear to other objects within a given system.

Summary

The other two are the discovery of non-Euclidean geometries and the rapid development of symbolic logic. For the first three decades of the century, the Kantian problematic was for the most part understood in pretty much the way Kant himself had thought of it. That is, necessity was understood as imperviousness to empirical revision, and substantiality was understood as a kind of epistemic non-triviality. What differentiated the rival philosophies of this period, primarily, were the different responses they gave to the discovery of non-Euclidean geometries. Here, three basic alternatives eventually came to be articulated.

The first of these, developed chiefly by Frege, attempted to preserve the basic Kantian judgement that both geometry and arithmetic are necessary. At the same time, however, it sought to distinguish two different kinds of necessity, one for arithmetic, the other for geometry. This it did in response to the discovery of non-Euclidean geometries, which it regarded as having revealed a fundamental asymmetry between geometry and arithmetic–namely, that though there are, at least at the level of conceptual possibility, alternative geometries, there are not alternative arithmetics. Arithmetical thinking was therefore thought to be a more pervasive part of rational thought than is geometrical thinking. Giving a proper account of this asymmetry then came to be seen as the primary duty of a philosophy of mathematics.

The second alternative, favoured by various of the early constructivists, modified the basic Kantian judgement that both geometry and arithmetic are necessary, and maintained instead that only arithmetic is necessary. It therefore also, like the first alternative, regarded the discovery of non-Euclidean geometries as having revealed a fundamental asymmetry between spatial geometry and arithmetic. Unlike the first alternative, however, it took this discovery to be best accounted for by reducing the truly mathematical part of geometry to arithmetic and rejecting its distinctively spatial part as being of an a posteriori rather than an a priori character. Arithmetic, on the other hand, since it admitted of no alternatives like those which non-Euclidean geometries constitute for geometry, was to be seen as arising from a source (viz., a very general kind of temporal intuition) which was wholly internal to the human mind and, so, a priori in nature.

The third alternative, developed by Hilbert, reacted to the discovery of non-Euclidean geometry in a different way still. It sought to put them to the test in order to determine how deeply into geometrical thinking they truly penetrate. The original non-Euclidean geometries were only shown to be independent of Euclid's axioms for plane geometry. They therefore left open the possibility that there are unknown and/or articulated features of the Euclidean plane such that, were they to be registered as axioms, the axiom of parallels *would* follow from them (together with the other axioms). They also left open the possibility that, even supposing the axioms of the Euclidean plane to be 'complete' in the sense just alluded to, the axioms needed to move from a description of the Euclidean plane to a description of Euclidean space are such that the axiom of parallels would follow from them (in conjunction with the other axioms), and would therefore not signify an independent feature of Euclidean *space*.

It was with the resolution of these matters that Hilbert was concerned. He therefore sought to produce a set of axioms so 'complete' that no axiom could be added to them without turning them into an inconsistent theory. In order to do this, he turned to a certain kind of continuity principle (viz., his so-called *Vollständigkeitsaxiom*) which, in Kantian fashion, he reckoned to belong to the domain of pure reason (or what Hilbert referred to as 'ideal' thought) rather than the domain of judgement (which Hilbert referred to as 'real' thought). This distinction between real and ideal elements in our mathematical thinking became the cornerstone of Hilbert's mathematical epistemology, both for arithmetic and, we believe, for geometry.

Hilbert did not, therefore, affirm the necessity of either arithmetic or geometry in any simple, straightforward way. Rather, he distinguished two *different* types of necessity operating both within arithmetic and geometry. One of these–that which attaches to the 'ideal' parts of arithmetic and geometry–he identified with the kind of necessity that attached to Kant's faculty of reason; a kind of necessity borne of the manner in which our minds inevitably work. The other, that applying to the so-called 'real' parts of mathematics, he saw as consisting in the presumed fact that all our thought assumes as a pre-condition the apprehension of certain elementary spatial features of simple concrete objects.

This, roughly, was Hilbert's complex conception of the a priority of mathematics. Clearly, it represented a significant departure from Kant's 'idealist' explanation of necessity as residing in the supposed inevitable tendencies of our minds to couch all experience in temporal terms, and all spatial experience in Euclidean terms. Clearly, too, it represented a significant departure from the ideas of the logicists and intuitionists. Its response to the discovery of non-Euclidean geometries (and such related discoveries concerning intuition of time as Einstein's theory of relativity) was not to posit an essential epistemological asymmetry between arithmetical and geometrical knowledge, as did (the original forms of) both logicism and intuitionism. Rather, or so it seems to us, it was to try and bring both under the authority of a single type of proto-geometric intuition concerning our apprehension of the elementary shapes or forms of concrete figures.

The 1930s witnessed the demise of Hilbert's Kantian programme at the hands of Gödel's theorems. Frege's Kantian logicism had earlier been vanquished by Russell's paradox, and the Kantian philosophies of the intuitionists, too, were under assault from both philosophical quarters (where their idealism was called into question) and mathematical quarters (where their ability to support a significant body of mathematics remained in doubt). Of the programmes of the early period, perhaps Russell's non-Kantian form of logicism remained most intact, being sustained chiefly through its embodiment of the recent advances in symbolic logic and its selection as the preferred philosophy of mathematics of the then-influential logical empiricist school.

Yet, even though the positive ideas of Kant's mathematical philosophy had largely been abandoned by the 1930s, his basic apparatus remained in effect until the early 1950s. It was only with Quine's attack on the analytic/synthetic distinction of the positivists that its influence, too, began to slip. The dominant trend in the philosophy of mathematics since then, if, indeed, there has been one, has been that of empiricism: an empiricism which denies that there is a difference in kind between the necessity of mathematics and that of the rest of our judgements; an empiricism which sees mathematics as governed by the same basic inductive methodology as governs the natural sciences; an empiricism which sees mathematical inference as reducible to logical inference. The ultimate adequacy of this overall empiricist approach as an explanation of the 'data' of mathematical epistemology, however, remains unclear. So, too, does the justification of its view of what these data are. Indeed, the problem of determining what are the data of the philosophy of mathematics is, I would suggest, among the more serious problems facing the subject as the century draws to a close.

3

Philosophy of Physics

Introduction

It seems that these three clusters of studies could also be found in the philosophy of chemistry. To disentangle what it is that is characteristic of studies in the philosophy of physics, I shall have to say something about what distinguishes physics from all other natural sciences. Nowadays, one would be reluctant to try to draw a hard-and-fast line between physics and the other sciences. But for the purposes of this article, a rough division might be effected as follows. Physics is the study of the most general properties of matter. In chemistry and biology the unique properties of particular kinds of matter are examined. With this rather vague prescription in mind, we can pick out as physics studies of such ubiquitous features of the material universe as its spatio-temporal structure, and of the common properties that every material being shares with every other, such as mass-energy and mobility. This way of distinguishing subject by scope is rather simplistic. It is only very recently that matter and radiation have been found to be mutually convertible and so to have common properties.

But the study of optics and the study of mechanics have always been or almost always been branches of physics.

It is worth emphasizing the antiquity of philosophical investigations of the science of physics. It has never been free of some philosophical content. In times of crisis questions about ontology and about method come to the surface. By a time of crisis, I mean a moment in the history of the investigation of the physical universe in which the best theories that we can construct according to the local criteria for identifying a good theory, are in apparently irresolvable conflict with one another while they are indistinguishable by reference to the results of observation and experiment. In these circumstances, philosophical speculation returns to the centre of the stage. Physics, more dramatically than any other science, develops through the interplay between philosophical analysis of conceptual foundations and what, at first sight, seems to be an independent scientific research programme. I shall illustrate this feature of the history of physics from time to time in the course of making my more detailed remarks.

Before I turn to sketching some particular examples of conceptual analysis, there is one further general distinction to be borne in mind. Since the days of Euclid's formalization of the science of geometry, mathematics has played a central role in the development of physics, but this role has not been uncontroversial. There is an important distinction between different ways of interpreting mathematical formalisms. Is the abstract mathematical representation of the laws of physics auxiliary or representational? Auxiliary mathematics consists of formal devices by which the knowledge of the physicist can be conveniently summarized and manipulated. I owe to John Roche a very simple example of auxiliary mathematics. This is the grid of lines of lattitude and longitude which geophysicists have laid over the earth. A more complex example is the system of deferent circles and epicycles with which Ptolemy calculated the ephemerides, the risings and settings of the heavenly bodies. We are not justified in assuming a priori that all the technical devices employed in some mathematical formulation of a law or theory have physical counterparts. The laws and principles of quantum mechanics can be expressed mathematically in terms of vectors in Hilbert space. But what could the physical meaning of the leading concepts of the Hilbert space representation possibly be? What physical sense could one give to the idea of a vector rotating in an infinite dimensional 'space'? Equally, it is a misunderstanding of the second Bohr theory of the atom to ask for the physical counterparts of the charged oscillators which replaced the planetary electrons of the earlier theory. On the other hand there are plenty of theories in physics, such as the Clausius-Maxwell theory of the behaviour of gases in which every element in the mathematical representation is taken to be the counterpart of some determinate feature of the physical system. Each variable in $pv = {}^1/_3\, nmc^2$ can be given a physical meaning in terms of the molecular model of gas.

I shall illustrate the three branches of the philosophy of physics by outlining some recent discussion of topics of perennial interest. To illustrate analytical and historical studies, I shall draw on problems of the interpretation of relativity theory. An alternative example could have been the concept of 'mass' which has also had a long and interesting development. It has been sharpened, sub-divided and differentiated in response to experimental and theoretical advances. To illustrate methodological studies I shall take some very recent work on the role of experiments in physics. I shall also outline the debates between realists and anti-realists, concerning the interpretation and role of theories in physics. Do they describe unobservable but real entities and processes or are they merely devices for predicting more phenomena? If the former, how could we ever know whether we are getting a better and better picture of the hidden world of causal processes if we can never observe it directly? To illustrate foundational questions I shall turn from the development of the dispositional treatment of the foundations of Newtonian matter theory to debates about the meaning of the leading concepts of quantum field theories. An alternative example could have been the discussions of the significance of the Bell inequality and the status of the EPR experiment.

Analytic and Historical Studies of Concepts

I shall divide the topics that fall under this heading into two broad groups. There are those which are concerned with the analysis of concepts which at first sight are taken to be independent of the particular theories within which they fall. So there are analyses of space, time, causality, property, etc., which, though influenced by particular theories in physics seem to be in some respects, independent of them. On the other hand, there are analyses of leading concepts such as mass, momentum, charge, force, etc., which it is hard to imagine being carried out independently of the particular theories in which they have, from time to time, been embedded.

These analyses, whether generic or specific, are usually conducted with respect to larger questions. So, for example, developing and arguing for analyses of concepts such as space and time, are part of long-running controversies between absolutists and relationists. Do the concepts of 'absolute space' and 'absolute time' make sense? Some relationists would argue that these hybrid concepts are logically incoherent. Discussions about the proper interpretation of the concepts of mass, charge, force and so on, are related to the broad issue of whether the physical properties of things are best understood as dispositions, powers and propensities.

In investigating the status of the kinds of concepts that I have mentioned, it is important to bear in mind that these analyses are relative to the state of physical science at the time that they are being carried out. Yet they bear upon such general questions as whether we should favour an absolutist or a relationist theory of space and time, or whether all physical properties are really dispositions. Debates such as that between Leibniz and Clarke over the nature of space and time, though set in the context of Newtonian physics, nevertheless have universal significance.

One cannot say that there is no philosophy of physics independent of the state of physical theory. The generality of the concepts involved sometimes allows us to investigate concepts and to develop arguments which transcend particular epochs in the history of the science itself.

At the beginning of this century relativity theory seemed to present a radical challenge to a well-established conceptual system for expressing spatial, temporal and causal relations. The community of physicists had become accustomed to the idea of an absolute frame of reference, though it was hardly ever necessary to invoke it in solving any actual problem in physics. According to the popular myth, relativity theory was something extraordinary and utterly radical. I hope to show that such a picture is far from an adequate portrayal of the way in which theories of space, time and space-time had evolved in the course of the development of post-Aristotelian physics. Among many deep questions that can be asked with respect to space and time is whether there is a relation between the spatial and temporal location of an experimenter or observer and the forms that experimenters or observers would use to express the laws of nature. Are they or are they not the same when an experimental apparatus is run at one place and time rather than another? Would the laws of nature seem to be the same if studied on a moving platform as on one which was stationary relative to some apparently fixed frame of reference? Or on one which was accelerating with respect to some other platform on which stood the apparatus with which the experiments had previously been done? Ultimately the deepest question would be–could we discover which system of bodies was moving or accelerating and which was *really* stationary by looking for indicative changes in the laws of nature?

There are two ways in which absolutist views on space and time have been challenged. Relativity theory challenges the idea that there is a privileged reference frame, absolutely at rest, to which all motions uniform or accelerating could be referred. There is also the 'relationist' challenge to that idea which can be mounted independently of the question about reference frames. Relationists believe, with Leibniz, that space and time do not exist independently of the material system of the world. They are among the relational properties of that system. Absolutists (lately called 'substantivalists' to distinguish them from those who favour ultimately a non-relativist position in physics) believe that the space-time manifold is a substance which exists independently of the material world of charges, forces and fields. My illustrative example concerns the absolutist/relativist issue, not the substantivalist/relationist debate.

Let me briefly sketch how one now sees the historical progression that leads to the contemporary interpretation of relativity theory. The idea that the forms of laws of nature are independent of spatio-temporal location is expressed in the technical notion of covariance of the law under a coordinate transformation. To take a simple example: changing the coordinates of a location in a Cartesian plane each by a fixed quantity is a transformation. If the coordinates were x and y before the transformation and are x-a and y-b after, it is as if we moved the whole reference frame a units to the right and b units upwards. If a law of physics has the same form before and after the transformation is applied to the coordinates, we say that it is 'covariant under the transformation'. However, that technical idea is a version of a more fundamental conception. It expresses the idea that the forms of the laws of nature are indifferent to (that is unaffected by) changes in location, epoch or relative velocity of the frame with respect to which they are studied. We can detect the very beginnings of the covariance or indifference to location idea in the writings of Nicholas of Cusa. Contrary to the Aristotelians, who believed that space and time had instrinsic structures, the laws of nature differing with the location in which they are studied within that structure, Cusa introduced a general principle of indifference. His elegant epigram ran as follows: 'the centre and the circumference of the universe are the same', or in other words, physical laws are indifferent to their location in space and also, he believed, in time.

The next step in freeing physical processes from the influence of space and time came with the work of Galileo. In a striking image, he asked us to imagine conducting experiments inside a ship on a smooth sea. He argued that it would be impossible to discover whether the ship was in motion or at rest relative to the sea, by experimenting in a closed cabin. The relative motion between ship and ocean would have no effect upon the results of our experiments. Physics would always be the same inside the ship, no matter what its uniform velocity was relative to the ocean. This is the principle of Galilean relativity. Newton certainly subscribed to this principle. The laws of mechanics were thought to be indifferent, or covariant, as we might say to the Galilean coordinate transformation. In that transformation we can change the mathematical expression for uniform motions by any amount we like, the mathematical equivalent of slowing down or speeding up the ship by a definite amount relative to the ocean, and the laws of nature will preserve their form when expressed in the new coordinates.

All went merrily until the development of a comprehensive set of laws for electromagnetism by Clark Maxwell. Voigt showed in 1891 that Maxwell's laws were not covariant under the Galilean transformation. This implied that it might be possible to find electromagnetic evidence of our real or absolute motion through some uniform and universal background. It seemed that, in principle, one could find one's way about and perhaps even determine one's velocity with respect to some absolute frame of reference. Since it was the electromagnetic laws which were not covariant under the Galilean transformation, perhaps an electromagnetic aether might serve as just the absolute background that physicists needed. This was the project of Michelson and Morley.

By the end of the nineteenth century, however, it had become clear that there was a coordinate transformation under which the laws of electromagnetism were covariant. This was the Lorentz transformation. The situation had now become very interesting. The Maxwell laws of electromagnetism are covariant with respect to the tranformation of Lorentz but not with respect to that of Galileo. The laws of mechanics are covariant with respect to the transformation of Galileo but not with respect to that of Lorentz. Both are covariant with respect to the Cusan transformation, but this was taken to be so obvious as not to be worth remarking. This was the situation that Einstein confronted.

Essentially, Einstein had to solve two problems. How to make a reasoned choice between electromagnetism on the one hand and mechanics on the other as the most basic physical science. His reasons for choosing the electromagnetic option have to do with the need he felt to preserve a thorough-going symmetry between the process of electromagnetic induction that occurs when a moving conductor cuts the lines of force of a stationary magnetic field and that which occurs when a moving magnetic field interacts with a stationary conductor. If we assume an electromagnetic aether, the processes will be different in each case. He thought this intolerable. So he chose to privilege electromagnetism by denying the necessity to postulate an aether. This choice served not only to eliminate the aether from physics but to elevate the Lorentz transformation to be the dominant principle of covariance. His second problem was to find a new form for the laws of mechanics so that they too would be covariant under the Lorentz transformation. If this could be done, physics would be unified. There would be one physics and all its laws would be independent of the place, moment and relative velocity of the material system in which they were tested. He found these laws and they are none other than the laws of the special theory of relativity.

But now we can see that there was a third problem. With respect to what spatio-temporal structure are the new laws of mechanics, together with the laws of electromagnetism, indifferent? Einstein did not solve this problem himself. We owe its solution to Minkowsky ([6.36]). In the Minkowsky manifold, space and time are not independent systems of locations and moments. There is a four-dimensional manifold in which physical processes are imagined to take place. Minkowsky coordinate systems, moving with uniform relative velocity with respect to one another, now become the frames of reference with respect to changes between which the laws of nature must be indifferent. Wherever and whenever an experiment is conducted in the Minkowsky manifold, the result should be the same.

General relativity simply extended the same idea one step further. Einstein pursued the project of finding a formulation for the laws of nature which would leave them covariant under the general coordinate transformation, including that between frames of reference accelerating with respect to one another. Corresponding to the Minkowski manifold of special relativity came the famous curved space of general relativity which represents a manifold, with respect to which the laws of nature are absolutely indifferent.

The story of relativity is the story of one programme progressively developed, acquiring a more and more sophisticated form as the history of physics has unfolded. It begins in 1440 with Cusa's *Of Learned Ignorance* and culminates in general relativity. At each step yet another candidate for the one absolute space-time manifold was deleted from physics.

There is a connection with the absolutist (substantivalist)/relationist debate. Clearly, if the laws of nature are indifferent to their presumed locations in manifolds of space or time or space-time, then those manifolds can play no role whatever in the physical sciences. Absolute space, absolute time and absolute space-time are redundant. Experimental proof of this redundancy by Michelson-Morley is not in fact part of the history of relativity theory. It was, so to say, a comforting result that confirmed Einstein in the wisdom of privileging electromagnetic laws and their properties in his programme of research.

Relativity has turned out to be a cousin of relationism. But the triumph of relationism is by no means the foregone conclusion. There are still some reasons for thinking that there may yet be a place for an absolutist space-time in the physical sciences. These have to do with the status of the space-time manifold as required by general relativity. It is argued that even in the absence of all matter-fields this manifold would still have a structure. Therefore, it cannot be just one of the sets of relations that order the material stuff of the world.

Exemplary cases of conceptual analysis of other physical notions have occurred throughout the history of physics. One notices in the Clarke-Leibniz controversy, as a footnote, a conceptual investigation into the notion of quantity of motion, in which the distinction between momentum and energy is foreshadowed.

Methodological Problems in the Science of Physics

From the earliest days of mathematical astronomy, the nature of physical theory has been a matter of perennial dispute. The arguments have turned on the balance that one can draw between epistemology and ontology. Clearly, in some sense, that which is known through perception, the observable, has some kind of privileged ontological status, even though we know that many of our ontological claims made on the basis of perception are disputable and have sometimes had to be revised. Physical theory, however, characteristically seems to refer to processes such as the orbiting of the planets, entities, such as subatomic particles and structures, such as the curvature of space-time, that lie beyond perception. What, then, is the status of our knowledge of these beings and what of their standing as existences alongside and perhaps as components of that which we can perceive? Anti-realists have tended to privilege the deliverances of the senses, particularly the sense of sight, both ontologically and epistemologically. Realists, more cautious epistemologically, have nevertheless been bolder ontologically, and have tried to find ways in which attributes of physical theories, particularly their power to engender new kinds of experiments, have been taken as grounds for interpreting them realistically, and thereby adding to our ontology, enlarging the list of kinds of things we believe to make up the world. However, sceptics have little difficulty in finding grounds for pressing the interpretation of physics back in the anti-realist direction. They ask how we can be sure of the truth of our laws, when we have very limited grounds for accepting them, the classical problem of induction. How we can be satisfied with theories when our grounds for accepting them are only their predictive and retrodictive powers, since it is easy to demonstrate that there are infinitely many theories with the same predictive and retrodictive power, the problem of under determination, first formulated by Christopher Clavius. Quite recently a new mood has spread among philosophers of physics. Diagnosing the source of the weakness of realism as a commitment to the principle that science aims to establish the truth of the propositions of physical theory, a new school of neo-pragmatist philosophies of science animated to some extent by a reading of the philosophy of Niels Bohr has appeared. These philosophers have argued that a realist interpretation of physical theory should be taken as doctrine that theories are good in so far as they give us just sufficient understanding of the unobservable to allow us to manipulate it. Manipulability is the concept which allows us to penetrate beyond the bounds of perceptibility.

There are some important consequences that flow from the new 'pragmatic' realism. In one respect it harks back to an eighteenth century conception of the logical status of physical properties. At that time, the popular philosophy of physics decreed that our knowledge of the physical world was a knowledge of the powers and disposition of otherwise unknown entities. We knew them through what they could bring about. In very much the same way, the Bohrian philosophy of physics asks us to consider the world in so far as it appears to us through the dispositions it displays in apparatus of our own devising. There is another consequence which follows from this point of view. The early ecologists developing concepts to understand the way in which species of animals are related to their physical environment have coined the concept of *umwelt*. The *umwelt* of a species is that region of the world which is available to the species by virtue of its biological endowments. The physical world is broader and richer than the sum of the *umwelten* of the animals which inhabit it. This concept was used to explain how it was possible for different species of animals to occupy the same physical environment. Each carved out its own *umwelt*. In a similar way, one could treat the worlds of physics as *umwelten* and the development of the physical ontology as the successive marking-out of an ever-changing *umwelt* for humankind with respect to the experimental apparatus and conceptual systems with which they explored it. So the new philosophy of science is realist in a new way. The world is that which we make available to ourselves and our knowledge of it is our knowledge of what it is capable of doing and being made to do by our manipulations.

The next question that obviously arises in relation to the move from anti-realist to realist conception of physical science concerns the nature of physical theory. If we think of physics in the anti-realism manner as the statistics of perceptual experiences or even the statistics of the performance of instruments, while all else is of no ontological significance and merely serves the role of intervening variables to carry us from one empirical statement to another, we would be inclined to follow the philosophies of science of Mach and Duhem. According to these philosophers theory performs only a logical role: an inference machine for Mach, a taxonomical system for Duhem. However, when one examines physical theory and considers the development of the theorizing in a certain field of enquiry, one is struck by the fact that the theories of physics do not seem to be about the world at all. They are at least, ostensibly, about models of the world. A model, in this sense, can be either an abstract representation (a homeomorph) or a paramorph, which is an analogue of something we do not yet know but believe to exist. A law of nature like pv=a constant, is an idealised description of the behaviour of an abstract version of a real gas. The corresponding theoretical proposition, $pv={}^{1}/3\ mc^{2}$, is a description of the molecular model of what a gas might be like. This view of theory has been about for more than three decades. It is now becoming popular again. A philosophical question that is immediately evident if we think about theory in this way, is how this view fits with realism. Thinking it through, the fit is rather nice. Having given up the idea that realism must be defined in terms of the truth of propositions but rather thought out with reference to the manipulability of objects to which theory guides us, the idea of theory centred on a model of reality is very attractive. Model and reality are, so to say, beings of the same sort, and we can consider the fit of a model to that of which it is a model in terms of similarities and differences. The problem of which similarities are to be counted as important is itself readily soluble by reference to the structure of the theoretical concepts which constitute the discursive part of the theory. Some properties will appear as essential and some as accidental. Similarities and differences gain their importance from the extent to which they are drawn from the real and nominal essences of the beings in question, that is from our ideas about their inner constitutions and from our criteria for assigning them to kinds on the basis of their observable properties.

The shift from positivism to realism to what one might call post-realism or neo-pragmatism also involves rethinking of the role of experiments. In the logistic account of science, whether positivistically conceived or developed in the fallibilist mode by Karl Popper, gives a logical account of experiment. An experiment is done in order to provide a rational being with a proposition of the form 'Some A are B' or alternatively, if it so turned out, 'Some A are not B'. The significance of the experiment is determined by logical relations between those propositions which describe the experimental results and the general hypotheses to which they are considered relevant. So the logical account requires us to accept either inductivism or fallibilism. In the former we would have to accept the force of some pattern of inductive reasoning, in which the result in some cases, those that have been studied, are generalized to all cases, a notoriously shaky inference. In the latter we would have to rely upon the fallibilist pattern, that while we can draw no certain conclusion from a positive result, a prediction that turns out to be false justifies us in rejecting the hypothesis from which it was drawn. Neither is satisfactory.

The evident mismatch between the logicist way of construing experiments with the neo-pragmatist view of science, suggests the possibility of conceiving of experimentation in a much broader fashion. If we are endeavouring to see by experiment how well our models match reality, then we should not think of experimenting as a way of producing propositions to stand in a logical relation to a theory. We should think of the performance of an experiment as doing work in the world directed towards producing a certain outcome under the guidance of the theory. The question is not whether the theory in use is true or false, but whether, read as a set of instructions, it enables us to do what we aim to do. There is a sense to be given to the verisimilitude of a theory, but not in the prepositional mode, not centred on truth. In particular we are concerned with manipulating the world as if it were well-matched to our model. Indeed, there are specific model-matching experiments, some of which have been of enormous importance in the development of science. One which I find particularly instructive is the Page and Townsend experiments, by which a model of the flow of fluids with respect to bounded media is matched against a subtle experimental revelation of what the structure of such fluid motion 'actually' is (that is what the motion appears to be by the use of an ultramicrosope). Interesting cases arise where our faith in the iconicity of our models is based only upon their power to suggest manipulations. It seems to me that our main reason for believing in the reality of the magnetic field is the set of effects we can bring about by procedures which, we believe, act directly on that field. Changes so induced then bring about effects we can observe, such as the swing of the pointer of a galvanometer.

The notion of 'model' which I have been assuming in this discussion so far is the familiar one of analogue. One system is a model of another if it is analogous in relevant ways. But there is another, connected, sense of 'model' which has also been prominent in the writings of some philosophers of science. In logic a model is a set of entities and relations which can be used to interpret an abstract calculus. If the formulae of the calculus, interpreted as meaningful sentences by the use of such a domain of entities and relations, are all true when used of that domain, then that set of entities and relations is a model for the calculus. In logic there is a calculus and a model is needed to give it meaning; in physics a model or analogue of reality is imagined and a theory is subsequently created by describing the model. At the end of the day, so to say, the relation between calculus, theory and model is the same in both cases. However, in the order of creation, logic and physics run in opposite directions.

Disputes about how physical theory should be presented can be found throughout the history of physics. They were particularly prominent in the sixteenth century when many philosophers treated the heliocentric and geocentric astronomies, many versions of which were on offer in the mid-sixteenth century, as alternative mathematical systems. Ways of answering the question as to which formal system was to be preferred were interestingly divided between those who thought that anti-realist criteria, like simplicity and logical coherence, were of prime importance and those who favoured realist criteria, like the ontological plausibility of the model of the solar system that the mathematical structure represented.

In the late eighteenth, through the nineteenth century, Newtonian mechanics became the focus of a considerable effort to rework its formal representation. This was in part animated by the discovery of Maclaurin's paradox, perhaps more justly attributed to Boscovich. Boscovich noticed that the grand Newtonian theory was internally incoherent, indeed, self-contradictory. The concept of action, 'force × time', which was essential to setting up Newton's third law, that in action by contact action and reaction are equal and opposite, required all such action to take place in a finite time. But the Newtonian ontology required the ultimate material particles to be truly hard, that is incompressible. It follows that all action by contact must be instantaneous, since the ultimate contacting surfaces cannot deform. Forces in instantaneous Newtonian impact would, according to the mechanical definition of action, be infinite. But there is no place for infinite forces in the Newtonian scheme. A variety of strategems were developed to try to resolve the difficulty. In general physicists in France tended to favour theories without forces, whereas the English and some of their continental allies tended to favour a mechanics without matter, the so-called dynamical interpretation.

Great advances were made in the mathematics of physics, in the course of the working out of these alternatives. The Lagrangian formulation, the Hamiltonian formulation and Hertz's immensely influential reformulation of mechanics were all attempts in one way or another to come to terms with the same fundamental problem.

But there is another kind of investigation which we could classify as foundations of physics. This is the project of finding the minimal or most elegant formal representation of a scientific theory. In these cases, the mathematician-philosopher is driven by an interest in the aesthetics of formalisation rather than by some metaphysical paradox, such as the one uncovered by Boscovich and Maclaurin. In the present century some very interesting developments have hinged on attempts to provide alternative formal representations in which the very foundations of a theory are revealed in a clear and unambiguous way. For example, the matrix theory of Heisenberg and the wave mechanical formulation by Schrödinger of the laws of quantum mechanics, though they were shown to be mathematically equivalent in some sense, were animated, in part at least, by ontological differences between their authors. Lucas and Hodgson sum up an enormous variety of ways of arriving at the Lorentz transformation. There are many different paths by which this important group can be arrived at. Sometimes exercises of this sort, the formulation of an alternative mathematical representation, are significant. But sometimes they seem to be little more than formal exercises, mere auxiliary mathematics.

The way that the results of experimenal manipulations enter into the record of physics as *phenomena* is exceedingly complicated. Goodings has analysed the paths by which the personal experience of the discoverer of a new phenomenon makes it available, conceptually and manipulatively to the community of scientists and ultimately to everyone. A key move in the transformation of personal experience into public phenomenon is the elimination of all traces of the human hand that was involved in the early occasions of its production. Goodings shows that the circularity of the motion Faraday demonstrated as a natural electromagnetic effect was extremely difficult to produce on the laboratory bench, let alone understand. To explain how this transformation is achieved Goodings introduces the idea of a 'construal'. It could be a way of describing, a picture, a diagram or anything by which a scientist, in interaction with others, makes the phenomenon available as such. In applying construals the great scientists of history transform complex chains of fragile steps into simple sequences of surefire manipulations. Faraday describes seventy-five steps in recording how he first produced circular electromagnetic motion in his own laboratory. The published description of the procedure mentions only forty-five steps. The final simplification to a mere twenty steps occurs in his set of instructions for anyone to reproduce the effect.

Foundational Debates

The indeterminacy of subatomic processes that had first appeared in experiments with electrons was eventually canonized rather than resolved in the mathematical theory of quantum mechanics. Quantum mechanics has been the source of one of the major conceptual problems which has beset physical science for the last seventy years. At the heart of Newtonian physics was an assumption of the strict causality of all physical processes and the determinate character of all physical effects. Quantum mechanics provides a formalism by means of which a description of the state of preparation of a system can be linked to predictions about the probabilistic distribution of the effects of certain treatments of that system. The theory provides no way in which determinate outcomes can be predicted from knowledge of the original state of the system. Here is the dilemma. Is this because the way we now understand the state of any physical system is actually complete? This would seem to imply that there are real propensities to vary the outcomes of identical operations performed on identically prepared systems, contrary to the ordinary notion of determinate causality. Or is there something missing in our knowledge of electrons and other subatomic particles, knowledge which would restore the determinate structure of physical theory? Perhaps there are 'hidden variables' which do behave deterministically.

Arguments about the viability of hidden variable theories are almost as old as quantum mechanics itself. Can we find a theory based on the assumption of the existence of a set of attributes which we can ascribe to subatomic particles and their states of preparation from the determinate mathematics of which we could recover the probabilistic results for quantum theory as it is now understood?

So far, the answer has been equivocal. It is now clearly understood that there is no way in which a theory employing the familiar classical concepts of momentum, energy and so on, could be formulated, to provide a determinate hidden variable theory. Every experiment so far conducted has only given stronger and stronger support to the 'Bell inequality', the mathematical condition that expresses the principle of no hidden variables. On the other hand, there have been hidden variable theories constructed using exotic concepts from which the existing quantum mechanical results can be recovered. However, they lack any serious degree of physical plausibility.

In quantum field theory, another and yet more fascinating conceptual problem has arisen. It is now some fifty years since the idea of expressing field interactions as the exchange of particles was first proposed. The idea has been, one might say, extremely successful, at the cost of the development of theories of incredible mathematical sophistication. The quantum theory of fields is now a very well-developed speciality in physics, but it leaves us with a tantalising conceptual problem. The particles which are exchanged in interactions, say, the photons that are exchanged in an interaction between two electrons, are not identical with the photons, the flux of which is the light with which we are familiar. These photons are virtual, that is, they exist in and only in the interaction, if they exist at all. Furthermore, as imagined, they have properties which differ from the familiar properties of the photon of light. They are like light quanta, but not quite like light quanta.

Recently, the idea of using the analogy between the light photon and the photon of quantum electrodynamics as a basis for developing theories of other kinds of fundamental interaction, the weak interaction, the strong interaction and even gravity has led to the proliferation of such 'virtual particles'. I think it would scarcely have crossed the minds of most physicists to ask about the reality of virtual particles, were it not for the use of the structure of reasoning through which the light quanta became the models for quantum electrodynamics. In quantum electrodynamics, the virtual photon is modelled on the real photon, if I may be permitted to put the matter that way. Then weak interaction particles, the w+ and w-, and z0 particle are modelled upon the virtual photon. They are all species of the same genus. Then, in a reversal of the reasoning which led to the conception of the virtual photon, the idea of a real w particle, or a real z particle seemed to be a natural development from the quantum field theory of the weak interaction. The programme for hunting the w's and z's was defined, and in the manner in which such events are achieved, they were eventually 'discovered'.

I believe that this pattern of reasoning which is characteristic of quantum field theory was at least in part responsible for raising the question of the reality of the intermediate vector particles which carry the forces of interaction. If there are real versions of these particles, then surely there is some sense to the reality of the particle as the physical bearer of the field.

A clear formulation of the idea that there is a distinctive set of properties, the behaviour of which defines the subject matter of physics first appears in the seventeenth century. At that time the perceptible attributes of material things were classified as primary or secondary depending on their relation to human sensibility. Those which existed only in the act of perceiving were classified as secondary. Those which were thought to exist independently of the perceptual capacities of human beings were taken as primary. Galileo seems to have so tightly tied the primary qualities to the science of physics as to make the one definitive of the other. Secondary qualities were marked by the way that they varied in quality, intensity and duration with the state of the human perceiver. Locke completed the philosophical treatment of the distinction by carefully analysing the relation that must be supposed to obtain between secondary qualities, such as the power of a body to induce a colour sensation in a human observer, or an observable change in another material body, such as the power of fire to melt ice, and states of material bodies by virtue of which they had these and other powers. He sharply distinguished between ideas and qualities. Ideas are mental, including, for instance, sensations of colour. Qualities are material, including the properties of coloured things. This distinction enabled Locke to come at the distinction between primary and secondary qualities by a different route from that followed by Galileo. Ideas of primary qualities resembled the qualities as they existed in the material world. But the ideas of secondary qualities did not. Red, as a perceptible quality, does not resemble whatever property it is that causes a human being to see an old Soviet flag as red in hue. Generalizing the theoretical use of the concept of primary quality, Locke took the qualities that 'in the material body' caused corresponding ideas of secondary qualities as just those which are central to the conception of matter as it is used in the science of mehanics. All this is tied together by the thesis that the quality in the perceived thing that corresponds to the idea of colour, say, that is the secondary quality itself, is nothing but a power, a power to induce the relevant sensation. What the word 'red' refers to in the thing that is seen as red is a disposition. But it is grounded in an occurrent state of the perceived thing. According to this metaphysical scheme that state must be some combination of primary qualities.

For the scientist-philosophers of the seventeenth century, physics was mechanics. It was the study of the primary qualities of material bodies. Friction, for instance, as a mechanical disposition, must be grounded in the atomic structures of the interacting bodies. Mechanics, the basic science, was based on an absolutist metaphysics. Locke's philosophical account of the foundations of physics required two main categories of concepts. One set of concepts were relational. Many of the qualities of material things are dispositions to cause perceptible effects in a human being or in other material things. Whether they are activated or not depends on the contingent existence of suitable targets for their activity. The other category of qualities was absolute. The properties on which the dispositions are grounded are primary. Primary qualities are so defined as to be independent of the relations between the material things which possess them and human beings or to anything else. Boyle sums them up as the 'bulk, figure, texture [arrangement] and motion' of the elementary material things or corpuscles. The word 'corpuscle' was preferred to the word 'atom' by the sophisticates of the period, since it left open the question of whether the constituents of matter that were elementary for chemistry or mechanics were truly atomic, that is, indivisible in principle. Though Newton listed many dispositions amongst the primary qualities of matter, he shared the assumption of his contemporaries that there were *some* absolute physical properties. In the second edition of the *Principia* the list of the mechanical properties of matter are a mixture of the occurrent and the dispositional. Newton writes of the 'extension, hardness, impenetrability, mobility, and inertia of the whole [body] which 'result from' the corresponding properties of the parts. To have inertia, says Newton, is 'to be endowed with certain powers'. Inertia appears in the list of primary, mechanical properties as the power to resist acceleration. But inertia is not mass in Newton's metaphysics. Mass is an occurrent property. It is that which grounds the disposition identified as inertia. To give mass its occurrent character Newton defines it as 'quantity of matter'. Since the mass per unit volume differs from material to material, a universal matter, serving as a basic common substance, would have to exist in different states of diffusion. In a substance of low density the matter is rarefied, while in a substance of high density it must be compressed. One physical scheme to accommodate this difference would be basic atoms in a void with more or less pores between them. There would be fewer such atoms in a given volume of a light substance than in a similar volume of one that was more dense. Newton seems to favour this account. The basic atoms would be full, and so of uniform density. Lacking pores they must be incompressible and impenetrable. I have emphasized above the problem this thesis raised for Newton's general mechanics.

Though most of Newton's primary qualities are dispositions, they are nevertheless absolute in one of the dimensions on which the notion of the absolute figures. In Rule III Newton asserts that they 'are to be esteemed the universal qualities of all bodies whatsoever' whether or not they are 'within the reach of our experience'. As primary qualities they are not relative to human sensibility, but by the same token they are a mix of the relational and the absolute. In Newton's scheme mass, inertia, extension and mobility must exist even in a body wholly isolated from all other material beings. It seems to follow from the definition of mass that Newton's physics includes at least two absolute properties of matter. The quantity of matter would seem to be unaffected by the presence or absence of other material things. Since the quantity of matter of a body is related to its spatial extension, and that to absolute space, it would seem that extension and mass are both absolutes in Newton's scheme.

If two globes, kept at a given distance one from the other by means of a cord that connects them, were revolved about their centre of gravity, we might, from the tension of the cord, discover the endeavour of the globes to recede from the axis of their motion.

By testing to see in which direction impressed forces bring about the greatest increase in that tension, we can find not only the angular velocity of the globes in absolute space but also the true plane of the motion with respect to that space. To suppose that a force will appear in the cord when the globes are set rotating, Newton must be assuming that the masses of the globes are unaffected by the absence of all other matter. The masses are absolute qualities. If mass is a quantity of matter, then indeed that assumption seems natural and also inevitable.

The generality with which Newton uses the notion of 'power' is evident in Query 31 of the *Opticks*.

And thus, Nature will be very conformable to herself and very simple, performing all the great motions of the heavenly bodies by the attraction of gravity which intercedes those bodies, and almost all the small ones of their particles by some other attractive and repelling powers which intercede the particles.

There is yet another root idea in Newton's conception that gravity cannot be a primary quality because it suffers 'intensification and remission of degree'. So there must be a more basic power, 'an agent acting constantly' which is the absolute, because non-relational element, in the physics of gravity.

Mach's criticism of Newton's metaphysics is usually presented as an attack on the assumption that mass is an absolute property. But Mach's argument develops in two steps. He first shows that mass is best considered as a relational property in his analysis of the basic laws of mechanics. The argument goes as follows: consider an impact of tensions. The body A is falling under gravity. When the string joining it to the body B, resting on a smooth surface becomes taut, A will decelerate and B accelerate. Since the string is taut at the moment of 'impact' the force decelerating A will be equal to that accelerating B, so Mach argues. Let that force be '*F*'. Then if the mass of B is *m*b, the mass of A is *m*a, and the accelerations *f*b and *f*a respectively, the equation of motion for the whole system is *ma.f*a =-*m*b.*f*b

Mass appears here, and in all other contexts of mechanics, as a ratio. In this case the ratio is equal to the negative inverse of the quotient of the accelerations. Mass and inertia are the very same relational disposition. Given this prior analysis, Mach's treatment of the experiment of the globes (and of the more complicated argument of the thought experiment we call Newton's bucket, which involves a refutation of the Cartesian conception of locally real motion) in which the assumption of the persistence of inertial properties into the isolated system of the globes is entirely consistent. It simply involves the generalization of the relationality of the mass concept to the components in the simple system of the impact of tensions to the structure and contents of the universe as a whole. Mach completes a trend that began in the sixteenth century, a trend to replace absolute versions of the properties of material things with relational properties. Not only are these properties dispositions, manifested only in the interactions between material bodies, they are also relational in the sense that they are not grounded in some intrinsic property of isolated material individuals but in their relations to all the rest of the bodies of the universe.

One can tie together the seemingly disparate mechanics of Newton (read relationally) with quantum field theory by the realization that there is a common structure to their deep ontologies. In what sense do *any* of the 'particles', actual or virtual, exist? It seems obvious that only a dispositional account of their manner of being makes any sense. By that I mean that our claims about the world-in-itself made on the basis of the experiments by which apparatus, has (permanently) a disposition to display itself in such and such a way in the behaviour of that apparatus. Or to put it in Popper's terms only the set-up has propensities to yield this or that phenomenon. The phenomena are ephemeral, but it is they which are particulate or wavelike or whatever it might be. In this treatment we have both of Bohr's famous principles, that of complementarity and that of correspondence. Complementarity because set-ups which exclude each other produce, as a matter of fact, disparate and complementary phenomena; correspondence because the state of an apparatus-world set-up can be described for a human community only in the terms made available in classical physics, the physics the concepts of which are paradigmatically defined by the things and events of the ordinary world.

Quantum mechanics has provided much controversy in philosophical interpretations. As it developed its theories began to contradict many of the accepted philosophies. However, all its mathematical predictions coincide with observations.

In most cases accepted philosophies are based on the everyday experience of the average human which is extremely limited as it does not include observation of ultra-small systems, or motion with high speeds, or experimenting with high energies, strong gravity, etc. Thus, common-sense "theories", "intuitions" or "feelings" cannot be relied upon when it comes to descriptions or explanations of the behavior of many systems and objects in nature.

The 18th century saw many advances in the domain of science. After Newton, most scientists agreed on the presupposition that the universe is governed by strict (natural) laws that can be discovered and formalized by means of scientific observation and experiment. This position is known as determinism. However, determinism precludes the possibility of free will. That is, if the universe, and thus the entire world, is governed by strict and universal laws, then that means that human beings are also governed by natural law in their own actions. In other words, it means that there is no such thing as human freedom (except as defined in compatibilism). Conversely, if we accept that human beings do have free will, then we must accept that the world is not entirely governed by natural law. Some have argued that if the world is not entirely governed by natural law, then the task of science is rendered impossible. However, the development of quantum mechanics gave thinkers alternatives to these strictly bound possibilities, proposing a model for a universe that follows general rules but never had a predetermined future.

4

Philosophy of Science in Today Scenario

Introduction

It is not that applied science evolves all by itself, as the application of knowledge for practical purposes certainly requires efforts, including research. But the research for any practical purpose need not, it was taken for granted, be a search for knowledge. To make this clear, it may be useful to contrast the classical, typically eighteenth century view with today's view: today we recognize within science not two but three categories; we recognize basic research in addition to the classical pure and applied research, where pure research is disinterested and applied research is the use of the fruits of pure research for practical ends; basic research is pure research directed at material which is not very interesting in its own right but which is expected to be very useful in practice. There is little doubt that today research claims prestige for itself because of its potential usefulness. That is to say, all research is claimed to be more-or-less basic. In the classical vein this was unthinkable, the value of science was deemed almost exclusively personal and research was deemed edifying.

Obviously, of the many thousands of citizens engaged in research proper, most are engaged in small tasks–which Thomas S. Kuhn has labelled 'normal'. And, he stresses, normal science is practical. He probably means by this that normal science is all practical, but let us admit that it can also be basic. The practical attitude to science is very modern; it is at most the result of the industrial revolution, and so nineteenth century at the earliest; more likely it is post-Hiroshima. Kuhn is a historian of science and so he should know the obvious fact that normal science in the eighteenth century was more for individual entertainment than for practical ends. This was not always so: anyone familiar even with the mere illustrations in the literature in the history of science in the eighteenth century will know that. This is reflected in the third edition of *Encyclopedia Britannica* of the early nineteenth century. The article 'Science' there is extremely brief, reporting that an item of knowledge belongs to the body of science if and only if it is certain. Though the article gives no instances, clearly, the best instances are either from logic and basic mathematics or from extremely common and undoubted experiences, though, of course, some high-powered scientific theories should count as well. Today, incidentally, it is generally acknowledged that of these seemingly most certain items, none is exempt from doubt and revision (except perhaps logic; this is still a contested matter). Next to that brief third-edition *Britannica* article on science is a long article on science as amusement, in which the contents of a famous popular eighteenth-century book (by Ozanam) is reported. We would recognize today the contents of this article as vaguely within the domain of high-school sciences, as it includes somewhat amusing experiences with mechanics, electricity, magnetism and the like. Probably these two articles were not conceived of together and they were put together by sheer lexicographic rules.

The picture which emerges from this description presents a concern with science which is pre-critical. It was at times purely intellectual, at time practically oriented, always with great implication for life in general, for daily life and for peace, but with hardly any concern for the problems and issues in the philosophy of science as recognized today. Today, many of the concerns of the field, epistemological, methodological and metaphysical, are traced back to writers of the classical era, especially David Hume and Immanuel Kant. There was a major difference between these two thinkers. Hume is typical of his class: he was a private scholar who was clearly concerned with the social sciences (politics and economics, in particular), whose contributions to the philosophy of science he himself saw as marginal and preparatory. Not so Kant, who was most uncharacteristic; he was a university professor, who was on the side of science, and who was reputed to be a polymath proficient in fourteen different branches (some of which he inaugurated as academic subjects, such as geography and anthropology). He was still primarily a philosopher, and even primarily a philosopher of science.

It is hard to examine this, quite generally received, assessment, since the expression 'the philosophy of science' is new. To repeat, traditionally the word 'philosophy' designated learning in general and empirical science in particular. After the defeat of the French Revolution, some fashionable reactionary philosophers swore allegiance to unreason. Other, more old-fashioned philosophers understandably attempted to distance themselves from the new advocacy of unreason, and one way they did this was by naming their own views 'scientific philosophy'. This name usually designated mechanistic philosophy; its adherents considered theology to be typically metaphysical and so they branded all metaphysics evil; this enhanced their claim for scientific status for their own, mechanistic metaphysics. This way the philosophy that upheld the traditional esteem of reason centred mainly on science and on reasonability in the moral life of the individual and the nation. It naturally tended to centre increasingly on epistemology, methodology and rational metaphysics as a main tool to combat unreason. The philosophers of unreason had–still have–their own philosophy of science, but this is scarcely recognized: the philosophers who defend reason against the attack on it from the advocates of unreason took a monopoly on science and its defence.

The philosophy of science thus evolved into a specific activity of philosophers of the rationalist persuasion–the activity of defending science against its detractors. This explains the poverty of the field today: today science has no worthy detractors to combat; and no dragons to slay, no heroic deeds.

Even the philosophers of science themselves are aware of this fact, as they defend science not only by singing its praise, but also by attempting to solve problems in epistemology and in methodology, and by seeking newer and better arguments to combat metaphysics with. They do this as a mere pious act, paying no heed to the possibility that the problems they pose are insoluble, at least insoluble as long as they are presented in the traditional manner and settings. They cling to the pre-critical, optimistic view of science in the face of the risks to the very survival of humanity which scientific technology has originated: they relegate these risks to the new field of the philosophy of technology (which is less than half a century old), as if their philosophy of science does not include the philosophy of technology and as if their philosophy of science does not credit science with scientific technology as a great achievement. It really is a cheap trick to admit to the field of the philosophy of science the praise for science as the source of the benefits from scientific technology and its great achievements, and to banish to the philosophy of technology the possible and actual ill-effects of the same scientific technology.

David Stove is exceptional. The efforts to solve the traditional problems of the philosophy of science, he says, are commendable even if these should turn out to be insoluble. For, he explains (in his book against those who have given up the traditional struggle, including Sir Karl Popper and Thomas S. Kuhn), the struggle is the ongoing defence of science and thus of traditional rationalist philosophy and thus of rationalism as such. This is a charming admission, but of a position that is obviously pathetic.

Public Relations for Science is Meaningless

In the year AD 1600 St Roberto Cardinal Bellarmino consigned Girdano Bruno to be burnt at the stake-allegedly because he taught that the universe is infinite, so that in all likelihood there exist other worlds like ours. Later on the said Saint issued an official threat to Galileo. Science was then rightly militant. Today science is triumphant; even the Church of Rome has recently admitted the superiority of Galileo's case over that which was contrived against him and officially endorsed there and then. However embarrassing this repudiation was, science became too strong to continue evading it. Today, science surrounds us and appears on all levels from the sublime, through the mundane to the abject.

In the sublime mood science is what Bertrand Russell called 'Promethean madness' and what Albert Einstein considered to be the scientific undertaking: 'tracing the Good Lord's blueprint of the universe'. In the mundane world of the modern industrialized metropolis, the impact of science on the intellectual, political, social and technical aspects of life is overwhelming; especially, the impact of scientific technology is so very prominent. The abject aspect of the impact of science on daily life has attracted a certain kind of philosopher of unreason, whose hostility to reason is expressed as a hostility to science, transformed into a hostility to scientific technology–on the grounds of prophecies of gloom and of apocalypse that should be blamed on scientific technology, the cause of the alienation of Modern Man. These prophets of unreason identify science with the foolish attempt to conquer and subjugate Nature and they are confident that Nature will soon avenge this treatment by devastation. They advocate the replacement of the Western harsh, indifferent attitude to Nature with a soft, intuitive, irrational, oriental attitude. This mixture invites very urgently the sifting of the grain from the chaff.

This quick survey of the impact of science on society that goes from the sublime to the ridiculous, has omitted the ridiculous. This dimension is normally absent: science is no cause for levity. The entertainment world is as much under the influence of science as any other component of our small universe, in its shaping of our tastes and opinions and values and in its stupendous media technology. But not as an object of hilarity; even as sedate entertainment it is almost entirely confined to the juvenile. Yet they raise in a fresh manner the question, what is science? The question seems to require an answer that is easy to apply to what we usually call science, including high-school science and nuclear physics and electronic engineering. This is an error: we may have a thing and the received model of that thing, and the two need not agree. In the literature on social anthropology, this is taken for granted. In that literature, the paradigm for the difference–between a thing and the received idea of it–is the difference between magic and the received idea of it: in every society that has been described by some anthropologists, there are magicians, and yet (unless we deem the scientists as powerful magicians as did Sir Francis Bacon in the early seventeenth century), we all agree that real live magicians never fitted the characterization of magic admitted in their society (except possibly contemporary modern society). Magicians like Merlin do fit the image of the magician, but they never existed (except perhaps among modern scientists). Do the modern scientists fit the image of science?

It is the image of science that is rather ridiculous, as it is put forward by the spokespeople for the public relations of science. This is not peculiar to science. The practice of public relations evolved unawares and uncontrolled as a part of the advertising world of the free market, where the shortest of the short-term interests govern, so that the most cynical opportunists set the tone. This is harmless enough when pertaining to the sales of soap, but not to the sales of the higher things in life, be they the arts, the sciences or religion. Whoever the individuals are who take upon themselves to express the social concerns of science, they are these days powerful individuals and they control the appointment of suitable individuals to the positions of spokespeople of science. The leading positions in this matter are well-paid professorships in the leading universities in the subject called the philosophy of science.

In brief, official philosophy of science, the philosophy of science boosted by the scientific establishment, is much less tolerable than the commercials on television which sell soap and other cosmetics. They are as remote from the Promethean madness of the search for the secret of the universe as eroticism is from the (intellectual) love of God.

If we grant this, and it is hard to deny it as we will soon have the occasion to notice, then we may begin by denying that there is anything more specific to science as such than to cosmetics as such. And the best characterization of science that can be given is in that vein: science is the Promethean madness, the attempt to trace God's blueprint of the universe, the search for the secret of the universe.

Science is a Cultural Phenomenon

There are some immediate, obvious objections to the view of science as a quest, and they centre on the missing object of the quest and on its trail, expressed in the following two questions. What will satisfy the scientific quest? Which way does one turn to be on its trail? These questions are reasonable and should be taken seriously, but they are presented as objections, and as objections, I will now show, they are residues from the pathetic public-relations department–by illustrating their immediate socio-political implications–on the assumption that there are no competitors within human culture.

The first objection is dominant in the semi-official literature on the matter: what exactly are we in search of? Are we in search for information or for knowledge? If for mere information, will any information do? If yes, why not be pleased with the information contained in primitive lore and in Scriptures? This series of questions looks so straightforward, but it is not. It begins well and degenerates: What exactly are we looking for? This is the right question even though obviously we do not know: we know what we look for when we look for a lost penny, but not when we look for a masterpiece while roaming in a foreign museum, much less when we seek the secret of the universe. The question is right, but we should not expect too much for an answer: anything remotely resembling a possible answer may be a tremendous excitement. But look where the series of questions ends. It ends with an insult to the competition. This is not serious; it is public-relations frivolity-especially since repeatedly the self-appointed public-relations spokespeople of science often find the quest formidable and even exasperating, so that they finally settle for the mere physical comforts that science-based technology has to offer to the modern world. (The leading public-relations spokesperson of science in the previous generation was Rudolf Carnap, the famous debunker of all speculations; his *magnum opus* was his *The Logical Foundations of Probability* of 1950; it starts with the formula for finding the truth about the world and ends by giving up the task, with the excuse that science is a mere instrument.) And so, when the circle is closed and science is praised as a mere instrument, the conclusion is not drawn but remains all the same: the mere physical comforts that science-based technology has to offer the modern world is superior to the primitive lore and to the Scriptures. This is hard to take seriously. Primitive lore and the Scriptures do not compete with modern science-based technology as conveyors of physical comforts, but they still are very interesting and deserve attention in many ways and on many levels. This is the end of the objection from the hostility to the idea of science as a quest from popular lore and the Scriptures: the quest is obviously not replaced by the study of popular lore and of the Scriptures; rather, the study is a part of the quest.

The public-relations spokespeople do not allow themselves to be dismissed so easily, and they respond with forceful objections to the dismissal here suggested: they want reason, namely science, to be the guide for life; they want science to offer but better technology and better education, and the two should go together (as the proper education for the next generation is essential for the technological challenges they will face), yet the competition will not agree. Admittedly taking the Scriptures as a science substitute is frivolous. Yet, however frivolous the competition is, its hostility to scientific technology and to scientific education must be taken seriously in the interest of the wellbeing of us all.

This rejoinder seems very serious and very responsible, but it is not. Responsibility will be served if the question of education and of the place of technology in the modern world be discussed not apropos science but apropos the design of a better education policy, the study of education and its purposes. And the same holds for the problems that are specific to high-tech society. Here we are discussing science, and as the search for the secret of the universe, not the implication this has for education and for the training for high-tech. True, science shares with other domains the search for the secret of the universe, including magic and religion. Do the public-relations spokespeople of science want to distinguish between the search conducted in a manner becoming science and in alternative ways? Or will the comparison suffice of the results of the search along different lines? Today it is agreed that the results tell the important tale: by their fruits ye shall know them. Do we know the difference in the results? Of course we do: even the most ignorant among the public-relations spokespeople of science have no difficulty in telling a magic text from a scientific one! To be precise, the semi-official literature of the public-relations spokespeople of science is not that advanced: only a few philosophers of science discuss magic proper in a manner which is up to the standard of current social anthropology, the scientific field of study which retains an exclusive claim over magic. Rather, the semi-official literature of the public-relations spokespeople of science is concerned to a large extent with the unmasking of items of pseudo-science as merely pseudo, namely, not the genuine articles they masquerade as.

Science Needs No Promoters

It should be granted that this is more challenging: the public-relations spokespeople of science can easily distinguish a genuine amulet or talisman from a page of a scientific paper, but lamentably all too often they cannot distinguish a phony page of a manuscript that deserves publication in a scientific journal from one that is not. This is why the public-relations spokespeople of science are never invited to act as referees for judging the merits of scientific research any more than the public-relations spokespeople of a financial concern will be asked to adjudicate on matters financial. This is why the public-relations spokespeople of science are so pleased and so proud when a scientist proper joins their ranks, even though they should know better. For, a scientist can contribute to the philosophy of science without joining the ranks of the philosophers, as many scientists often do. Hence, scientists become philosophers only as an admission of defeat as scientists, often after retirement. Max Born, the great physicist of the early twentieth century, who was also a somewhat less-great public-relations spokesperson of science, said that all able-bodied researchers should devote all their energies exclusively to science and permit themselves to turn to philosophy, if at all, only after retirement.

All this sounds rather evasive, yet it is the heart of the matter: analysing the means for distinguishing between the genuine research from the pseudo sounds a reasonable task, yet it clearly is very questionable, and probably it cannot be done: even scientists of the best repute are not very good at it. Proof: young Albert Einstein was deemed phoney by many scientists, and he was taken as suspect for well over a generation-while those who took him seriously debated hotly the question, was he right?

The great historian of physics, Sir Edmund Whittaker,

who was himself a serious scientist (he was the Astronomer Royal for Northern Ireland) was hostile to Einstein all his long life, and as late as in the mid-century, long after the heated debate had subsided, he declared that there never was an Einsteinian revolution and overlooked the heated debate completely. In a review of Whittaker's book Born said, I was there, I witnessed the heated debates and took part in them.

To return to the items that may be masqueraded as parts of science. These items are, among others, magic, theology and metaphysics. What is so pathetic about all the many studies dished out year in and year out by the public-relations spokespeople of science is precisely this: were they right, then there would be no regular problem in refereeing, and in the rare cases of such a problem, these very public-relations spokespeople of science should be the experts to consult. Such cases do not exist.

To be precise, one such case does exist: in the long history of the public relations of science, one such spokesperson was invited to speak as an expert on the matter at hand. It was the second 'monkey trial' so-called, the court case a few decades ago, in which a judge in Little Rock, in the State of Arkansas, USA, was called to adjudicate between the education department of that state and a religious sect which demanded that the official biology text books should include proper reference to scripture. It is a priori obvious that both parties were lamentably in the wrong: the education department was in error in proscribing such reference and the other party was in error in trying to bring in, not the Scriptures, but a certain dogmatic attitude. The judge had little choice but to side with the education department, simply because the public-relations spokespeople of the religious sect in question were even more inept than the public-relations spokesperson of science who was invited to speak for the department. He argued that the religious are dogmatic and the scientists are not. Even apart from the fact that many individuals are religious scientists, this is a naked falsehood: there are non-dogmatic religious sects and dogmatism is lamentably too common among scientists, religious and non-religious, and more so among science teachers. When it comes to the curse known as science-education inspectors, it seems that for them dogmatism is obligatory, though, as many obligations, it is at times not carefully observed. This is no complaint about the judge: he was facing in court dogmatism on both sides and had to choose the lesser evil. In that Arkansas court on that day, science appeared the lesser dogma and the lesser evil; but with the help of the public-relations spokespeople of science this will soon change-unless something is done about it. These public-relations spokespeople of science are not powerful at all, but they may do harm anyway, as they cover up some powerful evils: the more powerful science is, the more success it brings about, the more the dangers of its abuses, and unchecked it will be abused. This, after all, is the major lesson we learn from all science fiction, and Mary Shelley, H.G. Wells and Isaac Asimov spun yarns to let us absorb this lesson. Except that under the pious guidance of the public-relations spokespeople of science, readers of science fiction take the lesson to be mere fiction with no moral to it or as fiction with futile moral against the abuse of magic, not a real moral against the abuse of science.

We have arrived back at the claim of the self-appointed public-relations spokespeople of science that the evil of magic is in its masquerading as science, in its being pseudo-scientific. This is most parochial as an attitude: most magicians and theologians, even most metaphysicians, operated (and still do) in societies in which there is no familiarity with science so that they do not masquerade as scientific and so they do not qualify as pseudo-scientific. Even cargo cults, the magic rituals involving wooden copies of aeroplanes and other modern artefacts in the hope of inducing the gods to grant them to the worshippers, scarcely qualify as pseudo in any sense. To say of Moses the Law-giver and of Jesus Christ that their theology masqueraded as science defies the imagination. Only in response to the assertion of Maimonides, that Moses the Law-giver was a scientist, could anything like the charge of masquerading be launched-validly or not. *The* philosophers of science, however, that is to say, the semi-official public-relations spokespeople of science, are not interested in all this: they care little about societies overseas; they are here to advertise science here and this task includes the discrediting of the competition here. They therefore permit themselves at times to be agreeably tolerant to theology and to metaphysics-after proof is issued to their own satisfaction that the parties involved do not compete with science. Usually, that is, theology and magic are deemed competitors, and then the philosophers of science, that is to say, the self-same semi-official public-relations spokespeople of science, find themselves acting as bouncers for the exclusive club of science. The leading sociologist of science, Robert K. Merton, prefers the term 'gate keepers', as he deems it the less offensive of the two; it is more offensive, as will be clear when we find the answer, which should guide the bouncer, to the central question of the philosophy of science: who is and who is not a *bona fide* member of the club? What is science? Is there a quality of science that sets it apart from what the bouncers consider as the competition? For, clearly, science is open and gate-keeping makes it a closed club.

Science is not Super Magic

What is science? Science is a body of knowledge; science is what scientists do qua scientists; science is a tradition; science is any empirically involved research activity; science is a faculty in the university. All these answers are true and meet the question, yet they are highly unsatisfactory. Hence, the question was ill-put. Here it is in its proper wording: what is the essence of science?

This is a tricky question; without entering the hoary matter of the critique of essentialism we can re-word it: what differentiates science from......? Taking it seriously requires the study in depth of many competitors to science. The study of alien cultures is, of course, highly recommended and even the bouncers will not object to it unless it is done from the viewpoint of the competition. Yet controversy about alien cultures abounds in the scientific departments devoted to it, and consequently the task of characterizing science in opposition to them gets increasingly harder. Example: is Claude Levi-Strauss, who has created a revolution in the current view of myth and of magic, is he a bouncer or a competitor? He says he is a friend of science, a scientist indeed. Is he? The question is very difficult to settle and the anthropological literature is still struggling with it. Let us try to alter our strategy, then. Can we look at science rather than at the competition and find there some clear-cut characteristic that sets science clearly apart from all the competitors? If so, what is it?

This is the problem of the demarcation of science as semi-officially understood. The most traditional answer to it, to repeat, is that science is a body of theories, and what characterizes them is their certitude, our ability to prove their perfection and finality. The more modern answer is the theory that science is a prestigious social class which lends prestige to its ideas. These two answers are contested these days, though the first is advocated mainly by philosophers of science and contested mainly by sociologists and historians of science, and the second suffers the reverse role-we have here two groups of self-appointed public-relations spokespeople of science competing for the same territory. Let us take the first answer first.

In the twentieth century the impact of logic led to a shift on this matter. The scientific character of a sentence shifted: it was deemed not proof but provability. Now generally one cannot know if a sentence will prove true or false before it is compared with experience. So, a sentence was deemed not quite provable, but merely decidable; a sentence is decided if it is either proved or disproved. This doubled the number of entries: not only a proven sentence but also its negation is scientific, as the negation of a proven sentence is disproved, and the couple of sentences taken together is decidable. Now the claim was that though generally a sentence cannot be declared a priori provable, it was declared that every well-formed sentence is a priori decidable. The justification for the relaxation of the criterion of demarcation of science to the extent of letting the negation of scientific claims be scientific was the wish to corner the competition once and for all by permitting the competition to contradict science openly. If the competition does contradict science, they will put themselves to ridicule, and if not, we will be able to expose them as saying nothing. The very idea that one can permit the competition or not, and decide that they say something or not, shows that the advocates of this view took anti-science to be passé, that they were serving science in the supposition that it is winning anyhow. In any case, they met the surprise of their lives when they learned (from Kurt Gödel) that even in mathematics decidability is unattainable. In computer science it is at times an empirical affair: many tasks given to a computer for deciding the truth or falsity (called the truth value) of a sentence are performable, and demonstrably so; at times the demonstration is purely abstract, as when the time the task takes to complete is much too long. And some tasks are not known to be performable or not. And then, if such a task is given to a computer, then, if the computer finishes the task, it is performable and the truth or falsity of the sentence in question is decidable; but until the task is completed it cannot be decided whether the task will be completed soon or not.

At this juncture the story of contemporary philosophy of science gets much too involved. First, there are modified conceptions of the empirical character of science: the requirement from a sentence that may claim (empirical) scientific status is lessened by leaps and bounds. The exercise of the lessening of the requirement is curious: the input into it increases all the time, yet the output becomes less and less satisfactory, to the point that its own advocates are too unhappy about it to conceal their displeasure with it. Briefly, the idea of certitude is replaced by probability, by a limitation on the domain of the validity of the proof, and by the abandonment of the very concept of proof, which imports finality, in favour of the concept of relative truth. What all of these substitutes for the idea of decidability share is the following incredibly fantastic idea:

though a sentence is not usually decidable, its scientific character is. (In other words, though finding the truth or falsity of a sentence is not generally assured, the truth or falsity of the claim that it is scientific is easily assured.) This is a fantastic idea, since one way or another science is linked to truth, no matter how tenuously. Yet it is accepted upon faith. What accepting a sentence on faith means is not clear, but its political implication usually is: the society of the elect are known by their faith.

This is how the first answer, the idea of decidability, upheld by philosophers of science, slowly degenerates into the second answer, the idea that science is a social status of sorts, upheld by the sociologists of science.

Science is Public and Empirical

Our question has undergone some transformations. First we asked, what is science? This was replaced by, what is the essence of science? And this was translated to, what differentiates science from other intellectual matters? And this was narrowed down to, what characteristic is peculiar to science? And rather than go over the same exercise yet again we should translate this into the following, final wording: what is the specific characteristic of science?

There are two very generally accepted answers to this last question, what is the specific characteristic of science? This would be very comforting, except that the two answers do not overlap. The one is, science is public; the other is, science is empirical.

Take the public character of science first. The claim made here is that most intellectual activities are esoteric, closed to the general public, that entry is conditioned–whether on some natural gift or some specific preparation not given to all or both. Is this true? If so, then by what virtue do the public-relations spokespeople of science dare bounce people who wish to be or appear scientific? More than that: if science is open-exoteric-then why does science need public relations in the first place? It needs recruiting officers, talent scouts, instructors; but why bouncers?

What does it matter to science that some esoteric groups appear to be exoteric and other groups have esoteric reasons to oppose science? What does it matter to science, asked Einstein, if this or that church opposes it? If it is necessary to expose and unmask those who masquerade as scientists, is it not best to do so by examining the question, how open are their clubs? Perhaps the bouncers suggest that this is not such a good idea as it may deprive them of their jobs; if so then they are disqualified from debating this question because of a conflict of interests.

The second answer is that science is empirical. Now surely Sir Karl Popper is quite right when observing, as a matter of historical fact, that astrology and alchemy and even parapsychology, are empirical as well. The public-relations spokespeople of science are outraged by this observation, and they protest that the empirical evidence in question is highly questionable, and often it is simply lies. This complicates matters immeasurably by raising two tough questions. What evidence is not questionable? Are all scientific reports honest and all parapsychological ones lies? It has been reported that some people pose as parapsychologists and are liars; it has also been reported that some people are genuine parapsychologists and are not liars. Are the reported liars not simply pseudo-parapsychologists? Since people who falsely call themselves scientists are unmasked as pseudo-scientists, and science is free of responsibility for them, surely the same privilege should be granted to parapsychology! The question here is not who is a liar (this question belongs to sociology, to criminology, to cultural history), but, what grants a theory the right to scientific status, namely, what is the characteristic of science? Supposing it is claims to empirical character. Are we to alter this supposition in the light of criticism to say that it is the employ of scientific empirical evidence? This, surely, is hardly helpful, unless we know what makes evidence scientific. Whatever it is, two obvious, extreme answers are unacceptable. The one is that scientific evidence is true: history is full of (historically important) empirical evidence that is known to be false. The other is that empirical evidence is bona fide. For it is undeniable that some parapsychology is bona fide; indeed, some famous individuals whose contributions to empirical science is unquestionable were known parapsychologists. William Crookes is the standard example for that.

The matter of alchemy or of astrology is even more complex: historians of alchemy and of astrology tell us that the better practitioners of these activities were bona fide, and that some of them even contributed to what is now deemed chemistry and astronomy. And we have still not said whether all the bona fide empirical evidence should count as scientific. There is a vast literature, going back to the writings of Galileo Galilei as to this question. It is called the literature on theory-ladenness, and for the following reason: if empirical evidence is based on theoretical suppositions, then it may be false unless the suppositions are known to be true. Suppose they are known to be true. Suppose they are known to be true a priori? Then science cannot be said to be thoroughly empirical; assuming, as we often do, that no intellectual activity is utterly devoid of some empirical component, and empirical character ceases to be the differentiating characteristic of science. Suppose, then that the theocratical suppositions are known to be true on some empirical grounds. Then, are these free of theoretical supposition? If no, then the question returns full strength. If yes, then there is some empirical evidence based on nothing but experience. Can this exist? If so, do we have an instance of it?

Public-relations spokespeople of science hardly ever stay to hear all of these objections. Usually they or their seniors are in charge of (politically significant) discussions and they curtail them long before they are exposed that much. They have a strong technique to justify their impatience: no matter how abstract and distant from real life their discussion is, they sooner or later turn the complaint that their opponents are remote from real life. In real life, they intimate, science is successful. This success should be analysed. And profound analysis tells us that the success is predictive, i.e. it yields successful forecasts.

Thus, if we have no proof, we have systematic probability: the earthly success of science is too systematic to be merely accidental; rather, this systematic earthly success is due to the systematic success of science in its efforts to confirm its theories.

Science Power Worship is Hilarious

The discussion thus leads to the question, how come science is so systematically successful in its efforts to confirm its theories? What is the trick? Can it be learned? Can it be emulated by parapsychology? The answer must be, it can be learned, or else the success would not be so systematic. How can it be learned? There are two answers to this question, that of the traditional philosophers of science and that of the sociologists of science led by Michael Polanyi and his follower Thomas S. Kuhn. The one is exoteric, and so should be able to describe the formula that makes science an ongoing success; the other is esoteric and describes the knowledge of the formula an ineffable personal knowledge of the trade secret which is transmitted by master to apprentice. This is the worst aspect of the philosophy of science as currently practiced, as public-relations mulch: science is predictive success or it is nothing. If the worse comes to the worst, then scientists are better viewed as exoteric magicians who simply deliver the goods and no questions asked. But the trick is to take as much time as possible getting to the worst, and in the meantime perform the real function as bouncers. Let us review the discussion which leads to this blind ally and see clearly that it is but killing time, that the only serious, bona fide ideas involved in the time-killing activity are long dead.

There are two schools of thought in the establishment of the philosophy of science, inductivists and instrumentalists, so-called. Inductivism is the preferred view, as it suggests that scientific theories are probable, even if not provable. It is not clear what this probability of theories means, and, regardless of what exactly it is, it is not clear which evidence raises it and how. It will soon be shown that this is all sham. When inductivism is relinquished, its touchstone, the idea that the goodness of science is shown as it yields useful predictions, or probable forecasts, becomes more than a touchstone; it becomes the criterion of goodness: science, it is then suggested, is nothing but applied mathematics; its merit is practical. Consequently, it turns out, its merit is not theoretical but merely practical–it is merely instrumental.

Assuming for the moment that the value of science is nothing but true forecasts does not yield the conclusion that all true forecasts are desirable. The approval of true forecasts runs against the very well known very commonsensical facts, first, that some forecasts are terrible and are better not fulfilled and second, that a true forecast may mislead.

There is no question that this is the case, and the public-relations spokespeople of science are not in the least unaware of it. They do not deny this either. They only ignore it. What this oversight amounts to is clear: we are in control and there is no reason to fear that our forecasts are alarming or that we are misled by them. This is establishment talk. The world is threatened by destruction from pollution, from the proliferation of nuclear weapons, from population explosion and from the ever-increasing gulf between the rich nations and the poor nations. But there is nothing to worry about. All will turn out to be well.

Query: is this a scientific forecast or false prophecy? It is neither; it is cheap public-relations mulch. To see how unserious this mulch is one only needs see the low level of the current debates in the leading literature on the matter. The topic common to both inductivists and instrumentalists is the question, are there any items of empirical information free of theoretical bias? Is there any 'pure' evidence? Or is all evidence theory-laden?

The onus here is on the party that says there is 'pure' evidence: they should offer instances. There are none. The only candidates in history were Bacon's naive realism and Locke's sensationalism. Naive realism is refuted: the naive see the sun rise and set, and, to cite an example of Erwin

Schrödinger, the sun appears as not bigger than a cathedral, which means, given some simple trigonometry, that the distance between east and west is less than one day's walk. In an attempt to replace naive realism in view of the criticism from Copernicanism, Locke revived sensationalism, claiming that motion is not perceived. It is. Sensationalism is refuted, anyway, by myriads of experiments. This is the end of that discussion.

The next discussions concern theories. There is hardly a debate between the inductivists who ascribe to theories informative contents and the instrumentalists who deny that and read the theories as a mere façon de parler. Rather, each struggles with its own problems.

The theory of induction contains two competing sub-theories, which deal with the question, what kind of evidence confirms a given theory? They both violate the only rule of science universally endorsed within science since its enactment in the early days of the scientific revolution: both sub-theories do not confine their discussion to repeatable, (allegedly) repeated observations, but rather they refer to unique items of experience. In addition to this, each of these sub-theories is easily refuted by very simple arguments. A vast literature is devoted to these refutations in an effort to get rid of them; worse, still, as usual with public-relations spokespeople in a defensive mood, they do not state the difficulties they struggle with and so sound arcane.

The first of these two sub-theories of inductive evidence is the instantiation theory of inductive evidence: a scientific theory is confirmed by instances to it. What then, is an instance to a given theory? What is an instance to a theory of gravity? Anything falling? Decidedly not: a falling feather disobeys even Galileo's theory of gravity. What then counts? Rather than discuss gravity, the public-relations spokespeople of science discuss such generalizations as, 'All ravens are black', forgetting that when asserted, these are items of evidence, not theories. What, then, counts as an instance? Every item that does not contradict a theory is an instance of it, since theories can be stated as prohibitions: there exists no perpetual-motion machine, for example; no gas deviates from the gas-law equation, etc. And then every item that is not a perpetual motion machine instantiates the law of conservation of energy! It sounds very counter-intuitive to admit every non-refutation as an instance, since this invites all irrelevancies into the picture. This is known as Hempel's paradoxes (in the plural) of confirmation. The counter-intuitive character of this fact is taken to be powerful criticism of the theory, despite the obvious fact that the theory is anti-intuitionist and so its advocates should not be disturbed in the least by its counter-intuitive character. For, were it permitted to rely on intuition, then the intuition that the world is law-governed is strongest, and so it dispenses with the problem of induction ab initio. A vast literature is devoted to efforts to rescue the instantiation theory of induction from its (seemingly?) counter-intuitive character.

The second sub-theory of inductive evidence has for its background the musing that the function describing confirmation is a unique [!] function of both theory and all [!] extant evidence and of nothing more-or, if not uniquely determined, at least all such functions must [!] conform to the mathematical calculus of probability. The theory, if it can be called that, is that the desired evidence is that which renders a theory probable in accord with this musing. The musing has two advantages. First, it identifies the vague concept of probable hypothesis with the clear concept of conformity to the mathematical calculus of probability. Second, it offers a clear-cut estimate of probability–on the further musing that the probability of an event equals the distribution to which it belongs. Except that this musing has no room for distributions other than those offered by theories whose probability this musing should help us estimate.

But evidence does play a great role both in research and in practical life, the self-appointed public-relations spokespeople of science exclaim in exasperation. Indeed, this is so, and from the very start; what they have promised to expose us to is not this profundity but the answer to the questions how and why? They even overlook the more basic question, which is, does evidence play the same role in research as in practical life? It is more than reasonable to assume that the answer to this question is in the negative. The public-relations spokespeople of science take it for granted that the answer is in the affirmative. So much so, that they refuse to ask it–or to hear anyone who asks it. For, clearly, investigators, be they detectives or scientists-from popular fiction or from real life–do pay great attention to minute details, as they must, and then, when their search is concluded, they ignore most of the minute details and blow up the others. How else can the small details of scientific discovery grow so large as to cover the whole of our city-scape?

When researchers–detectives or scientists–follow a clue, they do so at their own risk. Hence, science is not as successful as it looks. Even in fiction detectives do lots and lots of legwork that ends up in blind alleys. But when successful, results have to be confirmed, and their confirmations have to be easily repeatable. When the success in question is scientific, it matters little to the practical world what these are.

When the success is claimed to be worldly, then there are legal standards for confirmation, that philosophers of science assiduously ignore. In medicine, for example, a claim for success has to be repeated in vitro, then in vivo on laboratory animals, then on human specimens under specified controls, and then proved satisfactory by some complex standards.

Is this not the inductive canon that the philosophers of science seek? No. It is not any philosopher's stone, but the real, human, limited, at times highly defective system. The established philosophers of science ignore it as it is no use to them in their self-appointed function as public-relations spokespeople of science, as self-appointed bouncers of the haughty club of science.

Popper's Critique of Inductivism is Overkill

Popper's critique of the instantiation theory of induction is simple: there are practices accepted in the scientific community concerning what counts there as confirmation, and these should be taken into account when a theory of confirmation is presented: not all instances confirm theories but, at most, those which were expected to refute it and failed. This is admitted obliquely by Carl G. Hempel, the chief discussant of the matter of instantiation and its afflictions, but not openly. Yet, he is not satisfied with the situation as he seeks a formal criterion for confirmation. He thus cannot fully admit Popper's (empirical) assertion that at most only failed refutations of a theory confirm it, as failure is not a formal criterion.

The more extensive criticism of Popper is directed against the identification of confirmation with some function abiding by the calculus of probability. This is surprising, since the probability sub-theory of confirming evidence and the instantiation sub-theory of confirming evidence are, of course, but variants of the theory of induction. (Indeed, the common way to dismiss the paradoxes is to dismiss most of the confirming instances as practically irrelevant by the claim that they scarcely raise the probability of the theory which, strictly speaking, they hardly confirm.) Perhaps he does so on account of its lingering popularity. For decades Popper presented criticisms of this idea, and it would have been dropped from the agenda, were the public-relations spokespeople of science able to exhibit some sensitivity to devastating criticism.

First, says Popper, confirmation cannot be probability as it reflects the force of the evidence and not the informative content of the theory prior to evidence. Therefore, at least confirmation should be probability increase, not probability. (This, he embarrassingly adds, resembles Galileo's announcement that gravity is proportional not to velocity but to its increase.) And probability increase is certainly not a function abiding by the formal calculus of probability. The point is easy to demonstrate. Here is Popper's demonstration.

Let us write 'P (h)=r' and 'P (h, e)=r' to denote absolute and relative probability in the usual way; suppose a theory hl is absolutely probable and some evidence e1 reduces its probability, whereas h2 is improbable yet some evidence e2 (which may be the same as e1 if you wish) raises its probability, but not much, so that

$P(h_1) \geq P(h_1, e)$,

$P(h_2) \leq P(h_2, e)$,

yet

$P(h_1, e) \geq (h_2, e)$.

Clearly, though h1 is more probable than h2 it is more confirmed by the evidence. The objection that this is impossible is groundless. Moreover, a model for it is easy to construct. Here is one.

Consider event E which is the next throwhof a die. Take the following cases:

h1 : E is not a 1.

h2 : E is a 2.

e: E is 1 or 2 or 3 or 4.

The more serious criticism of the identification of confirmation with probability is directed at the identification of probability with distributions. The probability of a hypothesis concerning a distribution cannot possibly be the same as the distribution it depicts, since we have competing hypotheses concerning a given distribution, and the sum of their probabilities is a fraction, but they can each ascribe a high distribution so that the sum of their distributions will exceed unity. Attempting to escape this criticism one may seek refuge in the preference for equi-distribution. This lands one in the classical paradoxes of probability. Attempting to escape this criticism one may seek refuge in the preference for confirmed distributions. This not only begs the question: it raises the paradox of perfect evidence: the evidence that fits a given distribution perfectly both raises its probability and keeps it intact-which is absurd.

How can one go on examining the defunct option that science equals probability? Only on the supposition that science is a success story and the public-relations spokespeople of science are convinced that the difficulties piled on their road to present science as a success story are marginal.

Is science a success story? Decidedly yes. What kind of success story? It is hard to specify exactly in all detail, but the first details are clear: science is a success story in that it needs no public-relations spokespeople and it is a success story not in their (vulgar) sense of the word.

Science is more than Scientific Technology

The vulgar view of science as success is the view of the scientist as a person with a powerful insight, a sort of a magician. Surprisingly, this view does not conflict with the view of science as esoteric, since it alleges that only scientific research is esoteric, not the fruits of science, which are for all to see. The idea that scientific research is somewhat mysterious does conflict with the inductivist idea that research, too, is open to all. This is the idea that science is open to a simple algorithm that can be mastered by everybody. This view of science is dismissed by Popper derisively as the idea of 'science-making sausage machine'. Under the influence of Einstein it is now generally rejected as too simplistic-by all except some zealous adherents to the original idea of artificial intelligence. Today, there is a vast and exciting literature on techniques to aid the process of developing ideas that may lead to discovery ('heuristic', is the Greek word for this, which was coined by William Whewell, the great nineteenth-century philosopher who was the first to criticize the idea that there can be a science-producing algorithm).

The idea of a science-producing algorithm proper was recently replaced by, or rather modified as, the idea of normal science, so-called, developed around 1960 by Thomas S. Kuhn. The popularity of his philosophy, if it can be called that, rests on his conception of science, normal and exceptional. The exceptional scientist is the leader who prescribes a paradigm, namely a chief example, and a normal scientist solves problems following it. This suggests that the real magic rests in the leadership, the scientific character of the enterprise they lead rests on the obedience of their followers, the normal scientists, and the problems the normal scientists solve are quasi-algorithmically soluble: they are not so simple that a computer or a simple mind can solve, but they are not so difficult as to defy solution.

It is easy to see the allure of this philosophy: it balances a few ideas that seemingly conflict with each other but which share the goodness of being both popular and useful for the celebrated self-appointed public-relations spokespeople of science; it presents science as assured but not without some expertise and hard work; it assures science its openness to a reasonable degree, so that the public-relations spokespeople of science can see a little of the mystery involved–just enough to advocate it but not enough to partake in it actively.

What is missing in the concoction is the mystery–not the alleged mystery of the leaders of science who cannot and would not divulge the secrets of their craft, but the unmistakable mystery that is the secret of the universe.

It is not that the self-appointed public-relations spokespeople of science are not willing to praise science as the big search; after all they will say anything to glorify it. But they will use the public-relations criteria to judge when it is advisable to praise the search as the intellectual frontier and when to present it in a mundane fashion. Except that they claim to be philosophers, and thus bolster each move with a principle, and thus render complementary compliments into contradictory credentials.

Let us see if this serious matter cannot be approached somewhat more seriously and without the tricks of the trade of the public-relations spokespeople of science.

Science is a Natural Religion

There is so much to do other than gate-keeping. Certain grounds may perhaps be cleared. Certain assertions should be endorsed as a matter of course or clearly dissented from, though, of course, we may also examine them in great detail if we wish. It should be conceded that traditionally science admits as evidence only repeatable evidence, though we may examine this characteristic in great detail if we wish. It should be admitted that traditionally science admits only items open to the general public, though we may examine this characteristic in great detail, too, if we wish. It should be admitted that some of the evidence which science traditionally admits as true is later on deemed false, but not overlooked; rather they are qualified and readjusted. Though we may examine all this in great detail, if we wish, we may want to know right now why these rules are deemed obligatory. The answer to this question is simple: it is taken to be the role of empirical science to explain known facts.

More should be stressed at once: the rules are introduced not as taboos but as reasonable commonsense ideas. It should be clear that one may break any of these rules, but openly and at one's own risk.

The paradigm is Max Planck, who took upon himself a most unusual research project and voluntarily and as much as he could ignored all items that he could not square with it. This was his own private road to quantization.

Past this we are ignorant, and it is advisable to admit ignorance in many areas and open them up for genuine research that may get the philosophy of science out of its recently acquired role as gate-keeper and bouncer and into a proliferation of researches. We do not know how empirical empirical science is, though we have the feel that some technologically oriented researches are much nearer to common experience than some speculative studies of first principles. We do not know how a research report is judged scientific and/or deserving of publication. We know that some erroneous criteria are used, and that much latitude is exercised in the matter; but more information is needed and more deliberation and experimentation. We do not know how much of science is empirical and how much is guided by general principles, by the culture at large and even by politics-international, national or of the local chapter of a scientific society or the local department in the university.

We simply do not know enough about how science intertwines with other activities, and we only have an inkling as to what is the minimum requirement for a society that wishes to allow it to flourish, namely, freedom of speech and of dissent and of criticism and of organization to that effect.

It is hard to say what other item, if any, is generally admitted as a basic tradition. This invites one to scout beyond the current horizon, and seek in the past some heuristic that might be helpful. And the point to start with should, perhaps, be the roots of the unbecoming hostility to metaphysics and to religion that is so characteristic of contemporary philosophy of science that induces its practitioners to undertake the lowly task of bouncers-with-a-sense-of-mission. The rise of modern science is the starting point, as the heritage from that noble period is in great need for revision.

Here, only one aspect of that period will be mentioned, the idea of natural religion. It is the idea that religion comprises a doctrine plus a ritual, that the doctrine is either revealed or natural to all thinking humans as such, and that ritual is prescribed in accord with doctrine. All of these items are nowadays known to be false, but let us overlook this for a while.

The idea of natural religion was that it is supplemented by revealed religion, not inconsistent with it. This cannot be admitted without some qualification on the religion under discussion, but here no specific religion is discussed. The idea of natural religious doctrine, natural theology or rational theology, so-called, is the proof of the existence of God. This proof is now dead. The idea of the natural or rational ritual is the idea of research as worship. This idea has a tremendous attraction to some researchers, including Einstein, and other researchers consider it silly. The main obstacle in this matter is not the item under consideration, but the idea of religion as belief. The involvement of belief as a central item in any religion is a very strong item of all Western religions–not of all Eastern ones. It also led to the idea that superstitions are prejudices, namely beliefs in ideas that are objectionable or at least not warranted. It is well-known that superstitious people are sceptical, not dogmatic as the philosophers of science describe them, though, of course, what they particularly lack is the ability to be critically minded about their guiding ideas.

Traditional philosophy of science took it for granted that the dogmatic and the superstitious share the errors of clinging to erroneous metaphysical systems, now better known as intellectual frameworks. It recommended not to endorse any unless it is proven. Then Kant proved that a proven intellectual framework is a set of synthetic propositions a priori proven. Then Russell and Einstein between them proved that such propositions do not exist, and the gate-keepers decided to oust all intellectual frameworks. These were reintroduced by social anthropologists and by the posthumous writings of Ludwig Wittgenstein, who spoke of them, somewhat enigmatically, as of 'forms of life'. As far as science is concerned, they were discovered by various historians of science of the Koyre school, and were then sanctified by those who identified them with Kuhnian paradigms. This is a gross error, of course, since the whole point of Kuhn's idea of the scientific paradigms is to prevent the conflict between the diverse scientific systems, especially the classical Newtonian and the modern. To use the jargon expression, he insists that paradigms are incommensurable. The whole confusion, and the bouncing that goes with it, will be cleared once we notice that intellectual frameworks do compete, and that science may both use some of them and be used as arguments for and against some of them.

Obviously, a researcher may consciously and clearly follow two different guiding ideas, employ competing intellectual frameworks–from not knowing which of them is true. Taking notice of this simple, commonsense fact will free the theory of scientific research from its obsession with rational belief. Current philosophy of science is fixated on the study of rational belief without any criticism of traditional ideas of belief in general and of scientific or rational belief in particular. The source of this idea was Sir Francis Bacon's superbly intelligent and highly influential doctrine of prejudice: the prejudiced cannot be productive researchers, since theories colour the way facts are observed, so that facts cease to act as refutations and as correctives of views, so that the prejudiced are blinded to contrary evidence and can only perpetrate their prejudices by endlessly multiplying evidence in their favour. This theory still animates the pseudo-researches of the self-appointed gate-keepers, even though it is amply refuted. The worst of it is that philosophy of science centres on the problem, what theory deserves acceptance, where acceptance means credence. Yet it is well known that we are unable to control our credence, certainly not to confine it to a simple algorithm. It is here that the roots of the erroneous view of science as a competitor of religion can be found and corrected. This is not to deny that scientific research can be a thoroughly religious affair, a dedication to the search of the secret of the universe. This is not to deny that the religious aspect of research is not obligatory either.

Once this is realized, the avenue is open to the study of science as a central item in our culture and to see the interaction of other items in our culture with science. It is interesting to view the philosophy of science as part-and-parcel of our culture rather than as an isolated item in philosophy. What isolates the philosophy of science from the philosophy of human culture in general is the idea of the gate-keepers that any item not quite scientific is beneath the dignity of the philosopher. This idea is not quite philosophical. Nothing human is alien to any philosopher-of science or of any other aspect of human culture.

The philosophy of science is concerned with the assumptions, foundations, methods and implications of science. In addition to these central problems for science as a whole, many philosophers of science consider these problems as they apply to particular sciences. Some philosophers of science also use contemporary results in science to draw philosophical morals.

Although most practitioners are philosophers, several prominent scientists have contributed to the field and still do. Other prominent scientists have felt that the practical effect on their work is limited: "Philosophy of science is about as useful to scientists as ornithology is to birds," according to physicist Richard Feynman.

Philosophy of science focuses on metaphysical, epistemic and semantic aspects of science. Ethical issues such as bioethics and scientific misconduct are usually considered ethics or science studies rather than philosophy of science.

Analysis is the activity of breaking an observation or theory down into simpler concepts in order to understand it. Analysis is as essential to science as it is to all rational enterprises. For example, the task of describing mathematically the motion of a projectile is made easier by separating out the force of gravity, angle of projection and initial velocity. After such analysis it is possible to formulate a suitable theory of motion.

5

Quantum Mechanics in Philosophy

Introduction

The structure of quantum mechanics may be divided into a mathematical formalism and a physical interpretation of that formalism. The mathematical formalism, which is the core of quantum mechanics, is a differential wave equation, the solution of which is usually termed the *psi* (ö) function; the wave equation was first developed by Erwin Schrödinger, who pursued Louis de Broglie's extension of the concept of wave-corpuscle duality from light to elementary particles of matter. The advantage of this formalism is that it yields, on a probabilistic basis, numerical values permitting a more complete mathematical description of microphysical states, including prediction of future states, than has any other formalism so far. The disadvantage of the mathematical apparatus of quantum mechanics is that the only widely accepted physical interpretation for it contradicts several of man's most basic intuitions concerning matter. Specifically, quantum mechanical computations, in contrast to the classical laws of the macrophysical realm, do not yield arbitrarily exact values for position and momentum coordinates of microparticles. According to the well-known uncertainty relation, the more exactly the position of a microparticle is known, the less exactly the momentum, and *vice versa.*

In view of the success of the mathematics of quantum mechanics for the derivation of useful physical values, the obvious question arises, What is the physical significance of the wave function? Can it be that matter is, indeed, undulatory? It is over this question of the physical interpretation of the mathematics of quantum mechanics that scores of philosophers and scientists have splintered their pens.

The evolution of quantum mechanical theories is a trail littered with unsatisfactory explanations. De Broglie originally proposed that matter is wavelike and that the waves described by quantum mechanics do not "represent" the system; they *are* the system. This explanation encounters enormous difficulties, far too complex to enter into here, but the nature of some of them may be indicated by noting that a literal acceptance of the physical reality of the wave function would involve such concepts as that of physical space with an almost unlimited number of dimensions. And most graphically, such an interpretation cannot explain satisfactorily the fact that a single micro-object upon impact on a sensitive emulsion leaves a spot, not the imprint of a wave front. Max Born originally suggested the alternative: Matter is corpuscular, and the wave function describes not the particles but our knowledge about them. This ingenious theory unfortunately runs into equally disastrous physical facts, which are best illustrated by reference to the now classic two-slit interference experiment. When particles are allowed to pass through two narrow slits in a barrier before striking a sensitive emulsion, the impacts form an interference pattern that can be explained only on the basis of the wavelike characteristics of microbodies.

The Copenhagen Interpretation, developed by Niels Bohr and Werner Heisenberg, resolves the contradictions of previous interpretations by postulating that no observable has a value before a measurement of that observable has been made. As Heisenberg declared, "The path comes into existence only when we know it." Thus, it becomes meaningless to speak of the characteristics of matter at any particular moment without empirical data in hand relating to that moment. It is senseless to speak of the position of a particle ("position" is a property of the corpuscular theory) without a measurement of position; it is equally unjustified to speak of the momentum (a wave property) without a measurement of momentum. This reconciling of classically incompatible properties by granting them existence only at the moment of measurement is usually called complementarity and is the heart of the most critical discussions of quantum mechanics.

Physicists and philosophers of science do not agree on a single definition of complementarity, but a satisfactory definition is the one just given that is, contradictory properties of a microbody are reconciled by granting these individual properties existence only at separate moments of measurement. Another formulation, one that evades the question of "existence" of properties but that nonetheless is commonly given, is to say that the quantum description of phenomena divides into two mutually exclusive classes that complement each other in the sense that one must combine them in order to have a complete description in classical terms. This latter view was the one accepted by Oppenheimer when he stated that the notion of complementarity "recognizes [that] various ways of talking about physical experience may each have validity, and may each be necessary for the adequate description of the physical world, and may yet stand in a mutually exclusive relationship to each other, so that to a situation to which one applies, there may be no consistent possibility of applying the other." It must also be added that even such early leaders in quantum mechanics as Bohr and Wolfgang Pauli did not entirely agree in their definitions of complementarity. The essential problem was the perennial one in the history of science: giving a verbal interpretation to a mathematical relationship.

Before World War II the views of Soviet physicists on quantum mechanics were quite similar to those of advanced scientists elsewhere. Russian physics was in many ways an extension of central and west European physics. The work of such men as Bohr and Heisenberg influenced scientists in the Soviet Union as it did everywhere. Indeed, Soviet physicists spoke of the "Russian branch" (*filial*) of the Copenhagen School, composed of a group of highly talented theoretical physicists, including M. P. Bronshtein, L. D. Landau, I. E. Tamm, and V. A. Fock. And yet behind this exterior of agreement with scientists everywhere on quantum mechanics (or, more accurately, disagreements similar to those everywhere), as early as the 1920's certain Soviet physicists were aware that dialectical materialism might some day be interpreted in a way that could influence their research. Lenin had, after all, devoted an entire book, *Materialism and Empirio-Criticism*, to the crisis in interpretations of physics and had particularly criticized the neopositivism of Ernst Mach, out of which much of the philosophy of modern physics grew. Lenin's assertion that a dialectical materialist must recognize the existence of matter separate and independent from the mind, while not inherently contradictory to quantum mechanics, could be regarded as at least uncongenial to the Copenhagen School's disinclination to comment upon matter in the absence of sensible measurement. And the extension of the concept of complementarity beyond physics to other realms, including ethical and cultural problems, by certain members of the Copenhagen School almost guaranteed some conflict with representatives of Marxism. As early as 1929 the leading Soviet philosopher at that time, A. M. Deborin, gave a lecture on "Lenin and the Crisis of Contemporary Physics" to the Academy of Sciences. But the first serious Soviet critique of the customary interpretation of quantum mechanics in a physics journal, rather than a philosophy journal, occurred in 1936, written by K. V. Nikol'skii. In the dispute that developed between Nikol'skii and V. A. Fock, a leading interpreter of quantum mechanics in the Soviet Union to the present day and originally an adherent of the

Copenhagen School, Nikol'skii called the Copenhagen Interpretation "idealistic" and "Machist," two appellations that were to be frequently utilized after World War II by Soviet Marxist critics. Nikol'skii's own view of quantum mechanics deserves examination for still another reason: It was a purely statistical approach, with only a few differences from D. I. Blokhintsev's postwar "ensemble" interpretation, which will be discussed in greater detail below.

With mention of Nikol'skii's "purely statistical" approach it is appropriate at this point to insert a few remarks on the concept of probability, which is crucial to any interpretation of quantum mechanics. Probability in quantum mechanics has been interpreted by different scholars in both epistemological and statistical senses. The statistical, or frequency, approach, used by Nikol'skii, was an attempted objective interpretation in which probability was seen as inherent in nature. On the other hand, a number of scholars have seen quantum mechanics, particularly through Born's original interpretation, as containing probability because of its epistemological assumptions, and have even discussed such peculiar things as "waves of knowledge." The distinction between these two approaches, often blurred in discussions of quantum mechanics, is absolutely necessary in attempting to decide whether a theory that is irreducibly probabilistic is also necessarily idealistic.

Fock's interpretation in 1936 of the physical significance of the wave function was essentially the same as that of the Copenhagen School, which combined Born's emphasis on the mathematical description of man's knowledge of the microworld with its own emphasis on the role of measurement; Fock stated in an introduction to a Russian translation of the 1935 debate of Einstein, Podolsky, and Rosen versus Bohr:

> In quantum mechanics the conception of state is merged with the conception of "information about the state obtained as a result of a specific maximally accurate operation." In quantum mechanics the wave function describes not the state in the usual sense, but rather this "information about the state."

The importance of this prewar position of Fock's lies in its subtle difference from his stated views after the war, when he was placed under heavy pressure to desert the Copenhagen School. Nevertheless, Fock's change in position was small compared to the swerves that occurred in the views of several other prominent Soviet philosophers and scientists.

The debate of the 1930's did not, however, leave a permanent imprint on Soviet attitudes toward quantum mechanics. Even many philosophers accepted much of the Copenhagen view. Early in 1947 M. E. Omel'ianovskii, a Ukrainian philosopher who with Fock and Blokhintsev completes the triumvirate whose views will be examined in detail here, argued a position on quantum mechanics close enough to the Copenhagen orientation to cause him intense embarrassment only a few months later. His 1947 book has become more interesting in recent years, since it represents a view that Omel'ianovskii later returned to and developed further. In this work, *V. I. Lenin and TwentiethCentury Physics*, Omel'ianovskii accepted much of the common interpretation of quantum mechanics. He recognized and used such terms as "the uncertainty principle" and "Bohr's principle of complementarity." He guarded against using these concepts in a way that might deny physical reality, as he said certain people (including Bohr on occasion) had done, but his major thesis in this book was a defense of the surprising but necessary concepts of modern physics against adherents of the determinism of Laplace, by then clearly outdated. Buried within Omel'ianovskii's arguments, however, one may observe, at least in retrospect, the core of his own interpretation of quantum mechanics and of his later criticisms of the Copenhagen School. Although he acquiesced in the vocabulary of Copenhagen, he emphasized that the correct interpretation of quantum mechanics began with a recognition of the peculiar qualities of microparticles, not with problems of cognition: "And so we have come to the conclusion that Heisenberg's uncertainty principle, like Bohr's principle of complementarity, is a generalized expression of the facts of the dual (corpuscular and undulatory) nature of microscopic objects." Thus, the uncertainty principle was not actually an epistemological limitation or a limitation of knowledge, but a direct result of the combined wavelike and corpusclelike nature of micro-objects, which was the *material* reason why classical concepts could not be applied to the microworld. In view of this material source of the phenomenon of canonically conjugate parameters, one could not expect ever to possess simultaneous exact values of position and momentum of elementary particles. For his recognition of the basic position of contemporary views on quantum mechanics, Omel'ianovskii was soon criticized severely, and eventually produced a second edition of his book, in which, most notably, he repudiated the principle of complementarity.

The most important event of the postwar years for Soviet scholarship was A. A. Zhdanov's speech on June 24, 1947, at the discussion of G. F. Aleksandrov *History of Western European Philosophy*, an event well known to historians of the Soviet Union. Only near the end of that speech did Zhdanov mention specific issues in science, and less than a sentence referred directly to quantum mechanics: "The Kantian vagaries of modern bourgeois atomic physicists lead them to inferences about the electron's possessing 'free will,' to attempts to describe matter as only a certain conjunction of waves, and to other devilish tricks."

Although Zhdanov's speech is now known as the beginning of the most intense ideological campaign in the history of Soviet scholarship, the Zhdanovshchina, the first few issues of the new journal *Problems of Philosophy* that appeared after the speech were surprisingly unorthodox. Evidently taking seriously the slogan of the journal "to develop and carry further" Marxist-Leninist theory the editors promoted vital discussions of several philosophic questions. In no field was this vitality more apparent than in the philosophy of physics; the second issue of *Problems of Philosophy* contained an article by the outstanding theoretical physicist M. A. Markov, a specialist in the relativity theory of elementary particles, which may well still be the most outspoken presentation of the Copenhagen point of view to appear in a Soviet philosophy journal since World War II. Just why Markov chose this moment, after Zhdanov's condemnation of Aleksandrov and during the tightening of ideological controls, to expose himself to criticism so extensively may never be known, but there are several hints available. Markov was a research scientist in the Physics Institute of the U.S.S.R. Academy of Sciences, the organization that in the past had most stoutly defended international viewpoints in science and that would do so in the future, incurring sharp criticism from political activists. It is probable that the theoretical physicists in the Academy, aware since the 1930's that, given the will, dialectical materialism could be used against prevalent interpretations of quantum mechanics, decided that the nascent ideological campaign meant that an official position on quantum mechanics was very likely to be imposed and felt that an early attempt to make that official position compatible with contemporary quantum theory was necessary. Markov probably knew well just how controversial his article would be, but hoped, first, that it would be vindicated, and second, that even if his point of view was rejected, the final compromise would be more palatable to the physicists as a result of his strong stand. Furthermore, Markov was able to capitalize on a feud among the professional philosophers. As the course of the debate illustrated, the chief editor of the new philosophy journal *Problems of Philosophy* was disliked by the old guard, which had published *Under the Banner of Marxism*, the major Soviet journal of philosophy from 1922 to 1944. The debate over Markov consequently contained many dimensions: It was an effort by the physicists to protect quantum mechanics, it was a volley in a feud among philosophers, and it was a decisive struggle over whether physicists or philosophers would have the ultimate influence on the philosophy of science in the postwar period.

Markov accepted modern quantum theory completely and agreed with Bohr's position in Bohr's debate with Einstein, Podolsky, and Rosen. Thus, Markov considered quantum mechanics to be complete, in the technical sense that no experiment that did not contradict it could yield results not predicted by it; and he consequently rejected all attempts to explain the behavior of microparticles on the basis of "hidden parameter" theories that would later permit restitution of the concepts of classical physics: "It is impossible to regard quantum mechanics as a classical mechanics that has been corrupted by our 'lack of knowledge.'" Such complementary functions as "momentum" and "position" simply did not have simultaneous values; to suggest that they did would mean contradicting quantum theory.

Not only was Markov's view on conjugate parameters typical of the Copenhagen School, but his approach to science bore few traits of dialectical materialism despite his initial quotations from Marxist classics. He asked that no statements be made that could not be empirically verified; he accepted relativity theory, including relativity of spatial and temporal intervals; he used the term "complementarity" without hesitation. To be sure, he affirmed that his view of science was "materialist" and criticized James Jeans and other non-Soviet commentators on science, but nowhere in his article did he make any effort to illustrate the relevance of dialectical materialism to science.

Markov maintained that "truth" is obtained from many sources; when we speak of knowledge of the microworld, which we gain with instruments, we are speaking about knowledge that has come from three sources: nature, the instrument, and man. The language we use to describe our knowledge is perforce always "macroscopic" language, since this is the only language we possess. The measuring instrument performs the role of "translating" the microphenomenon into a macrolanguage accessible to man. "We consider physical reality to be that form of reality in which reality appears in the macroinstrument." Thus, according to Markov, our concept of physical reality is subjective to the extent that it is formed in macroscopic language and is "prepared" by the act of measurement, but it is objective in the sense that physical reality in quantum mechanics is a macroscopic form of the reality of the microworld.

The role of the measuring instrument is one of the thorniest issues in quantum mechanics. Markov's view was essentially in agreement with that of the Copenhagen Interpretation, according to which the wave describing a physical state spreads out over larger and larger values until a measurement is made, when a reduction of this spread (wave packet) to a sharp value occurs. Such an interpretation does indeed imply that complementary microphysical quantities have no inherent sharp values but that such values instead result from, or are "prepared by," the measurement. The most imaginative attempt, no doubt, to illustrate the striking results of some interpretations of this view of microphysical measurements was made by Schrodinger: A live cat is imprisoned in a cage in such a fashion that it will be poisoned if a lightsensitive trigger is activated by a photon. In order to strike the trigger, however, the photon, whose position is mathematically described by a wave function spread through space, has to pass through a half-silvered mirror. If the emitter sends one photon toward the half-mirror at a certain time, say, twelve o'clock, and then no one looks at the cat until one o'clock, are we still to assume, Schrödinger asked, that the cat is neither dead nor alive until someone actually looks in the cage and thus "prepares" the cat in a state of life or death?

The cat paradox raises a basic question about the difference in quantum mechanics between relational aspects and subjective aspects. It is quite possible to interpret the cat paradox from a relational standpoint in which subjective considerations play no role. The moment of interaction between the microentity and the macroentity is not the moment when the observer looks in the cage, but the moment when the photon either was reflected by the half-silvered mirror or passed through it, poisoning the cat. As Hilary Putnam has pointed out, contemporary physicists explain the cat paradox by saying that all macro-observables always have sharp values, and therefore the poisoning of the cat (if it occurred) was in itself a "measurement." Similarly, the moment of measurement when a photographic plate is used to record a quantum phenomenon is the moment of exposure, not the moment a human being looks at the plate. Thus, although measurement theory is still an uncertain area in the philosophy of science, measurement can be made an essential part of quantum mechanics without necessarily including subjective considerations.

Whatever Markov's views on the cat paradox, his opinion that the instrument "prepares" the state of microphysical reality, together with his acceptance of the Copenhagen Interpretation in general, exposed him to criticisms from a number of quarters, ranging from dogmatic ideologues to ordinary physicists with hopes for the eventual replacement of the views of Bohr and his colleagues by an interpretation more agreeable to common-sense intuition. The Markov article very quickly became the occasion for a full-blown controversy, involving several dozen participants, on the nature of physical reality and the dialectical materialist interpretation of quantum mechanics.

The polemic began with the appearance of an article by A. A. Maksimov in the *Literary Gazette*, an unusual place for a commentary on the philosophy of science. The article, entitled "Concerning a Philosophic Centaur," contained very serious allegations against Markov. As the title indicates, Maksimov considered Markov a strange species, a creature combining Western idealistic views on the philosophy of science with professions of loyalty to dialectical materialism. The background of the article reveals that it was more than a mere alternative viewpoint on quantum mechanics. Originally a physicist, Maksimov was a veteran of the struggles in philosophy of the 1920's and 1930's, and had been a member of the editorial board of the journal *Under the Banner of Marxism* when it ceased publication in 1944. After the war he became a defender of the most dogmatic positions in the philosophy of science; he accused such scientists as Einstein and Bohr not only of committing grievous mistakes in the interpretation of science, but even of factual errors. Maksimov originally submitted his article to *Problems of Philosophy*, whose editor, B. M. Kedrov, disagreed strongly with it but decided to print it "for discussion purposes" along with a rebuttal. After Maksimov had already returned galley proofs to the journal, another article of his with an identical title and very similar content appeared in the *Literary Gazette*, prefaced with a condescending note from the editors about the questionable quality of the new Soviet theoretical journal *Problems of Philosophy.* Kedrov immediately canceled Maksimov's article in *Problems of Philosophy* and in the following issue accused both Maksimov and the editors of the *Literary Gazette* of bad faith. Why were the editors of the *Literary Gazette* carrying on a vendetta against another publication? Here again the facts may never be known, but any attempt to explain this very rare disagreement between editorial boards in Stalinist Russia must begin with the fact that the member of the board of the *Literary Gazette* responsible for philosophy was none other than M. B. Mitin, who had been one of the editors of the defunct *Under the Banner of Marxism* and was obviously discontented with the new philosophy journal that had appeared in the wake of Zhdanov's speech.

The central point of Maksimov's article was that Markov was a supporter of the Copenhagen Interpretation of quantum mechanics and was trying to whitewash this view with a few statements about its agreement with dialectical materialism. According to Maksimov, M. A. Markov, following directly behind Bohr, asserts that in physical experiments there is a mutual influence of the microworld and the instrument, which in essence can never be overcome. However, this argument in no way touches upon the basic epistemological question. Does microreality with such properties exist before the application of an instrument by man? M. A. Markov answers that it does not exist, but is "prepared" by the instrument.

After the appearance of Maksimov's article in the *Literary Gazette*, the editors of *Problems of Philosophy* published a discussion of quantum mechanics. A number of authors gave Markov strong support, revealing the numerous errors of Maksimov, and the noted D. I. Blokhintsev also took a fairly positive view of Markov's interpretation of quantum mechanics. Other critics, however, pointed out Markov's "anthropomorphism" in science, a result of his emphasis on the "observer" (L. I. Storchak) and his disregard of Party loyalty, or partiinost' (I. K. Krushev, V. A. Mikhailov). But the factor that determined the eventual disapproval of Markov's article was a decision, beyond any doubt promoted by the Party, to replace Kedrov as editor of *Problems of Philosophy* by D. I. Chesnokov. It is clear that Maksimov's attack on Markov played an important role in Kedrov's downfall. Maksimov *Literary Gazette* article was a clear criticism of Kedrov, and in a statement that appeared after Kedrov's dismissal, Maksimov commented:

> *Only a decisive rejection of the idealistic inventions of N. Bohr and M. A. Markov, only a decisive repudiation of the position taken on this question by the editorial board of the journal* Problems of Philosophy *can lead our philosophical organ out of this blind alley into which it attempted to lure several sections of our intelligentsia, those inclined to waver on the basic questions of Marxist-Leninist ideology.*

In a note in the third issue of 1948 the reformed editorial board of *Problems of Philosophy* admitted that the journal had not taken the correct position on quantum mechanics and particularly on Markov's article, which had "weakened the position of materialism." The article had contained "serious mistakes of a philosophic character" and was in essence a departure from dialectical materialism "in the direction of idealism and agnosticism."

In terms of personnel, the immediate casualty of the Markov affair was Kedrov, but in terms of the philosophy of science, the casualty was the principle of complementarity. The period from 1948 to roughly 1960 may be called, with respect to discussions of quantum mechanics in the Soviet Union, the age of the banishment of complementarity. Only a few scientists in this period, most notably V. A. Fock, attempted to include complementarity as an integral part of quantum theory.

This critical attitude toward complementarity after 1948 was made clear in an article by Ia. P. Terletskii that immediately preceded the final statement on the Markov controversy by the editors of *Problems of Philosophy.* Terletskii observed that Markov's article was actually an attempt to justify the acceptance of complementarity by maintaining that as a result of the role of measuring instruments as "translators" of reality, statements about microphysics often contradict each other. Such a view, thought Terletskii, was merely a restatement of Mach's opinion that scientists must describe nature in terms of sensations. A true dialectical materialist approach, however, showed, Terletskii continued, that the principle of complementarity was in no way a basic physical principle and that quantum mechanics could very well "get along without it."

The result of the Markov affair, then, was a victory for dogmatic ideologists. Maksimov, an ideologist, had triumphed over Markov, an active theoretical physicist in the Academy of Sciences. But it also became fairly clear that Maksimov was not capable of advancing an interpretation of quantum mechanics that held any chance of official acceptance. His articles on quantum mechanics revealed all too clearly his ignorance of the subject. And it was the same Maksimov who was simultaneously opposing not only Einsteinian relativity but even Galilean relativity, maintaining that every object has an absolute trajectory and that a meteorite inscribes this trajectory on the earth upon collision with it. Maksimov clearly represented pseudoscience, and his role in both quantum mechanics and relativity theory was a purely destructive one, isolating the "Machists" and "idealists" among Soviet scientists and winning a certain support for that service, but presenting no tenable alternatives to current interpretations of physical theory. As in the case of relativity theory, Maksimov soon lost his influence among Soviet interpreters of quantum theory. The period after 1948 was dominated instead by physicists and a small number of philosophers with some knowledge of physics, all of whom, however, were influenced by the atmosphere created by the Markov affair. Until approximately 1958 the major interpreter of quantum mechanics was the philosopher of science Omel'ianovskii, who drew upon the theories of the physicist Blokhintsev, advocate of the "ensemble" interpretation. Also important was Fock, who termed his interpretation a recognition of the "reality of quantum states." And a good many others, including A. D. Aleksandrov, Ia. P. Terletskii, B. G. Kuznetsov, as well as the foreign scholars Louis de Broglie, J. P. Vigier, and David Bohm, influenced the discussions of dialectical materialism and quantum mechanics. Soviet philosophers and scientists in the early fifties were restricted in their writings on quantum mechanics by certain reigning dogmatic interpretations of Marxist texts. For example, Marx in one passage of *Capital* had observed that "the properties of a given thing do not arise out of its relations to other things, but are only identified in such relations" This obscure passage, taken out of its context of economic analysis and employed in the analysis of physics, was interpreted by some Soviet philosophers as revealing the incorrectness of the Copenhagen School's emphasis on measurement (the relation of the microparticle to an instrument); these ultraorthodox writers believed that scientists should emphasize the inherent properties of the microparticle, not its relations with other objects. However, with the passing of Stalinism in the late fifties, other Soviet philosophers began clearing away these obstacles standing in the way of a genuine Marxist philosophy of science. An important book in overcoming the dogmatic interpretation of the above passage of Marx was A. I. Uemov s *Things, Properties and Relations,* published in Moscow in 1963. By the late sixties Soviet scientists and philosophers were able to defend materialism, as many of them without any question wished to do, without being hobbled by quotations torn out of the context of the writings of Marx and Engels.

An interpretation of quantum mechanics is a statement which attempts to explain how quantum mechanics informs our understanding of nature. Although quantum mechanics has received thorough experimental testing, many of these experiments are open to different interpretations. There exist a number of contending schools of thought, differing over whether quantum mechanics can be understood to be deterministic, which elements of quantum mechanics can be considered "real", and other matters.

This question is of special interest to philosophers of physics, as physicists continue to show a strong interest in the subject. They usually consider an interpretation of quantum mechanics as an interpretation of the mathematical formalism of quantum mechanics, specifying the physical meaning of the mathematical entities of the theory.

6

Origin of Life in Philosophy

Introduction

Most scientists and historians of science are very skeptical of easy associations of science and political ideology, and no doubt Waddington did not intend to imply a direct causal link here. In this rather casual comment in a book review he was opening the question of the possible influence of Marxism on the significant theories of life of the first half of the twentieth century, not attempting to answer it. There are many opportunities in the history of science for linking science and politics, but frequently upon close examination the links either dissolve or turn out to be much more complex than thought earlier. As we will see, there are very weighty pieces of evidence against the belief that Oparin and Haldane were applying Marxism in 1924 and 1929. Nonetheless, the question of the interaction of Marxism and biology in the twentieth century is a legitimate and important one.

At the outset it will be useful to compare what non-Marxist and anti-Marxist writers have done with two different developments in the history of science: the Lysenko affair and discussions of the origin of life. In neither case is it self-evident that Marxism as a system of thought was significant in the formation of the interpretations of biological phenomena known, respectively, as the Oparin-Haldane hypothesis and Lysenko's theory of inheritance. In both cases, however, the major actors, at times subsequent to the original development of the hypotheses, explicitly declared that Marxism was an important influence in their biological thought. Lysenko, Oparin, and Haldane all became vocal dialectical materialists. Yet if one mentions "Marxism and biology" to the average educated citizen of western Europe or America, he will think only of Lysenko. This tendency to explain an acknowledged calamity in science as a result of Marxist philosophy while assuming that a brilliant page in the history of biology had nothing to do with Marxism is a reflection, at least in part, of the biases and historical selectivity of anti-Marxist journalists and historians.

The important question still remains: Did Marxism have anything to do with the Oparin-Haldane hypothesis? Although this question cannot be answered at this time, and no doubt will long remain controversial, certain clarifications can be made. Before attempting these clarifications, I would like to make some general comments about A. I. Oparin and the issue of the origin of life. As a Russian, with an active life extending to the present day, Oparin is central to this study, while Haldane falls outside it. Oparin's initial work on this issue was prior to Haldane's entirely independent but similar approach; the British scientist graciously declared in 1963, "I have very little doubt that Professor Oparin has priority over me."

The question of the origin of life is one of the oldest in the history of thought. At almost all periods of time a belief in spontaneous generation was commonly held. This belief was by no means the property of one school of thought; Democritus, Aristotle, St. Augustine, Paracelsus, Francis Bacon, Descartes, Buffon, and Lamarck are only a few of those who expressed support of the concept, but within the frameworks of quite different interpretations of nature. With the development of microscopy the center of attention in discussions of the origin of life shifted to the level of the invisibly small. A famous debate in the 1860's between the French scientists Felix Pouchet and Louis Pasteur over the possibility of the spontaneous origin of microorganisms ended with a negative result that left the subject of spontaneous generation in disrepute for the remainder of the century. To be sure, a few writers such as H. Charlton Bastian continued to consider spontaneous generation possible. Friedrich Engels observed ironically that it would be foolish to expect men with the help of "a little stinking water to force nature to accomplish in twenty-four hours what it has cost her thousands of years to bring about." And most people did not notice that Pasteur himself commented in 1878, "Spontaneous generation? I have been looking for it for 20 years, but I have not yet found it, although I do not think it is an impossibility."

Much of the foregoing would seem to serve as an introduction to A. I. Oparin as a twentieth century exponent of spontaneous generation. Yet if one understands spontaneous generation to mean the sudden arising of a relatively complicated living entity whether an organism, a cell, or a molecule of DNA from non-living matter, Oparin has actually been an opponent of spontaneous generation. In his opinion, the belief that such an ordered entity as a cell or even a "living" molecule of nucleic acid could arise spontaneously is based upon "metaphysical materialism" and suffers from the same improbabilities (to be discussed below) as Pouchet's arguments.

Aleksandr Ivanovich Oparin has been a prominent biochemist in the Soviet Union for decades. He graduated from Moscow University in 1917 and subsequently became a professor there. He was closely associated with the Institute of Biochemistry of the U.S.S.R. Academy of Sciences, which he helped to organize in 1935 and of which he became director in 1946. In the same year he became a full member of the Academy of Sciences. In 1950 he received the A. N.

Bakh and I. I. Mechnikov prizes. He has worked on several different topics, including such practical ones as the biochemistry of sugar, bread, and tea production, but he is best known both in the Soviet Union and abroad for his theory of the origin of life. Over a period of almost half a century he has published numerous books, revised editions, and journal articles on this topic. As early as 1922 he delivered a talk on the origin of life to the Moscow Botanical Society; he subsequently published these views in a small booklet in 1924. Although frequently cited in the scientific literature, the 1924 publication was an exceedingly rare one and was not translated into English until 1967. Most references by English readers to Oparin's work have been to the 1938 and subsequent editions, which differed from his earliest publications in several ways that are interesting to the historian. These differences will be discussed subsequently. In recent years, Oparin's works have received international attention more rapidly; his 1966 *The Origin and Initial Development of Life* was translated into English by the U.S. National Aeronautics and Space Administration in 1968.

The most prominent contribution by Oparin to the study of the origin of life will probably be seen by historians as his reawakening of interest in the issue. The American biologist John Keosian wrote in his popular text *The Origin of Life*, "Oparin's unique contribution is his revival of the materialistic approach to the question of how life originated, as well as his detailed development of this concept." The British scientist J. D. Bernal commented in 1967 that Oparin's 1924 essay contains in itself the germs of a new programme in chemical and biological research. It was a programme that he largely carried out himself in the ensuing years, but it also inspired the work of many other people. Oparin's programme does not answer all the questions, in fact, he hardly answers any, but the questions he asks are very effective and pregnant ones and have given rise to an enormous amount of research in the four decades since it was written. The essential thing in the first place is not to solve the problems, but to see them. This is true of the greatest of all scientists.

Turning now to the question of intellectual and social influences on Oparin, it can be clearly established that from the early 1930's onward Oparin was influenced by dialectical materialism. The evidence is not only his frequent statements favoring dialectical materialism, but much more importantly, the very method of analysis of his later publications, which are permeated with an assumption of a process philosophy and a concept of differing dialectical levels of regularities in nature. All this is described in his works in the language of dialectical materialism. Oparin has spoken out so frequently on the relevance of dialectical materialism to theories of biological development that almost any of his publications of substantial length contains such statements. To be sure, there is the possibility that these sections of his writings are merely responses to political pressures, but if one reads in chronological order through Oparin's works published over many years at times of greatly varying political atmospheres, one can not avoid the conclusion, it seems to me, that dialectical materialism became an ever-increasing and substantial influence on his work. In 1953 Oparin wrote, "Only dialectical materialism has found the correct routes to an understanding of life. According to dialectical materialism, life is a special form of the movement of matter, which arises as a new quality at a definite stage in the historical development of matter. Therefore, it possesses properties that distinguish it from the inorganic world, and is characterized by special, specific regularities that are not reducible merely to the regularities of physics and chemistry."

And in 1966 in the book translated by NASA Oparin commented:

> *Regarding life as a qualitatively special form of the motion of matter, dialectical materialism formulates even the very problem of understanding life in a different way than does mechanism. For the mechanist, it consists of the most comprehensive reduction of living phenomena to physical and chemical processes. On the other hand, from the dialectical materialist point of view, the main point in understanding life is to establish its qualitative difference from other forms of motion*

> *of matter. Life finds its clearest expression (as a special form of the motion of matter) in the specific interaction of living systems organisms with the environment surrounding it, in the dialectical unity of a living body, and in the conditions of its existence.*

As I discuss in chronological order a number of Oparin's other publications on the origin of life, more specific effects of his dialectical materialist approach will come to light. This discussion should not be confused with a general history of Oparin's theories, which remains to be written and which would take into account in a much fuller way the development of biochemistry as a whole. In such a history dialectical materialism would play a less prominent role than in this discussion.

Nonetheless, it will always be seen, I think, as one of the important influences upon Oparin. If we turn to Oparin's 1924 booklet, however, we will find no mention of Marxism. Even more significantly, Oparin's analysis in this small book differed from almost all of his later works in containing no concept of "different levels of regularities," no statement that qualitatively distinct principles govern the movement of matter on different ontological levels. The Oparin of 1924 was a materialist (and here the influence of his political and social milieu no doubt enters in), but he seems to have been an old-fashioned materialist, one who believed that life can be entirely explained in terms of physics and chemistry. Compare, for example, his 1953 statement quoted above, in which he said that life is characterized by special regularities that are not reducible to those of chemistry and physics, to the following statement, which was contained in the 1924 booklet:

> *The more closely and accurately we get to know the essential features of the processes which are carried out in the living cell, the more strongly we become convinced that there is nothing peculiar or mysterious about them, nothing that cannot be explained in terms of the general laws of physics and chemistry.*

And in another place in the same work he commented, "Life is not characterized by any special properties but by a definite specific combination of these properties." Here again, the assumption is a reductionist one, although the "definite specific combination" allowed a bit of room for the later development of the concept of "special biological regularities" distinct from those of physics and chemistry, a concept that became fundamentally important to his work. Ironically, the Oparin who in 1924 campaigned against vitalism in the name of purely physicochemical explanations of life would in the fifties and the sixties defend the uniqueness of biological regularities against those molecular biologists who would try to explain life entirely in terms of the structure of a molecule of nucleic acid that is, in terms of purely physicochemical explanation.

The fact that Oparin did not possess a knowledge of systematic dialectical materialism in 1924 does not prove that Marxism had nothing to do with the timing or form of his expression. Russia in the early twenties was the home of a victorious revolution carried out in the name of Marxism. "Materialism" was one of the most popular slogans of the day, and most of it was of the rather elementary, mechanistic sort espoused by Oparin, not the more subtle dialectical materialism developed at a later day. Neither Engels *Dialectics of Nature* nor Lenin *Philosophical Notebooks* had yet appeared in print; these are the two books that more than any others have counteracted the severe reductionism of earlier materialism with the concept of qualitatively distinct realms of operation of natural laws. Oparin's materialism developed in parallel with the predominant philosophical views of his society. Communism probably had something to do with Oparin's 1924 statement, but not in the sense of a relationship between dialectics and theories about the origin of life; rather, Communism in Russia in the twenties provided the kind of atmosphere in which the posing of a materialistic answer to the question What is life? seemed natural. The Soviet Union of the twenties was an environment in which speculation about nature on the basis of materialist assumptions was not only welcome but very nearly inevitable. Oparin certainly had nothing to fear from the political and social milieu in expressing such views, and he made no effort to soften their impact. Similar viewpoints were still capable of causing quite an adverse reaction in Britain, as the reception of Haldane's similar views five years later showed; Bernal commented, "Haldane's ideas were dismissed as wild speculation." Haldane assured his readers in 1929 that his opinions (not entirely identical with Oparin's, but similar) were also compatible with "the view that pre-existent mind or spirit can associate itself with certain kinds of matter." Yet it was clear that Haldane did not share this vitalistic view, and his inclusion of it came from his hope that his scientific conception would not be rejected simply because "some people will consider it a sufficient refutation of the above theories to say that they are materialistic."

How close Oparin personally was to Marxist ideas at this early period is still unknown. From 1921 onward he was closely associated with the older Soviet biochemist A. N. Bakh, who was a political revolutionary and former *émigré*, and who published on Marxism as early as the 1880's. But until we know more about Oparin's philosophical orientation in the period 1917–24, not much more can be said about the influence of Marxist materialism on his 1924 publication.

It might be added parenthetically that a similar problem of interpretation exists in the case of Haldane. Like Oparin, he seems to have been most influenced by Marxist thought *after* he published his first fundamental work on the origin of life. Haldane's first article on the subject appeared in 1929, as already noted. As late as 1938 Haldane wrote, "I have only been a Marxist for about a year. I have not yet read all the relevant literature, although I had of course read much of it before I became a Marxist." Yet this does not prove, of course, that Haldane had not learned of Marxist ideas of development by 1929. Since the late twenties Haldane had been a leader of a group of Cambridge intellectuals who were very interested in Marxism. But there were other influences in the air, too, that were interesting to men like Haldane and relevant to the biological issue, such as the early process philosophy of A. N. Whitehead. Some writers have argued that Haldane's view of science grew out of a complicated interaction of reductionist biochemistry, the process philosophy of Whitehead, and Marxism. Whatever the ingredients and the proportions of the early intellectual influences on Haldane, a hand-in-hand evolution based on interaction seems to have occurred. It eventually resulted in his writing a volume entitled *The Marxist Philosophy and the Sciences*. A similar long-term intellectual development will be seen in Oparin, with the difference that his interests in Marxism continued to deepen throughout his career.

The task that faced Oparin in 1924 in biology was as much one of changing the psychological orientation of scientists as it was altering their research itself. He had to convince his readers that despite Pasteur's victory over Pouchet years before and the complete inability of scientists to produce even the most elementary living organisms in a laboratory, a materialist explanation of the origin of life was still worth the effort.

In retrospect, Oparin observed, we should not be surprised or fundamentally affected by the outcome of the Pasteur-Pouchet debate. Pouchet was indeed incorrect, but not because of his materialist assumptions. Even the simplest microorganisms, and certainly those that Pasteur and Pouchet observed, are extremely complex bits of matter; they possess an "extraordinarily complicated" protoplasm. How could Pouchet even suppose that such a highly differentiated form of matter could "accidentally" have arisen in a few hours or even days from a relatively formless mixture? To assume such an incredibly improbable occurrence was unscientific in the deepest sense, a violation of the principle of explaining nature in the simplest and most plausible fashion available. As Oparin observed,

> Even the simplest creatures, consisting of only one cell, are extremely complicated structures. The idea that such a complicated structure with a completely determinate fine organization could arise spontaneously in the course of a few hours in structureless solutions such as broths or infusions is as wild as the idea that frogs could be formed from the May dew or mice from corn.

How then could one begin to explain an origin of life on the basis of materialist assumptions? Only, said Oparin, by going back to the very simplest forms of matter and by extending the Darwinian principles of evolution to inanimate matter as well as animate matter. The "world of the living" and the "world of the dead" can be tied together by attempting to look at them both in terms of their historical development. Any finely structured entity, alive or dead whether a one-celled organism, a piece of inorganic crystal, or an eagle's eye seems inexplicable unless it is examined in historical, evolutionary terms. Pouchet failed because the specimens he thought would arise spontaneously microorganisms were already the ultimate products of an extremely long evolutionary history and can be brought into existence only through that chain of material development, not by side-stepping it.

In the section of his 1924 work subtitled "From Uncombined Elements to Organic Compounds," Oparin attempted to reconstruct the historic process that might have led to the origin of life, with the simple always preceding the complex. In order to follow this sequence, it was necessary for him to reject the customary thesis that all organic compounds had been produced by living organisms, a belief still widely held at that time despite the supposed synthesis of urea by Wöhler as long ago as 1828. To postulate that all organic compounds had been produced by living organisms was methodologically faulty, thought Oparin, since organisms themselves were obviously composed of organic compounds, and many of them far more complex than some of the products supposedly produced. Far better, he thought, to assume that at least some organic compounds antedated complete organisms and had been important to the origin of these organisms. An important stimulus to Oparin's thought on this subject was the theory of the carbide origin of petroleum advanced by the great Russian chemist D. I. Mendeleev many years before. Mendeleev had posed the possibility of the origin of the hydrocarbon methane by the action of steam on metallic carbides under conditions of high temperature and pressure, for example:

$$C_3Al_4 + 12H_2O \rightarrow 3\ CH_4 + 4\ Al(OH)_3$$

This inorganic source of methane would then have been followed by further transformations leading ultimately to petroleum. Oparin did not accept Mendeleev's hypothesis about the origin of petroleum (and it has not been accepted by geologists generally, although there have been a few attempts to revive it), but the idea provoked him to further thought about the inorganic origin of organic compounds. As late as 1963 Oparin continued to emphasize the importance of Mendeleev's idea to his original conception of the origin of life.

In order to provide the necessary temperatures and pressures and a source of energy, Oparin referred to the theory of origin of the earth that begins with the earth as an envelope of incandescent gas. Oparin maintained, "Only in fire, only in incandescent heat could the substances which later gave rise to life have formed. Whether it was cyan [nitrogen carbide] or whether it was hydrocarbons is not, in the final analysis, very important. What is important is that these substances had a colossal reserve of chemical energy which gave them the possibility of developing further and increasing their complexity." In the 1936 edition of his book, Oparin would tie this view of the origin of the earth to James Jeans's theory of planetary cosmogony, in which a star approaches the sun in such a way as to pull out by gravitational attraction a tidal wave of incandescent solar atmosphere. As we have seen in the chapter on cosmology and cosmogony, this theory came under heavy philosophical criticism in the Soviet Union in later years as "miraculous and improbable." Oparin later abandoned the Jeans theory of planetary cosmogony, finding other available sources of energy for the formation of complex hydrocarbons.

The prior non-existence of life was one of the necessary conditions for the origin of life, and consequently, now that life exists on earth it can not originate again, or at least not in the same way in which it first did. Oparin explained this conclusion quite graphically:

> *We have already seen in the last section, that at present what are lacking above all are those substances containing much chemical energy which are the only things from which life could develop and which, themselves, could only be formed at extremely high temperatures. However, even if such substances were formed now in some place on the Earth, they could not proceed far in their development. At a certain stage of that development they would be eaten, one after the other. Destroyed by the ubiquitous bacteria which inhabit our soil, water and air.*

Matters were different in that distant period of the existence of the Earth when organic substances first arose, when, as we believe, the Earth was barren and sterile. There were no bacteria nor any other micro-organisms on it, and the organic substances were perfectly free to indulge their tendency to undergo transformations for many, many thousands of years.

The posing of the prior non-existence of life as a necessary condition for the origin of life seemed somewhat more original in 1924 than it does today, since in the interim there has come to light a letter of Charles Darwin's in 1871 mentioning the same hypothesis. Several other scientists also seem to have mentioned this hypothesis in the late nineteenth and early twentieth centuries. In Oparin's later editions this seeming paradox was explained within the framework of dialectical concepts of natural law: On each level of being different principles obtain; therefore, the laws of chemistry and physics that operated on the earth in the absence of life were different from, and superseded by, the biological laws that qualitatively emerged with the appearance of life. In the case of man, the biological laws were transcended, in turn, by social ones.

Oparin continued his hypothetical scenario of the origin of life with a description of the way in which substances with complicated molecules form colloidal solutions in water. Such solutions are inherently unstable, however, and frequently form coagula or gels: "The moment when the gel was precipitated or the first coagulum formed, marked an extremely important stage in the process of the spontaneous generation of life. At this moment material which had formerly been structureless first acquired a structure and the transformation of organic compounds into an organic body took place. Not only this, but at the same time the body became an individual."

This emphasis on the arising of life in a liquid medium by means of the separating out of gels became one of the hallmarks of Oparin's views; once his general materialistic approach to the origin of life had been widely accepted, the gel theory or "coacervate theory" was often considered to be that part of his work that was unique to him. Consequently many current discussions of the validity of Oparin's views revolve entirely around the tenability of the coacervate theory.

The idea of life's arising in a sea jelly of some sort was not new, of course, having been a part of T. H. Huxley's bathybius hypothesis, but Oparin was to present it in a particularly plausible way. In 1924, however, Oparin did not even use the term "coacervate." All that was to come in later editions, after Oparin could make use of the research on coacervation carried out in the early thirties by Bungenburg de Jong. But both in the original 1924 work and in his subsequent publications, Oparin remained steadfast on the principle that life arose on a level of fairly large dimensions: the coagula, the gels, the coacervates, were all definitely multimolecular, and they all possessed a rather complex structure *before* they could be called "alive." After they became alive, a natural selection began that through the costs of survival resulted in increasingly viable and complex organisms.

From the philosophic or methodological point of view the transition from the "non-living" to the "living" is the crucial moment. Oparin has not been one to attempt rigorous definitions of "life," preferring to speak in metaphors or in terms of varying combinations of characteristics necessary for life, but it is clear that his opinion about the moment when life appeared has changed somewhat with time. In 1924 he described the moment when "the gel was precipitated or the first coagulum formed" and then observed, "With certain reservations we can even consider that first piece of organic slime which came into being on Earth as being the first organism. In fact it must have had many of those features which we now consider characteristic of life." This observation accorded with the reductionist, mechanist approach of the young Oparin, in which a simple physical process coagulation could herald a major transition. In later years, he would maintain that the first coacervate droplets were definitely not alive and that a "primitive natural selection" occurred among these non-living forms. The transition to life occurred after not only the more commonly named characteristics of life had appeared (metabolism, self-reproduction), but in addition, after a certain "purposiveness" of organization had been achieved.

One metaphor that Oparin used in his early statement remained constant throughout his works; this was his comparison of life to a flow of liquid. In 1924, he wrote that "an organism may be compared with a waterfall which keeps its general shape constant although its composition is changing all the time and new particles of water are continually passing through it"; in 1960, he commented, "Our bodies flow like rivulets, their material is renewed like water in a stream. This is what the ancient Greek dialectician Heraclitus taught. Certainly the flow, or simply the stream of water emerging from a tap, enables us to understand in their simplest form many of the essential features of such flowing, or open systems as are represented by the particular case of the living body." These metaphors, all based on the concept of the constant flow of matter in living organisms, involved Oparin in many discussions about whether relatively static entities sometimes considered alive (crystallized viruses, dried seeds) could be accommodated to his understanding of life.

If one shifts from Oparin's 1924 booklet to his 1936 major work (the first was approximately 35 pages in length, the latter, 270), a number of changes become apparent. The biochemist would notice the much fuller description of the initial colloidal phase and a subsequent section on the development of photosynthesis by the ancestors of vegetative organisms. The historian and philosopher would remark on

Oparin's growing philosophic awareness, his refinement of his definitions and his stated shift toward Marxist interpretations.

By 1936 Oparin could take advantage of the recent work by Bungenburg de Jong on "coacervation," a term used by de Jong to distinguish the phenomenon from ordinary coagulation. In solutions of hydrophilic colloids it is known that frequently there occurs a separation into two layers in equilibrium with each other; one layer is a fluid sediment with much colloidal substance, while the other is relatively free of colloids. The fluid sediment containing the colloids, de Jong called the coacervate, while the noncolloidal solution was the equilibrium liquid. Oparin emphasized the interface or surface phenomena that occur in coacervation; various substances dissolved in the equilibrium liquid are absorbed by the coacervate. Consequently, coacervates may grow in size, undergo stress with increasing size, split, and be chemically transformed. In discussing this active role of coacervates, Oparin was attempting to establish them as models for protocells. A "primitive exchange of matter" occurs between the coacervate and the equilibrium liquid, the beginning of that metabolic flow necessary for life, according to Oparin. To initiate life, however, Oparin said that it was necessary for coacervates to acquire "properties of a yet higher order, properties subject to biological laws." He had higher requirements for life in 1936 than in 1924, and his scheme now contained a phase of evolution of the lifeless coacervates.

Oparin's view of coacervate evolution depended heavily on the action of catalysts and promoters, the future enzymes of living cells. These chemicals greatly accelerated the transformations taking place in the coacervates. The "inner chemical organization" of these promoters of chemical reactions "became strengthened in the process of natural selection, insuring a gradual evolution." To be sure, this early "natural selection" was not identical with selection among advanced, truly living organisms, but nonetheless it did occur, and in the following way: Those coacervates containing superior catalysts grew more rapidly and split more often than the others and therefore gradually gained a numerical advantage. But speed was not the only requirement; even more important to Oparin was "a harmonious coordination of the velocities of the different reactions." The bound enzymes in the protoplasm of a living cell are released sequentially as the need for them arises; therefore, in order for cells to develop, a complicated orchestration of temporally separated operations was necessary. It was possible for some necessary reactions to occur prematurely and to decrease survival chances: "If the increase in the rate of a given reaction so affected the coordination between assimilation and degradation as to promote the latter, such an imperfect system would become mechanically unfitted for further evolution and would perish prematurely." Therefore, a very complex interweaving of separate operations gradually developed through natural selection.

In Oparin's 1936 scheme the transition from the non-living to the living was still not defined clearly. It occurred, he thought, when the "competition in growth velocity" was replaced by a "struggle for existence." This sharpening struggle resulted from the fact that the prebiological organic material on which the coacervates were "feeding" was being consumed. Ultimately, this shortage would lead to an important split in the ways in which organisms gained nourishment resulting in the distinction between heterotrophs and autotrophs but before that division occurred, the all-important transition to the biological level was reached. As the amount of organic material outside the coacervates lessened, the first true organism appeared. As Oparin described this moment:

The further the growth process of organic matter advances and the less free organic material remains dissolved in the Earth's hydrosphere, the more exacting "natural selection" tends to become. A straight struggle for existence displaces more and more the competition in growth velocity. A strictly biological factor now comes into play.

This new factor naturally raised the colloidal systems to a more advanced stage of evolution. In addition to the already existing compounds, combinations and structures, new systems of coordination of chemical processes appeared, new inner mechanisms came into existence which made possible such transformations of matter and of energy which hitherto were entirely unthinkable. Thus systems of a still higher order, the simplest primary organisms, have emerged.

It should be obvious from Oparin's scheme of development that he thought that heterotrophic organisms (organisms that are nourished by organic materials) preceded in time autotrophic organisms (those nourished by inorganic materials). Many scientists had earlier thought that the sequence was the opposite and assumed that carbon dioxide necessary for photosynthesis by autotrophic green plants was the primary material used in building up living things. Oparin found this thesis dubious. As evidence against it he cited the fact that heterotrophic organisms are generally capable of using only organic compounds for nourishment, while many autotrophic green plants "have retained to a considerable degree" the ability to use preformed organic substances for their nourishment. The significance here of the word "retain" is obviously one of time sequence; Oparin thought that all organisms had originally been heterotrophic, but that as the supply of organic food diminished, they split along two different paths of development. (This division is not, strictly speaking, the same as that between the plant and animal worlds, but is close to it, since green plants are largely autotrophic while all the highest and lowest animals and most bacteria and all fungi are heterotrophs.)

Oparin explained this scheme in a much fuller philosophic framework than previously. By 1936 he had read Engels *Dialectics of Nature*, and cited it in his footnotes, as well as the earlier-published *AntiDuuml;hring*. He commented that Engels had "subjected both the theory of spontaneous generation and the theory of eternity of life to a withering criticism." (In his 1924 work, "spontaneous generation" had still been a positive term, although he had thought the efforts to find it crude.) Oparin now thought that any effort to explain "the sudden generation of organisms" could rely only on either an act of "divine will" or "some special vital force." Such a view, Oparin observed, is "entirely incompatible with the materialistic world conception." On the contrary, "Life has neither arisen spontaneously nor has it existed eternally. It must have, therefore, resulted from a long evolution of matter, its origin being merely one step in the course of its historical development."

More indicative of essential changes in Oparin's thought was his shift away from mechanism. Crude materialism, the belief that all phenomena could be explained in terms of the elemental, was now one of the objects of his criticism:

Attempts to deduce the specific properties of life from the manner of atomic configuration in the molecules of organic substance could be regarded as predestined to failure. The laws of organic chemistry cannot account for those phenomena of a higher order which are encountered in the study of living cells.

Although Oparin now frequently cited Engels and thought that he had been remarkably prescient in his discussions of life, Oparin was also willing to interpret and modify Engels's formulations. When Engels said that "Life is a form of the existence of protein bodies," he did not intend to say, said Oparin, that "protein is living matter." Instead he meant that protein has hidden in its chemical structure "the capacity for further organic evolution which, under certain conditions, may lead to the origin of living things." This interpretation by Oparin fits well with his belief that life is not inherent in a structure, but instead is a "flow of matter," a process. Structure has a great deal to do with life, he thought, but to confuse it with life itself would be roughly like confusing a frozen stream of water with a flowing one. This emphasis on process, on "coordinated chemical reactions" rather than on determinate structure, would eventually involve Oparin in controversies with spokesmen from two quite different camps: the ultraorthodox dialectical materialists who wished to stick to Engels's literal word "protein" as the essence of life, and the new molecular biologists, who saw the essential features of life in the structure of nucleic acid and whose very terms of description "template," "code" carried the sense of the static.

It was the 1936 book, translated into English in 1938, that brought Oparin international stature. At this moment of first impact his primary message was still seen as the legitimacy of a materialistic approach to the study of the origin of life. Consequently, a number of writers who actually differed considerably with Oparin on details considered themselves in agreement. Haldane in his 1929 article, for example, hypothesized in contrast to Oparin a primitive earth atmosphere rich in carbon dioxide, described the first "living or half-living things" as "probably large molecules," and did not mention coacervates, coagula, or gels. These are important points of difference. Yet the hypothesis became known as the Haldane-Oparin (or Oparin-Haldane) one and even yet is frequently referred to in that way.

The regarding of all materialist theories of the origin of life as essentially the same was reflected in R. Beutner's comments on Oparin in his *Life's Beginning on the Earth*, published in the United States in 1938. Beutner learned of Oparin's work as his own manuscript was going to press, and was able to add only an addendum taking it into account. Beutner commented that "there is a close resemblance of Oparin's views on the origin of life, with those here evolved." Further, "Oparin draws the same conclusions which this writer tentatively developed in 1933, namely that 'life is just one of the countless properties of the compounds of carbon." (It should be clear from the foregoing discussions that Oparin decidedly disagreed with Beutner on this point.) And finally, said Beutner, the small difference between Oparin's and his approaches was "only concerned with the order of the essential events which preceded the appearance of life."

Beutner's reassurances notwithstanding, Oparin's approach was rather different from his, although both were, indeed, materialistic. Beutner was a reductionist, Oparin by 1936 a supporter of the concept of qualitatively distinct levels in nature. Much of Oparin's writing was based on the assumption that life is a multimolecular phenomenon, while Beutner flatly stated that "we know that a 'living molecule' is possible." (He was referring to Stanley's 1935 work with tobacco mosaic virus, a subject of much subsequent discussion in which Oparin would be deeply involved.) And the small "difference of order of the essential events" between Beutner and Oparin was a distinction that Oparin considered fundamental. On this issue Beutner wrote, "The essential feature of Oparin's hypothesis is that structural units with a remote resemblance to living organisms were *first* formed from the organic matter of the early ocean and that subsequently enzymes formed in them so as to enable them to assimilate substances from their environment. Our opinion, on the contrary, was that life-producing enzymes were the first to appear, through the action of electric discharges, without any structure around them as self-regenerating enzymes." Oparin believed that Beutner's fortuitous and lightning like appearance of complex and ordered substances capable of coordinating metabolism was merely another variant of spontaneous generation, a concept he could not accept. He felt that it ultimately led to a discrediting of materialism by failing to explain adequately the complexity of life. In contrast to Beutner, Oparin drew a sharp distinction between simple catalysts, perhaps composed of even one chemical element, and complex enzymes.

We must, therefore, conclude that the naturally occurring high-molecular enzymes of living cells are not individual chemical compounds but complexes made up of numerous catalysts and promoters. The tremendous power of enzymatic activity must be attributed exclusively to a favorable arrangement of components in this complex, which, of course, could not have arisen fortuitously but only as a result of a long evolution of living organisms. In this process ever new processes and promoters must have been gradually added on secondarily to the simple and not very active primitive catalyst.

Oparin's 1936 work remained substantially unchanged for twenty years. The 1941 edition contained few modifications; not until 1957 was a third, revised edition published, almost simultaneously appearing in Russian and in English. In the meantime the science of biochemistry was developing extremely rapidly. The new knowledge of molecular biology led to a union of biochemistry and genetics, culminating in the 1953 publication of the Watson-Crick model of the DNA molecule.

The relevance of molecular biology to theories of the origin of life was obvious to most observers in all countries, although just where the developments would lead was a subject of genuine debate.

The topic of viruses was particularly close to the question of the nature of life when viewed from the molecular level; viruses consist of nucleic acid (DNA or RNA) with a protein coat. The relevance of molecular biology to Oparin's work was to center, in part, on discussions of viruses. The most urgent questions could be stated in simple forms: Are viruses alive? If they are alive, in view of the fact that the simplest of them are essentially nucleic-acid molecules, was not Oparin incorrect in stating that life arose on the multimolecular level? Was not the first form of life a molecule of nucleic acid?

In the Soviet Union such issues were discussed in rather difficult circumstances, since the new union of biochemistry and genetics in world science had occurred at approximately the same time that Lysenko and his followers won control over Soviet genetics. Politically, Oparin and Lysenko were linked, however far apart they may have been in intellectual sophistication. Both had won favor from the Stalinist regime, both had built their careers within it, and around both there had arisen schools of biology that were officially described as "Marxist-Leninist" or "Michurinist." They benefited from the government, and they repaid the government in political praise and co-operation. Oparin was active in Soviet political causes in international organizations. As a high administrator in the biological sciences in the Soviet Union during these years he was involved in the perpetuation of the Lysenko school. From 1949 to 1956 he was Academician-secretary of the Department of Biological Sciences of the U.S.S.R. Academy of Sciences, a position that meant he exercised great influence over appointments and promotions at a time when these were keys to Lysenko's continuing power. The Soviet biologist Medvedev wrote in his history of Lysenkoism that in 1955 a petition directed against the administrative abuses of both Lysenko and Oparin was circulated among Soviet scientists. Oparin was for many years a supporter of Lysenko, praising him in print on numerous occasions. Nonetheless, as will be seen, Oparin struggled against several attempts by sympathizers with Lysenko to invade Oparin's field. Medvedev reported that in the final struggle with Lysenko, Oparin took a neutral position.

One of the low points of Oparin's intellectual career came in his praise in 1951 of the new cell theory of Olga Lepeshinskaia. Lepeshinskaia was a mediocre biologist of impressive political stature as a result of her membership in the Communist Party from the very date of its founding and of her personal association with Lenin and many other political leaders. In 1950, a year of great political pressure in the Soviet Union, Lepeshinskaia claimed that she had obtained cells from living non-cellular matter. Lepeshinskaia even maintained that she had obtained cells from noncellular nutrient mediums in as short a time as twenty-four hours. Her work won praise from Lysenko himself. It should be obvious from the past discussion of Oparin's views that he was skeptical in the extreme of all hypotheses that supposed the sudden appearance of a finely articulated entity from a less organized medium. Such an error had been made in the past, he thought, by all crude supporters of spontaneous generation. Engels had appropriately ridiculed those people who thought they could compel nature to do in twenty-four hours that for which thousands of years had been necessary. In principle, of course, such an event was possible, but Oparin thought only the best evidence, repeatedly substantiated, would have been convincing. He had in the past allowed room for the creation of life itself by man when his knowledge of biology became much more perfected. But in 1951, far from this moment, Oparin succumbed to the political pressures of Stalinist Russia and praised the "great service" of Professor Lepeshinskaia in "demonstrating" the emergence of cells from living noncellular matter even though Lepeshinskaia's evidence was rejected everywhere outside the Soviet bloc. He even agreed that this emergence of cells from living non-cellular matter was occurring "at the present time," although he had opposed such a view many times in the past. As we will see, not until 1953 did Oparin begin to resist these views in print. By 1957, however, he had returned to his flat opposition to spontaneous generation and to the sudden emergence of cellular forms in the manner described by Lepeshinskaia. Between 1953 and 1958 the supporters of Lepeshinskaia, and Lepeshinskaia herself, responded to Oparin's emerging remonstrances by, in turn, increasing their criticism of Oparin.

In the early 1950's some of the most aggressive criticisms of science being made in the name of dialectical materialism were actually crudely materialistic or even idealistic. Certainly Lysenko was incapable of sophisticated philosophic analysis of scientific theory. Oparin was criticized by ideologists who were close to Lysenko in their viewpoints. One of their objects of criticism was his opinion that although life had once arisen on earth, such an event would never again be repeated. Several militant ideologists felt that Oparin's view attributed to life a uniqueness that contradicted uniformitarian and materialist doctrines. These writers were somewhat similar to the materialists of the nineteenth century who felt that the doctrine of spontaneous generation was logically required by materialism. They called themselves dialectical materialists, but they ignored Engels's perceptive criticism of that form of spontaneous generation; the fact that Oparin also opposed spontaneous generation was evidence to them of his philosophical waverings.

In early 1953 Oparin commented on some of these criticisms. "Does life arise now, at the present time?" he asked. No doubt it does, he replied, because matter never stays at rest, but constantly develops ever-new forms of movement. But life is not arising now on the earth that stage of the development of matter has already been passed through here but instead on other planets in the universe. He admitted that his critics had made a legitimate point in noting that his books were entitled *The Origin of Life*, as if what happened on earth were the whole story.

Oparin's critics tried to find other similarities between his views and those of Jeans. Just as Jeans's "near-collision" of stars seemed to them to bestow an exceptional or miraculous character on the origin of the earth, so Oparin's establishing of very special conditions for the origin of life and his insistence that this event could never be repeated on earth seemed to attribute exclusive properties to the origin of life and, ultimately, to man. Oparin attempted to answer this criticism in his joint book with Fesenkov. The origin of life, the two writers said, was a perfectly normal development in the evolution of matter:

> *In its constant development matter pursues various courses and may acquire different forms of motion. Life, as one of these forms, results each time the requisite conditions for it are on hand anywhere in the Universe.*

But just because life is a lawful and normal development does not mean, they observed, that it should be seen everywhere. Those materialists who constantly seek to find evidence of the arising of life around them in order to illustrate the unexceptional character of life are ignoring the very genuine qualitative distinctions that exist in matter, and if carried to extremes, such views will lead to a form of hylozoism. Life must not be seen as an inherent property of matter, they thought, but as a special yes, exceptional form of motion of matter:

> *From the point of view of dialectical materialism life is material in nature, but it is not an inalienable property of all matter in general. On the contrary, it is inherent only in living beings and is not found in objects of the inorganic world. Life is a special, very complex and perfect form of motion of matter. It is not separated from the rest of the world by an unbridgeable gap, but arises in the process of the development of matter, at a definite stage of this development as a new, formerly absent quality.*

Just how rare is life in the universe? After a very long and detailed discussion of the physical requirements of life and the known characteristics of the universe, Oparin and Fesenkov came to the tentative conclusion that "only one star out of a million taken at random can possibly have a planet with life on it at some particular stage of development." But such a ratio should not be seen as bestowing anything approaching uniqueness upon life; indeed, the two distinguished scientists continued, in our galaxy there may be thousands of planets on which life is likely and "Our infinite Universe must also contain an infinite number of inhabited planets.

Oparin and Fesenkovs argument here was quite similar to that of astronomers in many other countries. From a methodological stand point it might be worthwhile to notice that by such calculations, given an infinite amount of time, even so rare an occurrence as Jeans's near collision of stars would occur with an infinite frequency. (Of course, some astronomers, referring to the red-shift phenomenon, would not grant infinite time to the universe.) But aside from philosophic considerations, there was significant empirical evidence against Jeans's hypothesis by the mid-fifties, and Fesenkov and Oparin, as professional scientists pursuing these problems, knew the evidence well.

In 1957 Oparin published the third revised and enlarged edition of his major work, with a more restricted title, *The Origin of Life on the Earth.* In this volume he attempted to answer a number of recent criticisms advanced against his system, and he incorporated much recent scientific evidence. His original 1924 booklet had now grown, thirty-four years later, to almost five hundred pages.

As in the past, one of Oparin's major points concerned his belief in the erroneousness of the concept of spontaneous generation. The book by Olga Lepeshinskaia, *The Development of Cells from Living Matter,* was in his opinion, "an attempt to rehabilitate Pouchet's experiments and thus to resuscitate the theory of spontaneous generation." Pouchet had expected microorganisms to be spontaneously generated; Lepeshinskaia also looked for spontaneous generation, but of cells from noncellular matter instead of complete organisms. Both had therefore sought the sudden appearance of order out of chaos, and such attempts are "foredoomed to failure." The formation of the "very complicated structure of protoplasm" that is contained in both microorganisms and the single body cell cannot have happened fortuitously, but rather as the result of a long evolutionary process.

Oparin applied a similar sort of criticism to those scientists who would propose the gene, a molecule, or a bit of DNA as the primordial speck of life. Each of these theories is materialistic and therefore commendable, said Oparin, in that it seeks a material basis for life, but it is also mechanistic in the sense of all theories of spontaneous generation: It takes as a starting point a bit of matter that is actually the end point of a long evolution and assumes that the story of life began there. Since no explanation for the origin of this coherent bit of matter is attempted, the whole interpretation, intentionally or unintentionally, acquires a mysterious aura.

By this time J. D. Watson and F. H. C. Crick had suggested the noted double-helix model of the macromolecule of deoxyribonucleic acid (DNA). It was also clear by 1957 that DNA is the inherited material of almost all organisms. The different sequences of the bases (adenine, guanine, thymine, and cytosine) in the nucleotides linking the helixes, and the varying amounts of bases in each organism, led to an almost astronomical number of possibilities of structural combinations. Thus, the DNA macromolecule appeared to be a code of life, varying for each species and indeed for each member of that species. Investigators were now beginning to speak of genes as "sections of DNA," and specialists in the origin of life began to suspect that the first bit of life was a DNA molecule.

Oparin considered the establishing of the structure of DNA to be an event of great importance, and described in detail, with the inclusion of diagrams, the achievement of Watson and Crick. But he was definitely opposed to the talk resulting from the work of molecular biologists about the "first living molecule of DNA." His argument was, at bottom, the same one he had used against spontaneous generation of organisms many years before. Referring to hopes for the appearance of complete microorganisms in infusions, Oparin had written then:

> *If the reader were asked to consider the probability that in the midst of inorganic matter a large factory with smoke stacks, pipes, boilers, machines, ventilators, etc., suddenly sprang into existence by some natural process, let us say a volcanic eruption, this would be taken at best for a silly joke.*

No longer, he acknowledged, did anyone expect the spontaneous generation of complete organisms, or even of complete cells; to stick to the metaphor, no longer did they look for the sudden appearance of the whole factory. But Oparin believed that those people who thought the story of life began with the fortuitous synthesis of DNA were making the same error; they did not pretend that a factory could suddenly spring into existence, but they acted as if they believed it possible for the *blueprint* of that factory to accidentally appear. Yet that blueprint (the molecule of DNA) contained all the information necessary for the construction of the factory; for such a body of coded information to suddenly appear was as wild as to assume the sudden materialization of the factory itself. In their emphasis on a molecule as the starting point of life, many scientists were ignoring the question that to Oparin was the most important of all: "How did the rigidly determinate arrangement of nucleotides in the DNA come into being?"

To Oparin, the synthesis of nucleic acids was a complex process of interaction on the supramolecular level, one requiring the presence of a complicated enzymic apparatus. This supramolecular approach was obviously fitted to his coacervate theory, although he recognized that other supramolecular mechanisms might also account for the origin of life. The important point was that the presence of more than one substance was required and that a hand-in-hand evolution of both nucleic acids and other substances had occurred: "Thus, on the one hand, the synthesis of proteins requires the presence of nucleic acids while, on the other, the synthesis of nucleic acids requires the presence of proteins (enzymes)."

This known interaction had already led biologists to speculate on the relative importance of nucleic acid and protein to the life process. In 1955 J. D. Bernal lectured on the origin of life at Moscow University and at the end of his presentation asked Oparin, "Which came first, nucleic acid or proteins?" Oparin's answer, presented in his 1957 book, provided a clue to much of his thinking:

> *This question reminds one somewhat of the scholastic problem about the hen and the egg. The problem is insoluble if we approach it metaphysically in isolation from the whole previous history of the development of living matter. Nowadays every hen comes from an egg and every hen's egg from a hen. Similarly, nowadays proteins can only arise on the basis of a system containing nucleic acids while nucleic acids are formed only on the basis of a protein-containing system. The hen and its egg developed from less highly organised living things in the course of their evolution. In the same way, both proteins and nucleic acids appeared as the result of the evolution of whole protoplasmic systems which developed from simpler and less well adapted systems, that is to say, from whole systems and not from isolated molecules. It would be quite wrong to imagine the isolated primary origin either of proteins or of nucleic acids.*

Many contemporary authors do not, however, follow this line of thought. They take the view that in the first place nucleic acids arose in some way and that at once, simply by virtue of their intramolecular structure, they were able both to synthesize proteins and to multiply themselves spontaneously. It is, however, clear from all our previous discussion that a hypothesis of this sort is in direct opposition to the facts as they are at present known.

Oparin saw a similarity between the current concentration upon the molecule of DNA as the primordial bit of life and the earlier concentration upon the gene. Furthermore, the view that the gene is a section of DNA molecule allowed room for the coalescence of the two approaches. In each case, however, valid research was being given what Oparin thought was an improper interpretation. "Life" to him was still a process, a flow, an interchange of matter, and could not be identified with any static form. Oparin thought that the new emphasis on DNA was the lineal descendant of earlier erroneous views such as: H. J. Muller's conception of the "random emergence of one successful gene among myriads of types of molecules"; T. H. Morgan's original "gene molecule"; C. B. Lipman's idea of the formation of a "living molecule"; R. Beutner's proposed "self-generating enzymes"; and A. Dauvillier's organic molecules with a "living configuration."

Many molecular biologists were willing to grant that Oparin's evolutionary approach to DNA had its merits, but they felt that his commitment to a definition of life as a multimolecular phenomenon led him to strained, if not absurd, positions on the issue of viruses. Viruses were discussed several times in Oparin's 1957 book and also at an international symposium on the origin of life that he attended in Moscow in August 1957. The topic also appeared in an important book published by Oparin in 1960.

The study of viruses has a great significance for an understanding of nucleic acids and protein synthesis; in viruses we can see with unusual clarity the process of replication that lies at the base of the life process. Furthermore, viruses seem to occupy a curious position, one often described as midway between the living and nonliving. By studying them, scientists have given their definitions of life the most rigorous tests. The most frequently examined virus has been the tobacco mosaic virus (TMV). Ever since W. M. Stanley succeeded in isolating TMV in a crystalline form in 1935, it has been regarded as one of the most suitable types of virus for study. The advantage of the TMV is not only that it can be crystallized and studied in a static form, but the undoubtedly related fact that while many other viruses contain a variety of substances-lipids, carbohydrates, nucleoproteins the TMV is relatively simple in structure, consisting of nucleic acid and a protein coating. In the mid-fifties several workers succeeded in separating the nucleic acid core from the protein. As two of the leaders in this field of research reported, "The nucleic acid alone can perform all the crucial functions of the virus, namely it can initiate infection and transmit the required genetic information."

Here then was a serious challenge to the earlier concepts of the nature of life, not only to Oparin's, but to those of all biologists. Stanley commented that in their efforts to distinguish the animate from the inanimate, "if viruses had not been discovered, all would have been well." But with viruses, researchers seemed to have come upon a form of "life" that could, at least in some cases, be crystallized and held static indefinitely, that in terms of size was smaller than certain molecules, that could grow and reproduce, and that had the ability to mutate during reproduction. Why not recognize them as fully "living" organisms? Some researchers did. Stanley said of viruses in 1957, "They are all, in short, by definition, alive." Others, including Oparin, believed that there were serious reasons for excluding viruses from the realm of the truly living. And Oparin was particularly firm in his opinion that neither viruses nor any other "living" form on the molecular level should be considered antecedent to all other living organisms. Such molecular forms were to him the products of life, not the producers. He felt that to regard them as the starting point of life would be to begin with the unexplainable and eventually to fall prey to metaphysical, mysterious interpretations of nature. His arguments will be considered in more detail below, after a few additional comments on the nature of viruses.

A clearer view of the action of viruses, and of the central role of molecules of nucleic acid in their function, can be gained by considering bacteriophages, those viruses that prey on bacteria. A particularly suitable example of a bacteriophage is the virus that attacks colon bacilli. These viruses first attach themselves to bacilli and then literally inject their interior nucleic acid molecules inside their hosts, leaving the protein coats outside. Once inside the cell walls of the bacteria, the viruses multiply until the bacteria burst, freeing the viruses for further conquests.

It is important to see that this phenomenon is not identical with the familiar forms of parasitic action in the biological world by which the parasitic organism gains sustenance from a host; there is a much more elemental and striking mechanism at work here. The virus is incapable of metabolic action by itself and, indeed, possesses none of the physiological mechanisms necessary for such action. It uses, instead, the mechanisms of the host to the ultimate degree, bringing to it only information suitable for gaining its goals. One might say that the nucleic acid is no more than a program for using an existing process for a different goal; it is as if an impostor came into an operating chemical plant with only a coded computer card under his arm and, by inserting that card into a central computer, redirected the flow of chemicals to yield a different product, one valued by him. Of course, this analogy, with its inclusion of man's intelligence and its somewhat mechanistic reference to a factory, has its weaknesses, but it may convey some of the strangeness of the situation.

Of Oparin's fairly extensive discussion of viruses, the aspect that most concerns us is his opinion on whether they are alive. He did not flatly declare in his 1957 book that viruses were not alive, but his argument certainly pointed toward that conclusion. It is indeed true, he remarked, that they can replicate themselves quite readily. But replication is not the same as life, he continued, since even inorganic crystals can replicate and grow. Furthermore, viruses cannot even replicate unless they are placed "inside" an existing life process. Oparin commented:

> *Nobody has succeeded in producing this so-called "multiplication" of virus particles under any other conditions or on any artificial medium.*

> *Outside the host organism the virus remains just as inert in this respect as any other nucleoprotein. Not only does it show no sign of metabolism but nobody has yet succeeded in establishing that it has even a simple enzymic effect. It is clear that the biosynthesis of virus nucleoproteins, like that of other proteins, is brought about by a complex of energic, catalytic and structural systems of the living cell of the host plant, and that the virus only alters the course of the process in some way so as to give specific properties to the final product of the synthesis.*

Although Oparin doubted that viruses were alive, he seemed to press that point of view less insistently in later publications. There were several potential avenues of compromise between the opposing points of view. Wendell Stanley, the man who crystallized TMV, even suggested at the 1957 conference, "Some may prefer to regard a virus molecule in a crystal in a test tube as a potentially living structure and to restrict the term 'living' to a virus during the time that it is actually reproducing. I would have no serious objection to this." But Stanley then went on to repeat his belief that viruses are alive, without stipulating the moment in time when they become so. It was unclear what position Oparin would take on the suggestion that viruses are intermittently "alive" and "dead."

As an author primarily committed to the elaboration and defense of a theory on the origin and evolution of life, Oparin's main interest remained not a precise definition of "life," but the description of a development in time during which life appeared at one point or another. The exact moment could be debated. His interest in viruses centered on whether they were in the main path of that development or in a branch. And he believed that the answer to the question was that they were in a branch; whether or not viruses are ever alive, they were not the primordial forms of life from which all the others developed. As he commented at the 1957 symposium:

> *Today I should like to formulate, in a couple of words, my own viewpoint which I have expounded and substantiated in my book. I assumed that what had arisen primarily, by abiogenetic means, was not the functionally extremely efficiently constructed nucleic acids or proteins which we can now isolate from organisms, but only polynucleotides or*

polypeptides of a relatively disorderly structure, from which were formed the original systems. It was only on the basis of the evolution of these systems that there developed the functionally efficient forms of structure of molecules, not vice versa.

In the opposite case one would have to conceive of evolution as imagined by Empedocles, who held that first there developed arms, legs, eyes and ears and that later, owing to their combination, the organism developed.

And in 1960 he came back to the topic by discussing the tobacco mosaic virus. He emphasized what happens when this virus attacks the cells of a tobacco leaf:

All that takes place is the constant new formation of a specific nucleoprotein with the help of the biological systems of the tobacco leaf. This means that the new formation is only possible in the presence of an organisation which is peculiar to life and consequently the first living thing was not a virus; on the contrary, viruses, like other modern specific proteins and nucleic acids, could only have arisen as products of the biological form of organisation.

In the last phrase of the above quotation Oparin was referring to the well-known fact that parasites frequently become simpler in organization as they become more and more dependent on their hosts and adapted to that ecological niche. All viruses are parasites. Oparin was therefore suggesting that although the coded nucleic acid in viruses is the evolutionary product of more sophisticated organisms, the viruses themselves are the ultimate results of parasitic "devolution." They have lost all but their genetic material itself; they are, so to speak, "escaped" bits of the genetic code, which reproduce themselves by using the metabolic processes of more sophisticated organisms. But, according to Oparin, they could never have arisen without the prior evolution of organisms with a metabolic capability.

The publication by Oparin that most clearly illustrated the refinement of his philosophic views was his *Life, Its Nature, Origin and Development*, published in Russian in 1960 and in English in 1961. And all this was written at a time of relatively little political pressure upon scientists in the Soviet Union, during years that were quite different from those of the late forties and early fifties, the only time when most non-Soviet observers have considered dialectical materialism to be significant in Soviet science.

The point to which Oparin returned again and again in the writing of this book is that dialectical materialism is a via media between the positions of frank idealists and vitalists on the one hand, and mechanistic materialists, exuberant cyberneticists, and supporters of spontaneous generation on the other. Dialectical materialism was indeed a form of materialism and therefore opposed to the idealistic view that the essence of life was "some sort of supramaterial origin which is inaccessible to experiment." But dialectical materialism was equally opposed to the view that all living phenomena could be explained as physical and chemical processes. To take the latter position would mean, said Oparin, "to deny that there is any qualitative difference between organisms and inorganic objects. We thus reach a position where we must say either that inorganic objects are alive or that life does not really exist." Dialectical materialism provides a means, Oparin continued, of accepting the principle of the material nature of life without regarding "everything which is not included in physics and chemistry as being vitalistic or supernatural." To dialectical materialists, life is a "special form of the motion of matter," one with its own distinct regularities and principles.

Life was to Oparin still a flow, an exchange, a dialectical unity: "An organism can live and maintain itself only so long as it is continually exchanging material and energy with its environment." In a living body, this exchange of material is "strictly co-ordinated in time and space" in an orderly way insuring self-preservation and self-reproduction. But metabolic exchange and reproduction were only necessary, not sufficient, conditions for life. A living organism must have, in addition, the characteristic of "purposiveness." This characteristic figured more prominently in this later work of Oparin's than it had in his earlier writings. He believed that purposiveness pervades the whole living world "from top to bottom, right down to the most elementary form of life." He recognized that his insistence on purposiveness as an essential feature of life had its dangers, since "in one form or another Aristotle's teaching about 'entelechy' has left its mark on all idealistic definitions of life." But Oparin believed that the purposiveness of the organization of life was "an objective and self-evident fact which cannot be ignored by any thoughtful student of nature. The rightness or wrongness of the definition of life advanced by us, and also of many others, depends on what interpretation one gives to the word 'purposiveness' and what one believes to be its essential nature and origin." He thought that dialectical materialists could avoid idealism by always studying this purposiveness in terms of its development, its origins. So long as purposiveness can be understood as a result of a historical interaction between the material organism being studied and its material environment, one need not fear idealism. It is only, Oparin said, when purposiveness is brought in from outside the boundaries of the material world, or is left so unexplained that such an origin seems implied, that biological explanations become idealistic. Hence, Oparin believed that the essential methodological guide through these dangers could be found in the writings of Heraclitus of Ephesus: "One can only understand the essence of things when one knows their origin and development." In this principle Oparin believed that dialectical materialism and Darwinism drew upon a common inspiration, one found in ancient philosophy.

Oparin maintained that his conception of life as a metabolic, purposive material process revealed the faults of a number of twentieth century views of life. Particularly common at the present time, he thought, were attempts to explain life on mechanistic, crudely materialist bases. These efforts were no doubt the results of the prior successes of science and should be praised for their materialism, but they so oversimplified biology that in the end they might harm its further development. The two best examples of modern mechanism, he said, were those interpretations based on cybernetics and molecular biology. The cyberneticists believed that their new approach was a "fundamentally new and universal road to the understanding of the very essence of life." Helpful as some of their achievements have been, Oparin continued, the cyberneticists were wrong in their belief in the unique superiority of the cybernetic approach. They, like the mechanists of the eighteenth and nineteenth centuries, believed that living things could be identified with machines. They did not take into account the differences between their models and the real thing with the same readiness that they did the similarities. And the differences, Oparin remarked, are very genuine: "However great the complexity or intricacy of its organization, an electronic calculating machine is still further apart in its nature from a human being than is, for example, the simplest bacterium, although this has not got the differentiated nervous system which the machine imitates so successfully."

What, according to Oparin, are these genuine differences between machines and living things? The machine is not metabolic in its composition, it is not constructed of living matter: "The composition of the living body is the very factor which determines its flowing character." And the flow, the metabolism, of each cell in the living body allows it to exist for long lengths of time, to repair itself. In contrast to this, said Oparin, the basic structure of the machine is static.

Another fundamental difference between man and machine, said Oparin, is the origin of its purposiveness. Machines have purposiveness, just as living organisms do, but it is placed in them by man. They will therefore always differ from the "truly living." In order to understand this interesting and debatable insistence by Oparin that life can only be understood in its origin, the following quotation from a science-fantasy that Oparin related is helpful. In this quotation not only will Oparin's emphasis on historical development as a key to understanding emerge more fully, but one will also see his concept of dialectical levels of regularities; for Oparin there exist distinct "physicochemical regularities," "biological regularities," and "social regularities." Only human beings display all three:

Let us imagine that people have succeeded in making automatic machines or robots which can not only carry out a lot of work for mankind but can even independently create the energetic conditions necessary for their work, obtain metals and use them to construct components, and from these build new robots like themselves. Then some terrible disaster happened on the Earth, and it destroyed not only all the people but all living things on our planet. The metallic robots, however, remained. They continued to build others like themselves and so, although the old mechanisms gradually wore out, new ones arose and the "race" of robots continued and even, perhaps, increased within limits.

Let us further imagine that all this has already happened on one of the planets of our solar system, on Mars, for example, and that we have landed on that planet. On its waterless and lifeless expanses we suddenly meet with the robots. Do we have to regard them as living inhabitants of the planet? Of course not. The robots will not represent life but something else. Maybe a very complicated and efficient form of the organization and movement of matter, but still different from life. It is impossible to grasp the nature of the "Martian robot" without a sufficient acquaintance with the social form of the motion of matter which gave rise to it. This would be true even if one were able to take down the robot into its individual components and reassemble it correctly. Even then there would remain hidden from our understanding those features of the organization of the robot which were purposefully constructed for the solution of problems which those who built them envisioned at some time, but which are completely unknown to us.

In this passage, Oparin's view of life emerges in a particularly colorful way. It is obvious that he would not accept a purely functional definition of "life." Less obvious is how he would meet the arguments of a functionalist. How, for example, would a man who meets such robots on Mars know that they are, indeed, robots? How would he know that he does not "have to regard them as living inhabitants of the planet," as Oparin put it? Surely such an explorer would expect extra terrestrial life, existing in conditions quite different from those of the earth, to have a quite different appearance from life he had already witnessed. How would he know to be suspicious even if what he saw displayed baffling characteristics? Evidently Oparin would answer that man might indeed make such a mistake, but upon further study he would probably begin to realize that the robots had a social origin, even if he never learned very much about that disappeared society.

The other form of mechanistic interpretation of biology that Oparin resisted in his 1960 book was that coming out of molecular biology. He believed that Erwin Schrödinger in his important and stimulating little book *What Is Life?* had both helped biologists and hindered them; by calling attention to the importance of molecular structure to genetics, he had helped spur one of the great scientific searches of the twentieth century, leading to the description of the DNA molecule. But Oparin thought that at the same time he had thrown at least some biologists back upon static approaches to life. It was Schrödinger who described life as "clockwork," reminding Oparin of eighteenth-century mechanism. And Schrödinger had written, "Now, I think, few words more are needed to disclose the point of resemblance between a clockwork and an organism. It is simply and solely that the latter also hinges upon a solid — the aperiodic crystal forming the hereditary substance, largely withdrawn from the disorder of heat motion."

Oparin believed that here Schrödinger was making a methodological error. He ignored the room for compromise contained in Schrödinger's phrase "largely withdrawn from the disorder.", which indicated that Schrödinger saw the hereditary substance as only relatively, not absolutely, static. But even if one admitted that the difference between the two men was a relative one, a difference remained. Oparin pointed out that he was not alone in believing that the mechanistic assumptions of molecular biologists were leading them astray. Erwin Chargaff, a biochemist from Columbia University, had warned against mechanism in a paper at the 1957 congress, when he asked, Is the cell really nothing but a system of ingenious stamping presses, stencilling its way from life to death? Is life itself only an intricate chain of templates and catalysts and products? My answer to these and many similar questions would be No; for I believe that our science has become too mechanomorphic, that we talk in metaphors to conceal our ignorance, and that there are categories in biochemistry for which we lack even a proper notation, let alone an idea of their outlines and dimensions.

And Chargaff had also seen, along with Oparin, that the mechanistic interpretations were leading to an improper emphasis on the nucleic acid molecule as the "patriarch" of life:

> *Thus, what started cosmically with beautiful and profound legends has come down to a so-called "macromolecule." If poetry has suffered, precision has not gained. For we may ask whether a molecule that merely provides for one cell constituent continually to make itself can teach us much about life and its origins. We can also ask whether the postulation of a hierarchy of cellular constituents, in which the nucleic acids are elevated to a patriarchal role in the creation, is justified. I believe there is not sufficient evidence for so singling out this particular class of substances.*

Thus, Oparin's denial of the primordial role of nucleic acid found sympathy in other ears and in other countries. An old-fashioned theist would have agreed with Oparin's and Chargaff's criticism of mechanism in biology, supplementing it with a story of divine creation. Chargaff's last statement above would fit very well with such an interpretation. But Oparin apparently would not have been troubled by this fact; since his dialectical materialism occupied a middle position between vulgar materialism and idealism, it followed that both the idealists and the dialectical materialists opposed mechanistic interpretations of life.

In October 1963 Oparin attended a conference on "The Origins of Prebiological Systems" at Wakulla Springs, Florida, sponsored jointly by Florida State University, the University of Miami, and NASA. At this meeting P. T. Mora, of the National Institutes of Health, submitted current theories of the origin of life, including Oparin's, to a methodological critique. He showed what has frequently been noted by philosophers of science, namely, that questions of singularity, or origin, are not in principle resolvable by experimental science. Thus, from the standpoint of strict logic and the methods of empirical science the question to which Oparin had devoted his life was not answerable. Mora was particularly critical of the application of the term "natural selection" (in the manner of Oparin) to nonliving systems.

Indeed, Mora said that the gap between physical science and biology is "too big to bridge." Consequently, Mora was extremely skeptical of attempts such as Oparin's to throw a bridge across the gap, and believed that it could be achieved only by committing methodological errors:

> *I believe that this accounting for the appearance of the first persistently self-reproducing unit in a prebiotic system is an unwarranted extension of the meaning of the word selection, used by Darwin in a valid, but different operational sense. Remember, the Darwinian selection and evolution concept was arrived at empirically, by observing the spectrum of living species. Of course, this Darwinian concept of selection is a necessary attribute of the living, which allows it to adjust to the changing milieu. But it is not a sufficient concept. Mutability can occur only when first there is present a persistent reproducibility. Thus, mutability should be considered as a consequence of the persistent reproducibility and not vice versa.*

Thus, two operationally different meanings of the word, selectivity one in the physicochemical sense and the other in the Darwinian sense, each of them valid when used in its proper context can give an illusion that the word may operate in a third sense. I believe this is the origin of the unwarranted speculation that some kind of molecular selection may account for the appearance of the first persistently self-reproducing, internally controlled, living unit.

What alternative explanation did Mora offer? He remarked that he believed that "we have to have a teleological orientation in our thinking." But when asked if he could describe this teleological orientation more fully, he replied, "No. I am very destructive in my attitude. I can't say anything conclusive and constructive and useful. I can only tell you that I consider that the impression of physics on scientists is so strong nowadays that it is difficult to think, without using the same concepts. In my mind, this almost parallels the situation which existed around the end of the scholastic period, when teleology was so overwhelming that considerations and approaches that physics now uses were very difficult even to think about."

Mora's presentation aroused considerable controversy at the Florida conference. He posed a very old and very important problem in the history of science. This issue is one of the fundamental problems of explanations of development, which can be simply stated in terms of Mora's thesis: We cannot obtain a higher degree of order or organization than that present in the interacting elements and the environment.

It was not Oparin but Bernal who took the responsibility for answering Mora at the Florida conference. Like Oparin, Bernal favored a materialistic, developmental explanation of the origin of life. He differed with Oparin by doubting the fundamental role of coacervation, favoring instead a process of clay mineral adsorption, but their two approaches both assumed the fruitfulness of trying to bridge the gap between the nonliving and the living. Bernal agreed with the validity of much of Mora's argument, and specifically, that questions of origin can not be explained on the basis of logic. They have instead, Bernal said, "a logic of their own." But, said Bernal, Mora draws a conclusion which is the opposite to the one which I would draw. The present laws of physics, I would agree with him, are insufficient to describe the origin of life. To him this opens the way to teleology, even, by implication to creation by an intelligent agent. Now both of these hypotheses were eminently reasonable before the fifteenth or possibly even before the nineteenth century.

But the real difference between Mora and Oparin-Bernal was not whether different laws exist on different levels. In fact, Mora believed in the existence of such different levels even more strongly than Oparin and Bernal, for he thought the gap between physics and biology was unbridgeable, and, therefore, the distinction between the two levels was absolute. Oparin and Bernal, on the other hand, saw the distinction as a relative one.

Oparin took up the question of the way in which the transition from one level to the next higher one occurs in his 1966 book *The Origin and Initial Development of Life*, translated by NASA in 1968.

In this work Oparin sketched out the "prebiological" state in greater detail, incorporating more of the recent evidence. He allowed more room for non-coacervate prebiological systems, pointing toward compromise with views such as Bernal's. The "coacervatelike" droplets, which had a complex and advanced organization but which are still simpler than "the most primitive living beings," Oparin now called protobions. The protobions went through a further evolution promoted by a process to which Oparin still insisted on giving the name "primitive natural selection." And in this section of his book Oparin cited Mora's criticism of his scheme at the 1963 Florida conference and attempted to answer it. He maintained that the "logic of its own" by which Bernal said the origin of life must be explained was, in fact, the logic of dialectics. As Oparin observed:

> *At present, a number of opinions have been expressed in the scientific literature on the competency of the use of the term "natural selection" only in respect to living beings. It is an opinion widely held among biologists that natural selection cannot be extended to objects which are not yet alive, and particularly not to our protobions.*

It is, however, erroneous to think that living bodies first originated and then biological laws or vice versa, that in the beginning biological laws were formulated and the living bodies arose.

Dialectics obliges us to consider the formation of living bodies and the formulation of biological laws as proceeding in indissoluble unity. It is therefore quite permissible to assume that protobions those initial systems for the formation of life evolved by submitting to the action not only of intrinsically physical and chemical laws, but also of incipient biological laws including also prebiological natural selection. Here we may cite an analogy with the formation of man, i.e., with the rise of a social form of the motion of matter which is even higher than life. As is known, this form took shape under the influence not so much of biological as of social factors, chiefly the labor of our ancestors, coming into being at a very early state of homogenesis, and improving more and more. Therefore, just as the rise of man is not the result of the operation of biological laws alone, so the rise of living bodies cannot be reduced to the action of only a few laws of inorganic nature.

In the above quotation Oparin's belief in a hierarchy of laws in nature is revealed particularly clearly; social, biological, and physicochemical laws all operate on different levels. The most difficult conceptual problem within the framework of Oparin's scheme is the transition from one realm of law to another. If one assumes, as Oparin did, that living matter evolved from non-living matter and human beings with their social life evolved from lower orders of animals, some method of explaining these transitions must be found. Oparin relied on a dialectical concept of the emergence of qualitative distinctions; he believed that "incipient forms" of laws of a higher realm could be found in the realm immediately below it. It is the sort of concept that has appealed to many thinkers in the past C. Lloyd Morgan's concept of "emergent evolution" was somewhat similar, but supplemented with the presupposition of God and it has a certain persuasiveness. Nonetheless, it should be admitted that Oparin's philosophy of biology suffered from a lack of precision in definition, on which critics such as Mora correctly centered their attention; furthermore, Oparin's emphasis on the irreducibility of biology to physics and chemistry and his increasing attention to "purposiveness" skirted ever more closely to the very real dangers of vitalism.

The Soviet philosopher I. T. Frolov recognized these pitfalls in his 1968 book on genetics and dialectics (discussed at the end of the previous chapter of this study) when he described the irreducibility of biology as more a result of man's incomplete knowledge than a characteristic of living matter itself. In Oparin's approach, living matter is inherently distinct from nonliving matter and cannot, in principle, be reduced to physics and chemistry. Frolov was less adamant.

It should be noticed that nothing in the philosophic system of a materialist absolutely requires him to believe that living matter on earth evolved from non-living matter. Materialists have usually supported this view since it seemed the best explanation for the origin of life on earth without reliance on a divine agent. But strictly speaking, there is another alternative available to the materialist; he can maintain that matter has existed eternally in the universe in both its living and non-living forms. Whether non-living matter actually evolves into living matter can then be left an open question without violating any assumptions of philosophic materialism. The life that exists on earth can be explained by saying that it resulted from the depositing of primitive organisms on the surface of the globe from elsewhere at some past moment in its history. Such a hypothesis is frequently called panspermia and has been posed in the past in various forms by such well-known scientists as Liebig, Helmholtz, and Kelvin.

A few Soviet scientists began to re-examine panspermia in the late sixties. A geologist, B. I. Chuvashov, wrote in 1966 in *Problems of Philosophy* that in his opinion life had existed in the universe eternally. He cited the recent criticisms of Oparin's application of the term "natural selection" to prebiological systems as one of the reasons for his dissatisfaction with Oparin's theory and his consequent interest in panspermia. Nonetheless, Chuvashov thought that nonliving matter may also occasionally develop into living matter, but perhaps only once in each planetary system or galaxy. Life was then spread throughout the neighboring area in the form of spores by meteorites and dust.

This view has been held by no more than a small minority of scientists in the Soviet Union; it enjoys similar modest support elsewhere. Some of the scientists now interested in panspermia have cited the presence of carbonaceous chondrites in the lunar soil samples brought back to the earth by the Apollo 11 expedition as support for their hypothesis. This evidence, however, is subject to differing interpretations and does not yet warrant conclusions.

7

Physiology and Psychology in Philosophical Science

Introduction

Surrounding Sechenov's views there soon grew up a controversy among the St. Petersburg educated public. The particular political and ideological scene of late-nineteenth century Russia influenced the course of the debate, with the radical intelligentsia usually, but not always, responding favorably to Sechenov's opinions and the government bureaucracy usually disapproving. In 1866 the book was prohibited for sale by the St. Petersburg censors, and Sechenov himself was threatened with court action for allegedly undermining public morals. Eventually Sechenov escaped trial, but the already existing link between materialism in science and radical politics was strengthened and made more apparent.

Although materialism was strong in Russian psychology before the Revolution, it by no means monopolized the field. Sechenov was thought of primarily as a physiologist, not a psychologist. Opposing the views of Sechenov and some of his pupils were not only the censors of St. Petersburg and representatives of the Church, but also many university professors of philosophy and psychology. Indeed, Sechenov was outside the mainstream of academic psychology in Russia. Nonetheless, the essential issues that he raised concerning the nature of the psyche and the relationship of the physiological to the psychological were hotly debated among Russian psychologists, physiologists, philosophers, and political activists in the last decades of the nineteenth century. The history of these debates is still insufficiently explored, but even a cursory examination reveals that some of the features of these polemics not only between materialists and idealists, but among members of each camp and of other groupings as well resemble discussions that have continued throughout the Soviet period.

The materialist tradition in physiological psychology was also strong in France; Sechenov's views have consequently been traced back to the French tradition through such men as the great French physiologist Claude Bernard, in whose laboratory the young Sechenov once worked. No doubt such an influence was genuine in Sechenov's development. One awaits further scholarship on the connections between Russian and west European schools of thought in this interesting area. Nevertheless, the originality of Sechenov seems clear; as Edwin G. Boring wrote in his *A History of Experimental Psychology*, Sechenov became the Russian pioneer in reflexology. We must, moreover, remember that he was far ahead of western European thought on this matter. His papers of 1863 coincided with Wundt *Vorlesungen über die Menschenund Thierseele, and 1870 was four years before Wundt Physiologische Psychologie. In no other country did reflexes yet seem to provide a means for studying cognition. Later, Pavlov read Sechenov's work and was, as a young man, greatly influenced by his argument.*

From the standpoint of the history and philosophy of science the greatest significance of Pavlov derives from his success in bringing psychic activity within the realm of phenomena to be studied and explained by the normal objective methods of natural science. In contrast to the introspective approach of many investigators of mental activity at the turn of the century, Pavlov's method was based on the assumption that psychic phenomena can be understood on the basis of evidence gathered entirely externally to the subject. He was not entirely original in his intention to proceed in this manner, but as a great experimentalist he was able brilliantly to combine this methodological assumption with unusual skill in devising and conducting experiments with animals. On the basis of these experiments he erected a theory of nervous activity that presented general principles aimed toward the eventual explanation of man's psychic activity on a physiological foundation.

Pavlov is, of course, best known for his theory of conditioned and unconditioned reflexes. Unconditioned reflexes, he said, are inborn forms of nervous activity and are transmitted by inheritance. Conditioned reflexes are acquired during the life of an organism and are based on a specific unconditioned reflex; conditioned reflexes are not normally inherited, although Pavlov believed that in some cases they could become hereditable.

In the classic case of the dog and the bell, the unconditioned reflex is the natural, inborn salivation of a dog in response to the stimulus of food. The conditioned reflex, salivation in response to a bell alone, is created by the prior repeated juxtaposition of the bell and the food. Pavlov further illustrated that "conditioned reflexes of the second order" could be created by using the conditioned response to the bell as a basis for the formation of yet another conditioned reflex to a third stimulus, such as a light. In the latter case, it must be emphasized that at no time was the original stimulus (food) combined with the stimulus triggering the second-order reflex (the light). In this fashion Pavlov was able to point to the quite indirect ways through association by which reflexes could be created. He believed that the psychic activity of man could be interpreted in this way, or at least on this foundation. This theory of the broad significance of conditioned reflexes Pavlov called the Theory of Higher Nervous Activity, and this phrase is a part of the standard terminology of Soviet physiologists and psychologists. Pavlov's view is supported by numerous contemporary non-Soviet scholars as well; their opinion is summed up in the words of a recent French biographer of Pavlov: "It is beyond doubt that many of our thoughts if not all are, or at least originate in, reflexes acquired from upbringing, education and culture, and strengthened by experience."

The inner structure of reflex action was described by Pavlov in terms of a "reflex arc," a term that would be the subject of much later discussion. The reflex arc had three links: the *affector* neurons, the *nervouscenters*, and the *effector* neurons. The original excitation caused in the sense organs by an external stimulus travels inward along a chain of affector neurons to the nervous centers; then another stimulus travels outward along the effector neurons to specific muscles or glands, causing a response to the original stimulus. The three links in this arc are sometimes described as sensor-connector-motor.

In the case of the formation of conditioned reflexes in man Pavlov believed that the nervous centers are located on the cortex of the cerebral hemispheres. "Temporary connections," an inclusive term embracing conditioned reflexes and other more rudimentary or fleeting linkages, are formed as a result of "irradiation" of stimuli reaching the hemispheres. In other words, stimulation is "generalized" in the hemispheres in such a way that other areas of the cortical region now react in the same way as that concerned in the original stimulus. Thus, the area of the cortex receiving nervous responses to light signals may be incorporated into reflex action originally based only on sound signals. As Pavlov wrote, "The fundamental mechanism for the formation of a conditioned reflex is the meeting, the coincidence of the stimulation of a definite center in the *cerebral cortex* with the stronger stimulation of another center, probably also in the cortex, as a result of which, sooner or later, an easier path is formed between the two paths, i.e., a connection is made."

By a process of training, inhibition, the reverse of irradiation, can also be illustrated. Physiologically, the area of the cortical region that has been irradiated is reduced by teaching the subject to discriminate not only between very different signals, such as sound and light, but between sounds of different vibrations. Thus, Pavlov was able to teach a dog to respond to a tempo of one hundred beats a minute but not to ninety-six, as a result of producing food only after the more rapid signal. After this process of inducing inhibition, Pavlov concluded that "the nervous influx produced by the stimulus is now communicated to only a very limited area of the cortical zone under consideration."

One of the most flexible concepts that Pavlov advanced, and one still exploited only to a rather small degree, was that of the "second-signal system," a feature unique to the psychic activity of man. Most of Pavlov's research was based on experiments with dogs, but in the latter part of his life he worked with monkeys and gorillas, and his interests were shifting more and more to what he considered the ultimate goal of neurophysiology the study of man. Man has fewer instincts than animals; Pavlov believed, therefore, that his behavior would be governed by conditioned reflexes to a much higher degree. Both animals and man can be conditioned in similar ways, but man, in addition, possesses the almost infinitely rich instrument of language. While animals responded to simple ("primary") signals or symbols (even a dog responding to a word command reacts to it in a fashion not dissimilar to the response to a bell or light), man responds to the meanings and incredibly rich associations conveyed by speech and writing ("secondary signals"). The language message that any one human subject receives will contain meanings and associations unique for him, given a message of even minimal complexity. And Pavlov saw the second-signal system as infinitely more complex than the primary one: "There is no comparison, qualitative or quantitative, between speech and the conditional stimuli of animals." Thus, Pavlov cannot be described fairly as a person who believed that human behavior can be reduced to the simple stimulus-response action of the noted experiments with dogs. He fully recognized that human beings were qualitatively quite distinct from other animals. But he believed, nevertheless, that human behavior is amenable to investigation on the basis of physiology, an assumption sensible and necessary in order for physiologists to investigate the human nervous system.

Pavlov's attitude toward psychology has been the subject of numerous inaccurate statements, many of which imply that Pavlov was opposed to the very existence of psychology. Pavlov did object to the concept of animal psychology, since he felt that there was no way for man to gain access to the inner world of animals. He was, further, deeply critical of what he considered metaphysical concepts presented in psychological terms. In his early years he was doubtful of the scientific validity of much that was presented as psychological research. As he grew older, and as experimental psychology steadily developed as a discipline, Pavlov became more and more kindly disposed toward psychology. In a speech given in 1909 Pavlov said:

> *I should like to elucidate that which might be misunderstood in these statements concerning my views. I do not deny psychology to be a body of knowledge concerning the internal world of man. Even less am I inclined to negate anything which relates to the innermost and deepest strivings of the human spirit. Here and now I only defend and affirm the absolute and unquestionable rights of natural scientific thought everywhere and until the time when and where it is able to manifest its own strength.*

But even in this statement affirming the right of psychology to exist, one can detect Pavlov's skeptical view of psychology. The last sentence implies a distinction between psychology and "natural scientific thought," which most psychologists would reject. And when Pavlov spoke of a fusing in the future of physiology and psychology, many psychologists thought that he was actually referring to an absorbing of psychology by physiology after the necessary progress in physiology had occurred. One must admit that Pavlov remained somewhat dubious about psychology as a science, although he was by no means so hostile as many later commentators have implied. Despite his frequent warnings against reductionism, his call for the study of the "whole organism," and his belief in the "qualitative and quantitative uniqueness" of man, Pavlov tended to see psychic phenomena, and especially the reflex arc, in somewhat mechanistic and elementary terms. This tendency was probably inevitable in the period when psychology was, indeed, heavily influenced by idealistic concepts and Pavlov had to struggle to establish his teaching on conditioned reflexes, now recognized as one of the great achievements of both physiology and psychology.

Pavlov was not a Marxist and did not defend his system in terms of dialectical materialism. For many years after the Revolution he stoutly resisted Marxist influences in educational and scientific institutions and even criticized Marxist philosophy. In the last years of his life, however, his views changed; he praised the Soviet government for its support of science, and he was impressed by the intelligence of individual Bolshevik leaders, such as Nikolai Bukharin. One of his pupils, P. K. Anokhin, a man whose views will be separately discussed, maintained that once in a conversation he tried to show Pavlov that his teaching about the contradictory but necessary effects of irradiation and inhibition was deeply dialectical and revealed the struggle and unity of opposites. To this observation Anokhin said that Pavlov responded, "There you are, it turns out that I am a dialectician!"

There are many aspects of Pavlov's thought that appeal to dialectical materialists. First of all, his primary goal, the explanation of psychic phenomena on the basis of physiological processes, is one that materialists have traditionally and understandably supported. His emphasis on the necessity to study organisms as a whole, "in all their interactions," rather than by isolating out one portion or one phenomenon has been praised by Soviet writers for being in agreement with the dialectical principle of the interconnectedness of the material world. His emphasis on the unique qualities of man, with his second-signal system, has been termed an understanding of the qualitative differences of organisms at different levels of complexity, based on the principle of the transformation of quantity into quality. His description of the human body as a system "unique in the degree of its self-regulation" has been seen both as a prefiguring of cybernetic concepts of feedback and as an understanding of the dialectical process of development.

Scholars in the Soviet Union, on the one hand, and those outside that country, on the other, frequently look upon Pavlovianism in different ways and almost as different things. Non-Soviet scientists often consider it a rather restricted body of experimental data and hypotheses concerning conditioned and unconditioned reflexes. To some of them his name is nearly synonymous with the mental picture of salivating dogs. Soviet scholars, on the other hand, see Pavlovian theory not only as this body of facts and conclusions, but also as an approach to nature in general and to biology in particular. Pavlov himself contributed to this latter understanding in a conversation with the American psychologist K. S. Lashley; when Lashley asked Pavlov to define the concept of "reflex," Pavlov replied:

> *The theory of reflex activity operates on three basic principles of exact scientific research: first, the principle of determinism, i.e., of a stimulus, a cause, a reason for every given action or effect; second, the principle of analysis and synthesis, i.e., of an initial decomposition of the whole into parts or units and then the gradual building up anew of the whole from its units or elements; finally, the principle of structure, i.e. the distribution of the actions of force in space, the timing of dynamics to structure.*

In reply to this statement by Pavlov Lashley observed that this definition of reflex was so general that it could be taken as the general principles of all science. But Pavlov stuck to his formulation, which is often quoted in Soviet discussions of the significance of reflex theory.

Some Soviet authors distinguish between the reflex principle in a philosophic sense and the reflex principle in a concrete, physiological sense, thus opening up considerable possibilities for recognizing certain limitations in Pavlov's teaching while retaining its methodological content. In 1963 F. V. Bassin, a Soviet scholar who called for much greater attention to the subconscious realm and pointed to certain elements of value in Freud's work at a time when this was rare among Soviet scholars, made this distinction; in his opinion, the most valuable aspect of Pavlov's work was the underlying idea of the essential dependence of biological factors on the environment. Bassin wrote:

> *That person who abandons the reflex theory in its philosophic sense abandons more than Pavlov's teaching: He abandons the dialectical materialist interpretation of biological processes in general. This is undoubtedly so, since the primacy of the reflex principle in its philosophic sense (i.e., the idea of the dependency in principle of biological processes on factors of the environment) is that basic, that most profound element that distinguishes us from the supporters of idealistic biology, with its emphasis on immanence, spontaneity, and consequently, the absence of the reflex principle in life processes. I mention this because it is necessary to see the difference between the reflex principle in its general philosophic meaning and as a concrete understanding of physiological structure.*

The history of psychology in Russia in the years after the Revolution is a very rich and contradictory story; since our center of attention in this volume falls on the years after World War II, it will be impossible to discuss the earlier period in detail. However, some of the features of the work of L. S. Vygotsky, who continues to be influential, will be considered below. More detailed discussions can be found in A. V. Petrovskii's *History of Soviet Psychology* (in Russian) or in Raymond Bauer's *The New Man in Soviet Psychology.*

Immediately after the Revolution members of several different schools of psychology could still be found in Russia. Those with the closest links to introspection and idealistic psychology were N. Lossky and S. Frank, both of whom lost their positions shortly after the Revolution. Another group was made up of experimental psychologists who had been heavily influenced by subjective psychology, but who moved after the Revolution to a position of neutral empirical psychology, hoping in that way to remain clear of the controversies. They included G. I. Chelpanov and A. P. Nechaev. A third group was made up predominantly of physiologists, such as V. M. Bekhterev, and hoped to reconstruct psychology on an objective, scientific basis. They usually doubted the validity of the term "psychology."

The first psychologist to call for an application of Marxism to psychology was K. N. Kornilov, a scholar with an interesting history in the discussions of the twenties and the thirties. At congresses of psychoneurologists in 1923 and 1924 Kornilov attempted to discern the operation of the materialist dialectic in his psychological research. He maintained that the dialectical principle of universal change could be seen in psychology "where there are no objects, but only processes, where everything is dynamic and timely, where there is nothing that is static." The dialectical principle of interconnectedness is illustrated, he continued, by the tendency toward "extreme determinism" in psychology, including the determinism of the Freudian school. This principle accorded well, further, with the views of Gestalt psychologists and with the emphasis on the importance of total patterns rather than discrete bits of experience. And a third principle the transition from quantity to quality by leaps is illustrated in many ways: color discernment, in which quantitative differences in frequency of light waves result in qualitatively distinct perception of colors; the concept of thresholds of perception, in which one senses change only after a considerable amount of quantitative stimulation of sense organs; and the Weber-Fechner law of weight and auditory discrimination.

Like Engels in his more enthusiastic moments, Kornilov seemed to see the operation of the principles of the dialectic on every hand. Not surprisingly, Kornilov was soon criticized for applying the dialectic in a "purely formal" fashion, for using it simply as a means of justifying research on which he was already embarked rather than as a methodology basically affecting the course of his work. He was particularly criticized for maintaining that "reactology" his term for his approach to psychology was a dialectical synthesis of the subjective and objective trends in Soviet psychology, one that would preserve a concept of consciousness, of the psyche, while at the same time utilizing the rich findings of the physiologists in the study of reflexes. One of his critics, V. Struminskii, observed that Kornilov was applying the materialistic dialectic to the world of ideas rather than to matter, combining intellectual currents in a Hegelian fashion: "This synthesis is not materialist, but a completely idealistic construction. The materialist dialectic does not concern itself with psychological currents as independent factors and magnitudes."

Despite Kornilov's attempt to identify the dialectic in his research, Marxism was not a major influence on his work. His effort to combine elements of subjective psychology and the newer physiological study of reflexes stemmed from his opinion that both possessed advantages. He thought that the physiologists and behaviorists were abdicating the responsibility of psychologists by occupying themselves exclusively with muscular responses. The traditional psychologists, on the other hand, were just as blindly ignoring the significant work of Pavlov, Bekhterev, and their followers. After 1923 Kornilov headed the Moscow Psychological Institute, where he worked with other scholars of later prominence, such as N. F. Dobrynin, A. N. Leont'ev and A. R. Luria. They were also in close communication with groups led by P. P. Blonskii and M. A. Reisner. All of these men at this time were experimenting eclectically with various currents in psychology in a manner that later became impossible because of ideological pressures.

In addition to reactology, the other major tendency in Soviet psychology and physiology at this time was the "reflexology" of M. Bekhterev. It contrasted sharply with reactology in its refusal to use subjective reports and such traditional terms as "psyche," "attention," and "memory." This school drew heavily on two different sources: the materialist tradition in Russian physiology stemming from Sechenov through Pavlov and Bekhterev himself, and American behaviorism. Bekhterev (1857–1927) had long before the Revolution maintained that every thought process, conscious or unconscious, expresses itself sooner or later in objectively observable behavior. On this basis he and his followers hoped to create a science of behavior. In the twenties their approach was so popular that the existence of psychology as a discipline was threatened. In the Ukraine in 1927 higher educational institutions replaced the term "psychology" with "reflexology" as a description of courses of study.

There was also in the twenties a genuine interest in Freudian psychology and much controversy over how well it fit with Marxist interpretations. It was by no means clear in this early period that Freudianism would become a pejorative term to Soviet Marxists. Part of the interest in Freud was simple curiosity; many of the articles in political and literary journals contained elementary descriptions of his work. Freud had not yet published his later, more speculative, works such as *Civilization and Its Discontents*, in which, in addition to some dubious psychological theorizing, there appeared a criticism of communism. To some Soviet writers Freud's teachings appeared as a victory of determinism, an end to free will. Writing in the major Marxist theoretical journal in 1923, the Soviet author B. Bykhovskii commented, "We conclude that despite the subjective casing in which it appears, psychoanalysis is at its foundation imbued with monism, with materialism and with the dialectic, i.e., with the methodological principles of dialectical materialism." And further, "I am not inclined to exaggerate the value of psychoanalysis. I know that it contains much that is controversial and that several of Freud's conclusions are arbitrary and one-sided. But I have tried to bring out the healthy kernel of psychoanalysis, and I think it is rather valuable and significant." Similar comments were made by such intellectual and political leaders as M. A. Reisner, A. P. Pinkevich, and Trotsky himself. But by the latter years of the twenties, discussion of Freud had shifted to open criticism.

Aside from the question of Freudianism, a new trend in Soviet psychology became discernible by the end of the 1920's. This trend stemmed from the rather widely held realization that with the defeat of the supporters of subjectivism and introspection in Soviet psychology, the greatest danger was now from the left from those militant materialists who hoped to swallow up psychology in a purely physiological understanding of mental activity. The defenders of psychology rallied around the concept of *psikhika* (psyche) and *soznanie* (consciousness) in what has been called a "great struggle for consciousness." This controversy, which ended in victory for the defenders of psychology and consciousness, bore many characteristics peculiar to the Soviet environment. It is well, however, to guard against the tendency of non-Soviet historians to look upon all events in the Soviet Union as *sui generis*, as irrelevant to intellectual history as a whole. These were years in which the validity of the concept of consciousness was being discussed in many countries. According to Boring:

> *The attack on old-fashioned analytical introspectionism was successful, and in the late 1920's Gestalt psychology and behaviorism found themselves practically in possession of the field. With their missions thus more or less accomplished, both these schools tended to die out or at least to lose their aggressiveness during the 1930's. Psychological operationism came in at this time to supplant behaviorism as a more sophisticated view of psychology, and the outstanding systematic issue in the early 1940's seemed to be whether the Gestalt psychologist could save consciousness, as observed in direct experience, for psychology, or whether the operationists would succeed in having it reduced to the behavioral terms which define the manner of its observation.*

Echoes of these changes in psychology internationally were reverberating within the Soviet Union. There, too, the criticism of introspectionism had been successful indeed, to the point of overkill resulting from the peculiar political instruments at the disposal of the Communist Party, such as increasing control of faculties and editorial boards. In the Soviet Union, as abroad, the crude mechanism of early behaviorism was being succeeded by a more sophisticated approach that, nonetheless, did not deny the achievements of the behaviorists.

In the Soviet Union there were other unique elements as well. The debates were increasingly cast in the terms of theoretical Marxism. Furthermore, the policy decisions of the Communist Party were beginning to have a direct influence on the course of the psychological discussions. The decision to embark upon a rapid industrialization program required great effort on the part of Soviet citizens and enormous will power. A psychology that left more room for voluntarism, for personal resolve and dedication, was welcome on this scene. This shift in Soviet psychology has been frequently discussed by previous authors, such as Raymond Bauer". In terms of Marxist theory the shift was explained on the basis of the Leninist "theory of reflection," which maintains that the mind, or consciousness, plays an active role in the process of cognition.

In the early 1930's the place of psychology in the Soviet Union became more secure, while Bekhterev's reflexology gradually lost its popularity. As we shall see in the cases of Vygotsky and Rubinshtein, Marxism was incorporated into psychological theory in a more sophisticated way. The increase in the subtlety of psychological theory was accompanied, perhaps surprisingly, by an increasing concern with such practical activities as industry and education. Industrial psychology, the scientific organization of labor movement (NOT), and psychotechnics all grew impressively. Educational psychology was also very important in the early thirties. The issue lying immediately behind the famous decree of the Central Committee of the Communist Party of July 4, 1936, "On Pedological Perversions in the System of the People's Commissariat of Education," seems to have been one of social class. The decree accused pedologists of attempting "to prove from the would-be 'scientific,' 'biosocial' point of view of modern pedology that the pupil's deficiency or the individual defects of his behavior are due to hereditary and social conditioning." It was the perennial issue of environment versus heredity and the practical question of how an educational system can overcome the deleterious effects of both. In the thirties in the Soviet Union a great effort was being made to achieve literacy among a backward population. From the standpoint of performing this monumental educational task, what was most needed were concrete suggestions in the field of elementary pedagogy, not theoretical and inconclusive discussions of the determining elements of intelligence. The American scholar Bauer seems quite correct in his observation that much of the criticism of the educational psychologists stemmed from the fact that they appeared, at least to their critics, to be "professionally more oriented toward finding an excuse than toward the development of a cure." From the standpoint of social reform, this was no inconsequential issue; similar controversies of great implication over the need to link theoretical analysis with practical reform could be found in the United States in the sixties. Academic social science frequently does become aloof to social needs, occasionally to an immoral degree. In the Soviet Union in 1936 the issue was resolved not so much by discussion from below as by political order from above.

The political atmosphere of the Soviet Union in the late thirties was grim, and the situation would be even worse immediately after World War II. The Stalinist system of control became firmly established. The great purges within the Communist Party eliminated several early defenders of innovative psychology. Soviet historians now openly admit that political controls did serious damage to many fields, including psychology. As M. G. Iaroshevskii, a Soviet historian of psychology, wrote in 1966: "The violation of Leninist principles in social life and in science hindered the development of the psychology of labor in the U.S.S.R. (the Psychotechnical Society and the journal Soviet Psychotechnics were abolished, and the teaching of psychotechnics in the universities ceased. In 1936 almost all laboratories concerned with psychotechnics and the psychology of labor were closed.)" Several pages later he continued in the same vein, "The criticism of pedology occurred in the complicated environment of the second half of the thirties and frequently was accompanied by a denial of all that was good in the work of Soviet scholars in pedology, and even in pedagogy and psychology, which had been developing in a very creative fashion."

Fortunately, important work had been done in the Soviet Union before these controls were imposed. A case is the achievement of L. S. Vygotsky, who did his research in the late twenties and early thirties. To the work of this significant Soviet psychologist we must now turn. Lev Semenovich Vygotsky (1896-1934) L. S. Vygotsky is one of the most important influences in Soviet psychology; in recent years his ideas have begun to spread outside the Soviet Union, particularly with the publication in English in 1962 of his *Thought and Language*. His influence is particularly remarkable in view of the fact that Vygotsky died of tuberculosis in 1934 at the age of thirtyeight; he rushed to completion some of his most important writings in his final illness. One of Vygotsky's best-known pupils, A. R. Luria, is supposed to have remarked many years later, "All that is good in Russian psychology today comes from Vygotsky." Luria dedicated his important monograph *Higher Cortical Functions in Man*, published in Moscow in 1962, to Vygotsky's memory, and remarked that his own work could in many ways "be looked upon as a continuation of Vygotsky's ideas."

Vygotsky has not always enjoyed the esteem of official circles in the Soviet Union, however. From 1936 to 1956 his writings were in disfavor. In 1950 Vygotsky's theories on the relationship of language and thought were contradicted by Stalin himself, as will be related below. Even in recent years, when Vygotsky regained his earlier popularity, he has frequently received a mixture of praise and criticism from Soviet historians of psychology. In 1966 A. V. Brushlinskii commented that Vygotsky underestimated the epistemological aspect of mental activity, but that nonetheless Soviet psychology was heavily in debt to him for his being the first to discuss in a detailed fashion the influence of sociohistorical factors on the human psyche. Vygotsky is now widely praised in the Soviet Union for this service, described as an important introduction of the Marxist approach to psychology. His works have been published and circulated widely.

In his introduction to Vygotsky *Thought and Language* Jerome Bruner sensibly warned non-Soviet scholars against interpreting Vygotsky solely in terms of Soviet conceptions of man; Vygotsky, he said, is an original who transcends ideological divisions. To identify some of the intellectual currents to which Vygotsky responded is not, however, the same as reducing him solely to those influences. In any case, there seems little question that Vygotsky was influenced by

Marxist philosophy, as he interpreted it. Non-Russian readers of his works may not believe that the influence of Marxism on Vygotsky was genuine, and for a very understandable reason; when Vygotsky's works were translated from Russian to English for publication in an abridged version in the United States, most of the references to Marx, Engels, and Lenin were omitted. Lenin disappeared completely. The translators believed, evidently, that the references to Marxism were extraneous to the scientific content of Vygotsky's writings and could be dropped without damage. As a result it is almost impossible for the historian of psychology without knowledge of the Russian language to understand the initial assumptions of Vygotsky's approach. In the original Russian, however, it is clear that Vygotsky attempted to show a relationship between his views on children's thought and Lenin's epistemology. He spoke of the "unity and struggle of the opposites of thought and fantasy" in cognition. He was, as we will see, critical of the epistemological dualism that he saw in Jean Piaget's theories of language, and in particular, Piaget's description of a child's "autistic" use of language. Vygotsky emphasized that a Marxist approach to language revealed its "external" or "social" origins.

One of the main problems to which Vygotsky addressed himself was the interrelation of thought and speech. His work on this topic has frequently been compared to that of Piaget. Vygotsky praised Piaget's work, calling it "revolutionary," but commented that it "suffers from the duality common to all trailblazing contemporary works in psychology. This cleavage is a concomitant of the crisis that psychology is undergoing as it develops into a science in the true sense of the word. The crisis stems from the sharp contradiction between the factual material of science and its methodological and theoretical premises, which have long been a subject of dispute between materialistic and idealistic conceptions."

Piaget in his early work postulated three stages in the development of the modes of thought of a child: first, autism, second, egocentrism, and last, socialized thought. In the first, or autistic, stage the child's thought is subconscious and is directed toward self-gratification. The child does not yet use language and has not yet adjusted to the existence of other persons, with their desires and needs. He is not susceptible to the concept of truth and error. In the last stage, socialized thought, the child has adapted to reality, tries to influence it, and can be communicated with through language. He has recognized laws of experience and of logic. In the intermediate stage, egocentrism, the child "stands midway between autism in the strict sense of the word and socialized thought." He uses language, but only to himself; he is thinking aloud. Thus, all three stages constitute a scheme of the development of the thought of a child based on the assumption that "child thought is originally and naturally autistic and changes to realistic thought only under long and sustained social pressure."

Vygotsky accepted much of this description by Piaget of the individual stages of child development, but he rejected the direction of flow of the underlying genetic sequence. As Vygotsky described it:

> *The development of thought is, to Piaget, a story of the gradual socialization of deeply intimate, personal, autistic mental states. Even social speech is represented as following, not preceding, egocentric speech.*

The hypothesis we propose reverses this course. We consider that the total development runs as follows: The primary function of speech, in both children and adults, is communication, social contact. The earliest speech of the child is therefore essentially social. At first it is global and multifunctional; later its functions become differentiated. At a certain age the social speech of the child is quite sharply divided into egocentric and communicative speech. (We prefer to use the term communicative for the form of speech that Piaget calls socialized as though it had been something else before becoming social. From our point of view, the two forms, communicative and egocentric, are both social, though their functions differ.) Egocentric speech emerges when the child transfers social collaborative forms of behavior to the sphere of inner-personal psychic functions. In our conception, the true direction of the development of thinking is not from the individual to the socialized, but from the social to the individual.

And thus Vygotsky arrived at the concept for which he is best known, the "internalization of speech":

> *Piaget believes that egocentric speech stems from the insufficient socialization of speech and that its only development is decrease and eventual death. Its culmination lies in the past. Inner speech is something new brought in from the outside along with socialization. We believe that egocentric speech stems from the insufficient individualization of primary social speech. Its culmination lies in the future. It develops into inner speech.*

Since Vygotsky believed that egocentric, and ultimately, inner speech stemmed from primary social speech, occurring through a process of internalization, it was necessary for him to explain the source of the mental states in the earliest stage, the autistic stage of Piaget. What about this child who has not yet "internalized" any part of primary social speech, who has not yet learned to speak at all? Can he think? It becomes obvious that if Vygotsky were to grant that such a child thinks, then he must find quite different roots for thought and speech. And this he did. According to Vygotsky, thought and speech have different genetic roots and develop according to different growth curves that "cross and recross," but "always diverge again." There is a "prelinguistic phase in the development of thought and a preintellectual phase in the development of speech." A crucial moment, explored by William Stern, occurs when the curves of development of thought and speech meet for the first time; from this time forward "speech begins to serve intellect, and thoughts begin to be spoken." Vygotsky believed that Stern exaggerated the role of the intellect as a "first cause of meaningful speech," but he did agree that "Stern's basic observation was correct, that there is indeed a moment of discovery" when the child sees the link between word and object. From this point on, thought becomes verbal and speech rational.

The source of prelinguistic thought is, thus, separate from the source of speech. Prelinguistic thought has a source that is similar to the embryonic thought of some species of animals, while speech always has a social origin. Vygotsky saw a clear tie here with Marxist analysis:

> *The thesis that the roots of human intellect reach down into the animal realm has long been admitted by Marxism; we find its elaboration in Plekhanov. Engels wrote that man and animals have all forms of intellectual activity in common; only the developmental level differs: Animals are able to reason on an elementary level, to analyze (cracking a nut is a beginning of analysis), to experiment when confronted with problems or caught in a difficult situation. It goes without saying that Engels does not credit animals with the ability to think and to speak on the human level.*

At this point in Vygotsky's analysis, his critic is likely to chastise him for reductionism, for drawing too crude a similarity between man and animals. But Vygotsky felt that the answer to such criticism lay in emphasizing the qualitatively new characteristics that emerged after the lines of thought and speech crossed, after the child's great discovery referred to by Stern had occurred. According to Vygotsky, the stage that followed this inter-section was not a simple continuation of the earlier:

> The nature of the development itself changes, *from biological to sociohistorical. Verbal thought is not an innate natural form of behavior but is determined by a historical-cultural process and has specific properties and laws that cannot be found in the natural forms of thought and speech. Once we acknowledge the historical character of verbal thought, we must consider it subject to all the premises of historical materialism, which are valid for any historical phenomenon in human society. It is only to be expected that on this level the development of behavior will be governed essentially by the general laws of the historical development of human society.*

Thus, Vygotsky developed for the explanation of the inter-relation of thought and language a scheme that contained a high degree of inner consistency and arrived eventually at

Marxist conceptions of social development. Thought and language have different roots thought in its prelinguistic stage being tied to the biological development of man, language in its prerational stage being tied to the social milieu of the child. But these two categories become dialectically involved once the link between them occurs, when the child perceives that every object has a name; from this point onward, one cannot speak of the separateness of thought and language. The internalization of language causes thoughts to be expressed in inner speech; the effect of logic on speech results in coherence and order in oral communication.

Several aspects of Vygotsky's scheme remained unclear. For example, he drew a parallel between the prelinguistic thought of a child and the mental activity of animals, such as chimpanzees. Yet Vygotsky of course granted that, physiologically, there are genuine differences between the brain of a child and that of a chimpanzee. Nonetheless, to what extent those differences result in a qualitatively different sort of prelinguistic thought in the child was not clear in his writings. Within his conception the socio-historical influences conveyed in language surpassed the biological superiorities of the human brain in accounting for the distinctions between man and animal. As a materialist and monist, Vygotsky agreed that the very sociohistorical factors that he emphasized also had their material causal sources, back in the biological development of man. He quoted Engels's descriptions of the influence of the use of tools upon the development of man. In the final analysis, then, the different roots of thought and language were only relative, not absolute. In the life of the individual human, however, the roots were distinct, and it was here that Vygotsky put his emphasis.

Vygotsky's opinion that language and thought have different roots and that "prelinguistic thought" exists in the early life of a child directly conflicted with Stalin's teachings on linguistics. Stalin wrote in *Marxism and Linguistics*:

> *It is said that thoughts arise without language material, without the language shell, in, so to speak, a naked form. But this is absolutely*

> *wrong. Whatever the thoughts that may arise in the mind of man, they can arise and exist only on the basis of language terminology and phrases. Bare thoughts, free of the language material, free of the "natural matter" of language do not exist. Only idealists can speak of thinking as not connected with the "natural matter" of language, of thinking without language.*

A clearer contradiction by highest authority can hardly be imagined, if one remembers not only Vygotsky's identification of separate sources of thought and language, but also his assertion that "there is no clear-cut and constant correlation between them." Consequently, the rebirth of interest in Vygotsky's writings occurred only after Stalin's death. The ruler of the state had dictated an interpretation of Marxism that the Marxist scientists and intellectuals of the country failed to perceive.

Piaget did not become familiar in detail with Vygotsky's criticisms of his early work until 1961 or 1962, when the abridged translation of *Thought and Language* was made available to him. At that time, twenty-eight years after Vygotsky's death, he published his *Comments on Vygotsky's Critical Remarks*. A discussion of this courteous and interesting reply — in which Piaget granted many essential points to Vygotsky, but retained a number for himself unfortunately can not be included here.

At this point I would like to shift attention to the postwar period and, particularly, to the person of S. L. Rubinshtein, the Soviet scholar who in the last thirty years has exercised the greatest influence in questions concerning philosophical interpretation of psychology and physiology. First, however, it is necessary to describe the ideological pressures upon Soviet psychologists and physiologists in the immediate postwar period. The physiology session of 1950 was one of the most important events of this sorry epoch in Soviet scholarship. From an intellectual standpoint the 1950 conference is much less interesting than that of 1962, when deStalinization had revived Soviet physiology and psychology, and consequently, the later session will receive more attention in this chapter. For those persons more interested in the 1950 meeting and its immediate results, an English language version of the proceedings is available, as well as several other accounts.

In the period immediately following the war Soviet scientists in many fields initially hoped for a relatively relaxed ideological atmosphere. We have already seen something of the efforts made by several scientists in physics and genetics to return to the more tolerant conditions of the 1920's. This effort failed; immediately after A. A. Zhdanov's death in 1948 controls in several scientific fields, including physics, genetics, cosmology, structural chemistry, and physiology, were tightened. The causes of this ideological campaign are very difficult to identify; the personal characteristics of Stalin seemed to be the most important factor, although the strained international situation and the availability of levers of control in Soviet society were also important conditions permitting Stalin to exercise extraordinary influence on science and scholarship. During the years 1948–52 conferences on science and ideology were held in a number of different fields at which political pressure was exerted on scientists; the "Pavlov" session on physiology and psychology occurred June 28–July 4, 1950, and was sponsored jointly by the Academy of Sciences of the U.S.S.R. and the Academy of Medical Sciences of the U.S.S.R. Unfortunately, the English translation of the speeches given at this conference, published in 1951 in Moscow as "Scientific Session on the Physiological Teachings of Academician Pavlov," does not contain the speeches by P. K. Anokhin, I. S. Beritov, L. A. Orbeli, and others. These speeches were critical of the official position.

In the inaugural address Sergei Vavilov, president of the Academy of Sciences and brother of the deceased geneticist Nikolai Vavilov, indicated that the function of the congress was to return to established Pavlovian teachings; he thus implied what was already known, that there would be no genuine effort at the congress to seek new understandings of the difficult problems of physiology and psychology on a materialist basis. Coming almost two years after the genetics conference where Lysenko and his

Michurinist school were officially established as the representatives of the only correct approach to genetics, the physiologists and psychologists were quite aware that the outcome of their conference was predetermined. Vavilov gave the official diagnosis of the state of Soviet physiology and psychology when he commented in his opening statement:

> *There have been attempts not too frequent, happily at an erroneous and unwarranted revision of Pavlov's views. But, more frequently, the ideas and work of researches have not kept to the highroad, but wandered into byways and fieldpaths. Strange and surprising though it may seem, the broad Pavlov road has become little frequented, comparatively few have followed it consistently and systematically. Not all our physiologists have been able, or have always been able, to measure up to Pavlov's straight forward materialism. The time has come to sound the alarm. Our people and progressive humanity generally, will not forgive us if we do not put the wealth of Pavlov's legacy to proper use. There can be no doubt that it is only by a return to Pavlov's road that physiology can be most effective, most beneficial to our people and most worthy of the Stalin epoch of the building of Communism.*

The Soviet physiologists who came under the heaviest criticism at the conference were P. K. Anokhin, L. A. Orbeli, and I. S. Beritov. I. P. Razenkov, vice president of the Academy of Medical Sciences, charged that Anokhin, one of the Soviet Union's most distinguished physiologists, "has been guilty of many a serious deviation from Pavlov's teachings, has had an infatuation for the fashionable, reactionary theories of Coghill, Weiss and other foreign authors." The attribution of pejorative meaning to the term "foreign authors" was typical of the chauvinistic temper of these Stalin ist years. Anokhin's thoroughly materialist and scholarly approach to physiology will be considered in a separate section of this chapter. After retreating somewhat from his earlier modifications of Pavlov's teachings, which he saw as somewhat obsolescent although fundamentally important in their time, Anokhin returned in the late fifties and early sixties to the views for which he was criticized in 1950. To the extent that dialectical materialism had an intellectually interesting relationship to Anokhin's work, that relationship must be sought in his writings before 1950 and after 1956. The intervening period was a dark one, not only for Soviet physiology and psychology, but for much of Soviet scholarship.

In a history of Soviet psychology published in Moscow in 1967, a work described by an American psychologist as "pioneering" despite its faults, A. V. Petrovskii told of the "dogmatism" in Soviet psychology following the 1950 session. Petrovskii observed that in the early 1950's there was a strong tendency toward the "liquidation" of psychology entirely, replacing it with Pavlovian physiology. This "nihilistic attitude" toward psychology, Petrovskii continued, was reminiscent of the reflexological currents of the early twenties, when the legitimacy of psychology had also been doubted, but:

> *If in the twenties the negative attitude of the reflexologists and behaviorists toward psychology could be largely explained though not justified by the objective need to criticize the vestiges of subjectivism in psychology, by the beginning of the 1950's the idea of "liquidating" psychology could not be based on any principled considerations whatsoever.*

Thus, the current major Soviet historian of Soviet psychology condemned Stalin ism in his field in strong terms. To be sure, he dodged the issue of the extent to which the events of the fifties were the responsibility, not only of Stalin, but of the system that permitted him to exercise such power.

Despite the political and ideological pressures of these years Soviet physiology and psychology continued to live and to develop. With better conditions after Stalin's death these fields moved forward once again. One of the best illustrations of the survival ability of Soviet scholars and of their continuing intellectual vitality in the face of great obstacles is found in the person of S. L. Rubinshtein.

Sergei Leonidovich Rubinshtein (1889–1960)

One of the lifelong goals of Sergei Leonidovich Rubinshtein was to give a theoretical analysis of the nature of consciousness and thought on the basis of dialectical materialism. His attempt in this direction was obvious in his writings over a period of many years, from his 1934 article entitled "Problems of Psychology in the Works of Karl Marx" to his 1959 book on the principles of psychology, in which he commented that his interpretation was heavily influenced by the "dialectical materialist understanding of the determination of psychic mental phenomena."

Rubinshtein was important in the formation of contemporary Soviet attitudes toward psychology. In 1942 he founded the department of psychology at Moscow University. Around him in the psychology section of the Institute of Philosophy of the Academy of Sciences of the U.S.S.R., which he headed from 1945 to 1960, there grew up a whole school of investigators of the relationship of the psychological to the physiological within the theoretical framework of Marxism. His advanced textbook *Foundations of General Psychology*, published in several editions, was the most authoritative voice in the field for Soviet graduate students. The first edition, published in 1940, received a Stalin Prize. His later work, *Being and Consciousness*, is regarded at the present time as a "deeply creative Marxist work" and has been published in many countries and languages, including Chinese; in 1959 it received a Lenin Prize.

From all these official honors one might think that Rubinshtein was an ideological hack, a mere apologist for Marxism. He was not. Possessing a broad-ranging and subtle mind, he produced even in his relatively elementary 1934 article "what is regarded as the first adequate Marxist theory of motivation and ability." During the worst period for scholarship in the Soviet Union, the Stalin ist years immediately after World War II, Rubinshtein came under heavy criticism for his "objective, non-Party" approach to scholarship and for certain of his theoretical formulations. He bent under the pressure, but he did not break. Though he sharpened the point of his ideological pen, he was still the sort of person who, at the 1947 discussion of the ideological failings of Aleksandrov *History of Western European*

Philosophy, would make a plea for the study of formal logic. This was at a time when formal logic was being displaced by dialectical logic, a campaign with much political support from Party followers. In the 1960's, Rubinshtein emerged again as the most prominent theoretical voice on the knotty problems of the nature of consciousness. He published three books on the topic in 1957, 1958, and 1959 respectively. Upon his death in 1960 the editors of *Problems of Philosophy* honored him with a necrology, which observed that his work would long continue to be of value to psychology.

Although certain themes for example, the definition of "consciousness" run through almost all of Rubinshtein's works, there was something of an evolution in his views, a slight but perceptible change that does not seem entirely explainable as a result of political pressure. He was a psychologist, not a physiologist, and his first works are deeply psychological in tone. As time went on, however, he moved more and more toward physiology, maintaining that only with a recognition of the material basis of mental activity could one proceed to an analysis of its most difficult problems. This shift from psychology to physiology is interesting when compared to the contrasting case of Anokhin, a dialectical materialist physiologist, but who, as time went on, became more and more aware of the importance of psychology. In Rubinshtein's case, political influences are undoubtedly relevant, since in the fifties he was criticized in a major Party journal for ignoring Pavlovian physiology; in the later revisions of his work he took this criticism into account. But his last works, published in the more relaxed late fifties, continued to stress the importance of physiology in such a way as to suggest that he had sincerely modified his own views, not just bowed to pressure. In the late fifties he developed several new formulations of the psychophysical problem, which he considered more successful than his earlier attempts and which his contemporaries have also judged more successful.

In the 1946 edition of *Foundations of Psychology* Rubinshtein began with an emphasis on the significance of the subjective to psychology. Here he was still defending the hard-won ground of the late twenties and thirties, when psychology's very right to exist and the legitimacy of the term "subjective" had been questioned. Rubinshtein noted that there are no such things as "sensations" or "perceptions" alone, but only a *person's* sensations and perceptions. This was to him, in 1946, the very ground on which psychology begins, its first principle: "The characteristic of 'belonging' to an individual, to a subject, is the first characteristic trait of all that is psychic. My experiences are presented to me differently, in another perspective, as it were, than to any other person. The experiences, the thoughts, the feelings of a subject are *his* thoughts, *his* feelings; these are *his* experiences, a piece of his very life, his flesh and blood." The difference between this 1946 view and his later writings is more one of emphasis than contradiction, since in the earlier writing Rubinshtein recognized, in addition, the psyche as a "reflection of objective reality, existing outside and independently from it." Nonetheless, what one says first is important in judging his intellectual priorities; Rubinshtein in the forties spoke first of the individual; in the fifties, he spoke first of objective reality. In the later years his effort to defend psychology as a discipline was replaced more and more in his priorities by an effort to give a theoretical definition of the psyche, one that was materialistic and monistic and yet assumed a permanent place for psychology as well as for physiology. It was a delicate, subtle, and perhaps inherently inconsistent operation.

In 1946 Rubinshtein's theoretical terminology for the relationship of the psychic to the physical was that of a "dual correlation" (a formulation he would later modify). As he observed:

> *Every psychic phenomenon is, on the one hand, a product, a dependent component, of the organic life of the individual, and on the other hand, a reflection of the external world surrounding it. These two aspects, always produced in the consciousness of man in unity and mutual interpenetration, appear here as experience and knowledge. The element of knowledge in consciousness especially emphasizes the connection to the outside world, which is reflected in the psyche. Experience is, first*

of all, a psychic fact as a piece of the very life of the individual, a specific manifestation of his individual life.

Rubinshtein attempted here to distinguish between the objective and the subjective, to tie the objective world to "knowledge" and the subjective world to "experience." Experiences were to him the "subjective aspect of the life-course of a personality." Experiences would be the particular realm of study of psychology.

Rubinshtein already realized that his distinction here could not be maintained in a rigorous fashion. Every "experience" was also connected to the objective world, but differed from "knowledge" in the degree to which the subjective exercised influence: "An experience is an experience of something and, it follows, knowledge of something. It is called an experience not because the second aspect knowledge is absent, but because the vital, the personal, aspect in it is predominant." Therefore, "experience" and "knowledge" are not pure categories, but are based on the degree to which either the subjective or the objective is governing. Every experience will contain an aspect of objective knowledge, and every new piece of knowledge gained by an individual will contain a bit of subjective experience. But Rubinshtein felt that the distinction was still a valid one. In experience, subjectivity was predominant; in knowledge, objectivity governed. This distinction led him to a definition of "consciousness" as a unity of experience and knowledge, a unity containing both subjective and objective elements."

This analysis has already brought us to the heart of the psychophysical problem and opens up almost endless possibilities for further comment. No problem is more difficult to solve within the terms of either materialism or idealism. But before pausing for these considerations, I, would like to proceed to Rubinshtein's criticisms in 1946 of both "psychophysical parallelism" and "psychophysical identity," and to his reformulation of the problem in the later 1950's.

So far, Rubinshtein's analysis reminds one of psychophysical parallelism, with "knowledge" being the series of physical or real events and "experience" being the series of psychological events. But he firmly rejected such parallelism:

> The theory of psychophysical parallelism is based on two errors. It is mistaken to oppose dualistically psychic and physical phenomena as two series of events entirely separate from each other; it is also mistaken to propose that between these two series there is a simple correspondence in the spirit of the old localization theories, according to which there is a sharp correspondence of the psychic to the nervous cell. The unsoundness of this view is illustrated by all contemporary evidence from experimental and clinical research.

Not only did Rubinshtein reject the theory of an unconnected parallel series of psychic and physiological events, but he also disavowed what he called the theory of mutual interaction. In the latter interpretation, psychic and physiological events are seen as a parallel series that *do* influence each other. Support for this interpretation may be found in the fact that physiological changes in an organism often are connected with changes in psychic behavior, or in the opposite phenomenon in which psychic processes, such as strong emotions, often have physiological effects. But Rubinshtein observed that the theory of mutual interaction was based on the same erroneous assumption as the theory of psychophysical parallelism namely, the dualistic belief that the psychic and the physical are two entirely distinct kinds of phenomena. It then follows that these two sets of phenomena either influence each other (mutual interaction) or remain entirely separate (parallelism). Rubinshtein concluded:

> The struggle of the supporters of the theory of mutual interaction against the transformation of man's consciousness into an "epiphenomenon" deprived of all real significance in principle may be completely justified, but they then proceed on the basis of the dualistic premises of the theory of interaction to the completely unjustified notion that psychic forces act externally, as it were, on the flow of physiological processes.

It was the identification of specifically "psychic" forces, entirely outside the realm of objective reality, that caused Rubinshtein to continue looking further for a solution of the psychophysical problem.

There would seem to be a logical exit left for Rubinshtein, the "theory of identity" that is, the reduction of either the psychic to the physical or the reverse, the reduction of the physical to the psychic. The latter was clearly unacceptable. It led directly, he said, to phenomenalism or even open spiritualism. Starting out with an assumption of the existence of an objective, physical, and material reality, Rubinshtein could hardly end up with a reduction of the physical world to psychic events.

Rubinshtein's theoretical exits seemed to be closing down rapidly. Within the theory of identity one path remained, the reduction of the psychic to the physical. It is the solution that many people outside the Soviet Union assume has been adopted there. It fits well with the popular conception of Pavlovianism. Yet Rubinshtein found it no more palatable. Its goal of explaining the psychic in terms of physics and chemistry was to him an unrealistic and impossible one. It was an example of vulgar, reductionist materialism, a failure to recognize the qualitative differences of matter on different levels. Further, it seemed to leave little or no room for psychology, a discipline that had struggled for its existence in the Soviet Union and that Rubinshtein firmly supported.

What was left? Rubinshtein advanced what he called the principle of psychophysical unity:

> *The principle of psychophysical unity is the basic principle of Soviet psychology. Within this unity the materialistic bases of the psyche are determining, but the psyche retains its qualitative specificity; it is not reduced to the physical properties of matter and is not converted into an ineffective epiphenomenon.*

Rubinshtein attempted, therefore, to work out a position in which there is between the psychic and the physical a unity that allows each to retain its specific characteristics. Consciousness is neither one nor the other, but both. This unity is one of contradictions, resulting in a sort of complementarity between the psychic and physical properties of consciousness, a complementarity that parallels the wavelike and corpusclelike properties of light particles.

Rubinshtein's 1946 formulation was not successful, and reading his writings of that time, one thinks that he realized it. It papered over apparently irreconcilable differences. When physics attempted to escape the dilemma of quantum mechanics by simultaneously attributing wavelike and corpusclelike properties to light, it was not quite destroying itself in the process (though it might have seemed that way). Materialists could (and did) adjust to this strange concept of physics by speaking of relativistic "matter-energy" instead of matter alone, and observing that both waves *and* particles, and all combinations thereof, would be matter-energy. But a materialist could hardly say that consciousness is both "psychic" and "physical" and leave it there. He needed an equivalence principle here, too, but he did not have one conveniently at hand. If "psychic" is a category, either it must in some way be equated with matter-energy, or one must abandon the view that only matter-energy exists, thereby destroying a fundamental assumption of dialectical materialism. Rubinshtein knew that the only solution lay in linking the psychic to some form of matter hence his statement above that "the material bases of the psyche are determining" but he was extremely vague on this linkage. He recognized his solution as "unfinished," and called for further attacks on this "difficult assignment."

Although Rubinshtein's 1946 position was vulnerable, the criticisms that were made of him by V. Kolbanovskii in the Party journal *Bol'shevik* in September, 1947 were lightweight intellectually. Most of the criticism was based on Rubinshtein's alleged insufficient political militancy, his failure to criticize adequately psychological theories advanced in western Europe and North America, and his lack of Party spirit.

One of the theoretical criticisms advanced against Rubinshtein stuck, however, and in 1952 he revised his position, abandoning in the process his principle of psychophysical unity, which supposedly described the "dual correlation" of the psychical and the physical. As he wrote:

> *Materialistic monism means not a unity of two sources–the psychical and the physical but the presence of a single source, a material source in relation to which the psychical is a product.*

This was the beginning of Rubinshtein's revision of his theoretical position, the final result of which was his 1957 monograph, *Existence and Consciousness: Concerning the Place of the Psychical in the Universal Inter-Connections of the Material World*. In the latter work he arrived at a stronger materialistic formulation of the psychophysical problem. Thus, Rubinshtein's revisions seemed to have been more a result of his own awareness of the inadequacy of his earlier position than of the superficial criticism he received during Stalin's last years.

In order to understand this analysis, it is necessary first to see some of the assumptions on which it was based and then proceed to its details, including the extensions of materialism that it contained.

Rubinshtein's approach was based on a rejection of the "classic" argument for cognitive idealism. As he wrote:

> *The basic argument of idealism is the following: In the process of cognition there is no way for us to "jump out" of our sensations, perceptions, and thoughts; this means that we can not attain the sphere of real things; therefore, we are obligated to recognize that the very sensations and perceptions themselves are the only possible objects of cognition. At the basis of this classical argument of idealism lies the thought that in order to attain the sphere of real things, it is necessary to "leap out" of the sphere of sensations, perceptions, and thoughts — and that, of course, is impossible.*

This line of argument assumes what it is trying to prove. It assumes that sensations and perceptions are only subjective constructs, external to things themselves, to objective reality. But actually objects participate in the very origin of sensations; sensations, arising as a result of the influence of objects on the sense organs, on the brain, are connected with objects in their very origin.

Rubinshtein was correct in stating that this standard argument for idealism assumes what it is trying to prove–that is, it assumes that sensations and perceptions are something other than material reality and therefore must be escaped from in order to approach reality. Since that escape cannot be accomplished, the argument goes, one must accept sensations and perceptions as objects of cognition themselves, and can define them as ideal forms if one wishes. But what Rubinshtein did not explicitly say (although it was implicit in his argument) was that *his* argument also assumed what it was trying to prove. The person who believes that sensations and perceptions are meaningful only as forms of material reality has also made an unprovable assumption. He has as much right to his assumption as the idealist does to his, but he can not justifiably maintain that he has "proved" his case while the idealist has merely assumed what he pretended to prove. All of this merely restates the writer's earlier opinion that the option between materialism and idealism is a matter of philosophic choice, not a matter of logical proof. As the parallel postulate in geometry is the starting point from which several geometries can be constructed, depending on the assumptions made, the mind-body problem is the point from which several philosophies can be built, depending on the assumptions made. The genuinely difficult problem in the case of epistemology is not what can be proved and what can not, but the dilemma presented by the occasional grounds for choice among philosophic assumptions, recognized as such. If science had to wait at every point for rigorous proofs, it would not proceed far. The best form of materialism could be constructed on a few principles openly recognized as unprovable assumptions for which there are, nonetheless, persuasive arguments. Rubinshtein never stated that he was proceeding on such a basis, but what he did say was perfectly reconcilable with such a position.

To Rubinshtein, then, sensations and perceptions do provide an entree into the real material world. He described an epistemology of interaction, of praxis. In 1946, he had rejected the theory of mutual interaction on the basis that it assumed separate interacting series of psychic and physical events. He still, in 1957, rejected such a theory; his new epistemology of interaction was based on the premise of the interaction of an internal material brain with reflections of external material objects. Two totally separate series of events were not interacting, since both series were based on matter. Thus, what he earlier called experience, or subjective, was to be unpeeled in layers, like an onion, and revealed as also based on objective reality, on matter.

The general philosophic framework from which he approached the problem was one in which the universe is an interconnected material whole. It is an age-old concept, one with similarities to many older systems. As Rubinshtein described this universe:

> *All phenomena in the world are interconnected. Every action is an interaction; every change of one entity is reflected in all the others and is itself an answer to the change of still other phenomena acting upon it. Every external influence is refracted by the internal properties of that body, of that phenomenon to which it is subjected. It was not for nothing that Lenin wrote: ". it is logical to propose that all matter possesses a property essentially similar to sensation, to the property of reflection."*

This property of reflection is expressed in the fact that every thing is affected by those external influences to which it is subjected. External influences condition even the very internal nature of phenomena and are, so to speak, laid up in it, preserved in it. On the strength of this, all incident influences, all influencing objects, are "represented" or reflected in an other objects. Each phenomenon is in a certain degree "a mirror and echo of the universe." At the same time, the result of this or that influence on any entity is conditioned by the very nature of the latter; the internal nature of phenomena is that "prism" through which single objects and phenomena are reflected in others.

This expresses the fundamental property of existence. On this conception is based the dialectical materialist understanding of the determination of phenomena in their interaction and interdependence.

In this interesting and ambitious passage Rubinshtein based himself on concepts already existing within Soviet dialectical materialism, but he presented them in a more complete and speculative form than usually found. The statement that every phenomenon is in a certain degree "a mirror and echo of the universe" derives directly from Marx; commenting on the physiological function of eyes and ears, Marx commented that "these are the organs that tear man away from his individuality, converting him into a mirror and an echo of the universe." However, whether Marx would have extended this limited statement concerning man's sense organs to the broader generalizations of Rubinshtein is by no means clear. The concept of "reflection" that runs through the passage is, of course, derived from Lenin, as Rubinshtein quoted to indicate. Here again, a small comment was expanded into broader meaning. Lastly, a principle of determinism, of universal causation, is also obvious in the passage.

This formulation of causation in an interconnected world became important to Soviet psychology, however. To Soviet theorists as well as to others it has a certain speculative persuasiveness. Rubinshtein's "prism," the internal state of an object through which external influences are refracted, was frequently cited. The editors of *Problems of Philosophy* commented in 1960: "The position defended in *Existence and Consciousness*, in which external causes act through internal conditions, has an essential meaning for the whole system of scientific knowledge."

The higher the level of the evolution of matter, the more weight the internal factors have: "The 'higher' we rise from living organisms to man the more complicated is the internal constitution of phenomena and the greater is the share of the internal conditions compared with the external." Consciousness in man is that form of material reflection in which internal factors play a greater role than in any other form of reflection. Rubinshtein's position on the nature of psychic activity now unfolded. Psychic activity, he said, is both an activity of the brain and a reflection of the external world. Therefore, psychic activity has two different aspects the ontological and the epistemological. The ontological aspect of the brain is its existence as a nervous system, a material object of great complexity currently being studied by physiologists. The epistemological aspect of psychic activity derives from the cognitive relationship of psychic phenomena to objective reality. Rubinshtein believed the distinction between these two aspects to be relative rather than absolute. While the epistemological aspect of psychic activity is dominated by connections with the outside world and the ontological aspect is primarily determined from within, it should always be remembered that the brain itself is also, in the end, a result of the influence of the external world. Thus, there is a difference of causal time scales here. The brain is a product of the external environment acting over the entire period of natural history. It is a material brain of great organizational complexity formed by natural selection from matter of simpler organization. But ontologically, it exists as a completed entity at any given point in time that cognition takes place, and its internal constitution "refracts" the reflection of external reality, which is also material. Thus, the interaction that occurs is material in origin on both sides, and both sides are products of objective reality, but being formed at different times, in different places, and in different ways, they interact.

The ontological statement that psychic activity is a function of the brain must never be taken, Rubinshtein warned, to mean that it is a function determined entirely from within, by its cell structure. If such a mistake is made and psychic activity is looked upon as entirely a function of the brain or of the sense organs, then this function will inevitably come to be considered an expression of the states of these organs and lose in that fashion its links to the outside world. This result, said Rubinshtein, would be physiological idealism: "The brain is only an organ of psychic activity, not its source."

The epistemological aspect of psychic activity, as advanced by Rubinshtein, was a form of Leninist reflection theory:

> *The theory of reflection can be roughly described as follows: One senses and perceives not sensations and perceptions, but things and phenomena of the material world. By means of sensations and perceptions one cognizes things themselves, but sensations and perceptions are not these things but only their forms; sensations and perceptions cannot be directly posed in place of things. One can not speak as frequently is done about sensations or perceptions as ideal things existing apart from all material reality in the ideal world of consciousness, similar to the way things exist in the material world. The sensations and perceptions are forms of the object. Their epistemological content does not exist independently of the object. In this way the dialectical materialist theory of reflection firmly excludes a subjective understanding of the psyche.*

An element of Rubinshtein's arguments that now emerges in a crucial way is the nature of sensations and perceptions, which he called "forms of objects existing in the material world." We know from his previous argument that these forms are influenced both by the brain ("refracted by internal conditions") and by objective reality ("reflected within the brain"). We also know that in the case of the brain, the "highest" form of matter, the internal factors exercise more influence than in any other example of reflection we can point to in the universe. Does this mean that the internal factors exercise *more* influence than the external ones? How can we ever know anything certain about the external world? How do Rubinshtein's forms differ from Plekhanov's "hieroglyphs," criticized by Lenin, and do they differ from the forms of *Bildtheorie* psychology?

It is clear already that Rubinshtein did not agree with the presentational epistemology favored by some simple Marxists (and the interpretation that many people, Marxist and non-Marxist, have put upon Lenin's theory of reflection). Sensations and perceptions are not direct "pictures" of external objects, but "forms" of them that are refracted–that is, modified–in the process of transmission. But Rubinshtein also said that he was opposed to representational epistemology, to *Bildtheorie,* in which sensations are considered to be not faithful pictures of reality, but merely representations of it. This position is interesting, since Rubinshtein's "refraction" or "reflection" does not seem far, on the face of it, from representational epistemology. The root of his opposition to representational epistemology seems to have been his belief that it would not allow man to get to the "essence" or "truth" of things, or even to know that he was making approaches to it; at most it would give him some obscure "indicators" of reality. Rubinshtein wanted more. He thought that not only must one recognize that the external objects influence the mental reflection of reality, but that this influence is the most important factor affecting that reflection. As he commented, "In the process of perception the form depends first and foremost on the [external] object, the form of which it is. But the form is not a dead imprint of the object."

Rubinshtein even rejected the "two-factor theory," a thoroughgoing praxis, saying that dialectical materialism dismisses at its very base the theory of two factors, according to which reflection is determined on the one hand by the subject and on the other by the object. Reflection psychic activity and its product is not determined by the object by itself, nor by the subject by itself, nor by one plus the other. Psychic activity is defined by the reflected object.

When Rubinshtein said that psychic activity was not determined by the "subject plus the object," he apparently meant that subjective and objective factors were not of entirely unknown or equal relative influence. Thus, he assigned weights to the internal and external factors of the cognitive process, which resulted in the following delicate and speculative balance of influences: Internal factors are very important in the formation of perceptual forms of reality, more important than in any other case of the universal phenomenon of reflection, but they are still less important than the external factors, the links to objective reality existing outside the brain. Thus, man's knowledge is still in an important sense a faithful reproduction of external reality. Man confirms the truth about that objective reality in practice, when his formulations are either proved or disproved by actual results.

The causal sequence involving consciousness is not, Rubinshtein believed, from consciousness to external reality, but from external reality to consciousness: Therefore the question, How do perceptions make the transition from forms to things? is an incorrectly posed question. "Man does not exist because he thinks, as Descartes put it; he thinks because he exists."

In his assignment of relative weights to internal and external factors Rubinshtein had added another unprovable though fruitful assumption to his system. A strict materialist could accept an explanation of the process of cognition in which the internal factors played *more* of a role than the external ones without contradicting himself, so long as he added to this explanation the belief that the internal factors were also material and had, in their turn, been caused by external influences during the process of evolution. The verification of truth through practice could still play the same role as in Rubinshtein's scheme. Rubinshtein, as we have seen, accepted this evolutionary understanding of the brain.

His addition of a weighting scheme was gratuitous, but reassuring within a tradition that preferred an epistemology in which there occurred as faithful a transmission of information from objective reality to consciousness as possible.

Rubinshtein still believed that the term "subjective" is a legitimate one. "Subjective" was to him a term used to indicate that every aspect of psychic activity displays characteristics unique to the person concerned. Every sensation, every thought, was subjective in this sense. The value of the word "subjective" was not destroyed, in Rubinshtein's opinion, by the fact that these subjective qualities of the individual had, in turn, their objective origins. To deny these origins would be to absolutize subjectivity, to fall prey to a thoroughgoing "subjectivism": "The major way to overcome subjectivism is not in its denial, but in a correct understanding of the subjective as a form of manifestation of the objective." At any one point in time, every person is influenced by both subjective and objective factors, although the human race over the period of its whole history has been influenced only by objective ones. Out of this evolutionary process is created two causal chains that interact with each other. The product of the interaction is consciousness. Psychology studies this interaction. Physiology studies the brain as an organ of the human body.

To a person who is willing to pay the price of some speculation in order to arrive at a conception of consciousness, Rubinshtein's scheme possessed strong points. It advanced a more sophisticated conception of consciousness than previously found within the tradition of materialism. To be sure, it possessed weaknesses, the most obvious of which was connected with the oldest problem in philosophy, pushed back into a more remote recess, but still there. The mind-body problem emerged now around Rubinshtein's "forms" (*obrazy*) of reality. He defined sensations and perceptions as forms of external material reality. Were these forms material themselves? Is psychic activity, consciousness itself, material? Rubinshtein said no; the forms were "reflections" of objects, not objects themselves. In this way, he said in 1959, "psychic activity is ideal as a cognitive activity of man, and the term 'form' of an object (or phenomenon) is an expression resulting from this recognition." Thus, the dialectical materialism of Rubinshtein contained a category of phenomena called "ideal" that was different from "material." Was this not a surrender of his assumption of monistic materialism? Not at all, he maintained. "The key to the solution of the problem is the fact that, to use a phrase of Hegel's that was specially noted by Lenin, one and the same thing is both it itself and something else, since it appears in different systems of connections and relations." Relying on this Hegelian principle, Rubinshtein maintained that in an epistemological sense the psychic is ideal, while in an ontological sense it is material. The ideal element is precisely the "forms" of reality. Rubinshtein stoutly affirmed, "We are convinced that the recognition of the idealness of psychic activity does not convert it into something spiritual, does not withdraw it from the material world." Many of his critics remained unconvinced. As Rubinshtein observed shortly before his death, "more and more frequently people are affirming that the psychic is material. The partisans of this point of view, which has recently received a certain currency in our philosophic literature, are shutting themselves up inside the ontological aspect of the problem and do not take the trouble to correlate it with the epistemological aspect." We shall hear more from these "partisans of the materiality of consciousness," who were indeed speaking loudly in the philosophic literature of the late fifties and the sixties.

The 1962 Conference

The most interesting event concerning Soviet Marxist philosophy and psychology in the sixties was a great conference on the subject held in May, 1962 in Moscow. This all-Union Conference on Philosophic Questions of Higher Nervous Activity and Psychology was convened jointly by the Academy of Sciences of the U.S.S.R., the Academy of Pedagogical Sciences of the Russian Republic (R.S.F.S.R.), and the Ministries of Higher Education of both the U.S.S.R. and the R.S.F.S.R. More than one thousand physiologists, psychologists, philosophers, and psychiatrists participated, coming from all over the Soviet Union. The reports and debates of this conference were published in a volume of 771 pages. Buried in this record are many sharp differences of opinion. It is the best single source for an understanding of the philosophic issues in Soviet physiology and psychology since the passing of the Stalinist era.

The resolution that was approved at the conclusion of the conference inevitably involved such compromises among various points of view that it reveals much less than the debates themselves. Nonetheless, the resolution did reproduce the general tone of the conference. The statement noted that the physiology of the nervous system, like many other facets of biology and psychology, was going through a special period of development in which it was coming ever closer to the physical and mathematical sciences. New methods of experimental research electrophysiology of brain structures and nerve formations on the cellular and subcellular levels; use of computers; statistical methods;

the theory of information; and cybernetics all were leading to new understandings of physiology and psychology. Cybernetics in particular seemed promising. The task of Marxist psychologists and physiologists was to find a way of incorporating these new and valuable sources of knowledge into their disciplines without falling prey to crass materialism on the one hand or idealism on the other.

In the conference reports and debates themselves, two issues emerged as those of the greatest importance. The first was: In view of all this remarkable new knowledge about physiology and information systems, is the reflex approach advocated by Pavlov still valid? The second question was the old one: How now must we define the term "consciousness"? Most of the debates centered around these two questions – legitimate issues of interest to psychologists and physiologists in all countries. On both issues the conference broke into contrasting points of view.

Validity of the Reflex Approach

The question of the validity of the reflex approach arose in several different forms: around discussions of the significance of Pavlov, the usefulness of the term "reflex arc," and the meaning of the phrase "higher nervous activity." The most energetic critic of the Pavlovian concept of reflexes was N. A. Bernshtein. Bernshtein's opinion that the Pavlovian teaching had become quite obsolescent in the light of modern science was supported by, among others, N. I. Grashchenkov, L. P. Latash, I. M. Feigenberg, M. M. Bongard, and, more indirectly, P. K. Anokhin. Opposing these speakers, in the most direct fashion, were E. A. Asratian, L. G. Voronin, Iu. P. Frolov, A. I. Dolin, N. A. Shustin, A. A. Zubkov and V. N. Chernigovskii.

All these speakers acknowledged Pavlov's immense stature in the history of psychology and physiology. The difference of opinion centered not on his past significance, but on the continuing fruitfulness of his approach. Some of the disagreements were semantic: he defenders of Pavlov tended to describe his views in a very broad, methodological fashion; the new critics looked upon Pavlovianism in a way similar to most non-Soviet physiologists and psychologists–that is, as a stimulus response approach to nervous behavior. Yet underneath the misunderstandings and heated arguments was a real issue: Was the dialectical materialist understanding of physiology to be tied to the name of Pavlov, or would it acquire other means of identification? This was a question of authentic concern. As one of the supporters of the traditional point of view, V. N. Chernigovskii, observed: "We know that there is a whole group of young people who are skeptical about a whole series of principles of [Pavlov's] teaching on higher nervous activity. I call this group the Young Turks." Yet the Young Turks were in many cases not so young.

N. A. Bernshtein thought that a revolution had occurred in physiology since the beginning of the second quarter of the twentieth century, requiring the modification of many traditional physiological concepts but, given the necessary attention, permitting a new and superior interpretation of life within the traditions of dialectical materialism. The most important element of this revolution, he maintained, was cybernetics. He agreed that cybernetics had some dangers, particularly in the form in which its foreign founders expressed it, but he thought that if this new subject were placed on "the correct methodological rails," it could bring invaluable assistance to the study of biology in general and physiology in particular.

The most important contribution of cybernetics to the problems concerning Bernshtein was the possibility that it presented of explaining on a materialist basis the process of goal-seeking. An organism, seen from the cybernetic point of view, has a definite goal of action; Bernshtein spoke of a "physiology of activity" to distinguish it from the "simply reactive" physiology portrayed in Pavlovian reflex theory. This goal of action must be analyzed carefully, he said.

The goal of action in other words, the result that the organism is striving to attain is something that must be achieved but that still does not exist. Therefore, the goal of action is a reflection or model of the necessary future, coded in one way or another in the brain.We should notice that the concept of the reality of such a coded brain model the extrapolation of the probable future creates the possibility of a strictly materialist interpretation of such concepts as purposefulness, advisability, etc.

To speak in a metaphor, we can say that the organism constantly plays a game with nature surrounding it a game whose rules are not defined and whose course of progress, "conceived" by the opponent, is unknown.

In contrast to Pavlovian theory, which Bernshtein characterized as assuming "an equilibrium between the organism and its surrounding milieu," the new conception of life processes assumed "an overcoming of that milieu," a surmounting of the environment. The activity of organisms was directed, he thought, not at simple self-preservation or homeostasis, but at movement toward "a specific program of development."

Bernshtein was quite aware of the dangers of his formulations, which soon resulted in his being criticized at the conference as a teleologist, but he believed his critics were simply ignorant of modern science. He thought that physiologists had been very slow in adjusting to the full implications of the concepts of probabilistic laws in nature. Many of the physiologists of the Pavlovian school, he implied, still dreamed of explaining the human body as a "reactive automat" with its actions rigidly determined in a way that was thoroughly predictable once sufficient facts had been collected. But these orthodox determinists were actually crippling modern materialism, Bernshtein indicated, by tying it to outdated concepts:

Of course, the form of behavior of a reactive automat is more obviously deterministic than the behavior of an organism that is constantly forced to make active choices in stochastic conditions. But the discarding of the concept of an organism as a reactive automat, existing "because" of the stimuli that affect it, is by no means a retreat from scientific determinism in the broad sense; that this is so should be clear from the fact that the shift from describing phenomena by means of single-valued functions to its description by means of the theory of probability does not mean a retreat from a position of strict science.

The possibility of many-valued functions in biological phenomena was also very attractive to Grashchenkov, Latash, and Feigenberg. Their report, "Dialectical Materialism and Several Problems of Modern Neurophysiology, Neurophysiology," centered on this issue. They, too, believed that the old conception of the structure of the reflex was "incapable of explaining the observed physical facts." But they thought that Pavlov's system contained a great deal more flexibility than some of his critics believed. The key to many-valued functions in physiology was in the past experience of the organism, as they said Pavlov himself had indicated in his writings on reinforcement. This emphasis on the past, on the genetic approach, was a traditional characteristic of Marxism and should be easily accepted.

Grashchenkov, Latash, and Feigenberg described how many-valued functions distinguished a biological organism from a "simple automat": "One and the same physical stimulus can evoke different reactions, and in turn, one and the same reaction can evoke different stimuli. This phenomenon depends not so much on the inborn structure of the nervous system as on prior individual experience."

Grashchenkov and his colleagues were of the opinion that the concept of prediction on the basis of past experience was the premise of many schemes being currently proposed by Soviet physiologists; Anokhin's "acceptor of action," Bernshtein's "physiology of activity," E. N. Sokolov's "nerve model of a stimulus," and several of I. S. Beritov's views on the physiological structure of behavior all stemmed from such a premise. The characteristic feature of all these hypothetical "predictive structures" was the probabilistic nature of prediction: "Of all possible predicted results the one is chosen which has the highest probability." And Grashchenkov and his friends observed, "The fact that in the process of evolution organisms have developed a mechanism for probabilistic prediction should not be surprising." Such an ability was essential to survival. Furthermore, rather than contradicting determinism, it broadens it by showing that "the final result of dynamic reactions is determined by the information flowing into the brain and by the past experience of the organism."

V. S. Merlin, of the Perm Pedagogical Institute, also thought that the new concepts of probability were very fruitful in physiology. He maintained that a given "nervous-physiological process" does not necessarily give rise to a given "psychic process." In the old days, Merlin continued, such a view would have seemed unacceptable for a materialist, but now that the full significance of quantum mechanics has been realized by materialists, it becomes apparent that laws of probability are fully acceptable as "causal," and therefore there is no reason that a similar approach cannot be taken in psychology and physiology. This would obviously leave room for a much less strictly determined psychology.

But there was still such a thing as "too much room." Grashchenkov, Latash, and Feigenberg guarded against carrying the new concept of probability all the way to a belief in complete spontaneity in psychic phenomena. The distinguished Australian neurophysiologist J. C. Eccles was frequently criticized at the conference for using the uncertainty of quantum mechanics as a means for postulating a realm of action for "mind" as distinct from matter. In his *The Neurophysiological Basis of Mind* Eccles had given considerable credence to the view that "mind could control the behavior of matter within the limits imposed by Heisenberg *Principle of Uncertainty*." The dialectical materialists rejected such a view as being based on the assumption of mind-body duality.

The issue of spontaneity is one that has emerged in several of the discussions described in this book. It also appeared in Grashchenkov's analysis of electroencephalograms (EEG's). He and his colleagues were a little disturbed that some researchers described the EEG patterns as "random and spontaneous." Several physiologists had even said that there is no physiological significance that can be attributed to the EEG. Grashchenkov felt that such an observation would imply a discarding of even probabilistic determinism for a complete, crass indeterminism. There was, he and his co-authors said, a discernible "functional organization" to the EEG, which was obvious in the differences in EEG patterns of states of sleep, narcosis, asphyxia, attention, alarm, and so on. Rather than being "spontaneous," EEG characteristics probably derived from one of two possibilities: Either these rhythms were a direct reflection of the processes of self-regulation, including the vegetative ones; or they were the manifestations of a sort of "undirected searching operation, an oscillation through a range of values, a sounding out of the possible initial displacements of functions subject to regulation."

In sum, then, Grashchenkov, Latash, and Feigenberg agreed with Bernshtein in attempting to modify traditional Pavlovian conceptions, but they were more careful in observing the pitfalls of such an approach.

Another issue very much debated was the continuing validity of the term "reflex arc." Bernshtein believed that the concept of a reflex arc was a part of obsolescent "classical" reflex theory of the first quarter of the twentieth century; he suggested the term "reflex circle." Grashchenkov, Latash, and Feigenberg were equally unhappy with the concept of an open-order reflex arc, but suggested instead what they called a "cyclical innervational structure." Still another speaker, V. N. Miasishchev, of the Bekhterev Psychoneurological Institute in Leningrad, proposed that the reflex be considered a "spiral." He maintained that this model "quite obviously follows the Leninist formula of development. It is a concept that is correct both philosophically and scientifically." He was referring here to Lenin's statement that the approach of the human mind to reality is not a "mirroring," but instead an approach that is "split in two, zigzag like."

Each of these proposed modifications had the same goal: to indicate that nervous activity is based on a continuous flow of feedback (afferent) signals that serves as the source of information for constant corrective signals. This inward flow of information also changes the very structural nature of the corrective mechanism itself by increasing the store of past experience "deposited" in it. In a sense, the corrective mechanism "manufactures itself" from this store of information. The Soviet interpreters saw in this approach a way of uniting social history (the past history of the individual) and natural history (the inherited characteristics of the species) in a single materialist explanation of behavior.

These critics of the traditional reflex approach were soon themselves the objects of considerable disapproval. The defenders of Pavlov accused the Young Turks of simplifying Pavlov's views by equating his concept of the reflex with the mechanistic one of Descartes. E. V. Shorokhova and V. M. Kaganov, for example, said that Bernshtein regarded reflexes as purely physiological phenomena, ignoring Pavlov's view that they were both physiological and psychological. Shorokhova and Kaganov continued that Bernshtein's concept of the "physiology of activity" contained a definition of reflex that had not changed since the days of Sechenov and was limited in the same sense as the one current in "modern west European physiology." They wished to discard this "atomistic" view but retain the term "reflex" as a description of a phenomenon that was "internally cybernetic."

Other speakers were even more skeptical of Bernshtein and his supporters in their comments. Ia. B. Lekhtman of the Lesgaft Institute of Physical Culture declared:

> *The archaic interpretation of N. A. Bernshtein of arbitrary actions as spontaneous acts of the nervous system automatically raises our doubts. We have before us an obvious and unequivocal indeterminist interpretation of arbitrary actions, which was long ago refuted by materialist science.*

Such interpretations strengthen the ancient fable of free will, he continued, and "if the ideas of cybernetics nourish this fable, then all the worse for cybernetics."

A. A. Zubkov of the Kishenev Medical Institute carried the critique farther. Referring to Bernshtein's statement about the way in which an organism overcomes its environment in accordance with a specific program of development, Zubkov asked, "Just what is this 'specific program of development'? Who drew up this program and put it into living matter as one would put it into a cybernetic machine? There's something here that smells of Aristotelian entelechy."

The approach of L. G. Voronin, Iu. P. Frolov, and E. A. Asratian, old supporters of the Pavlovian school, was to deprecate the originality of people like Bernshtein, Grashchenkov, and Anokhin. Asratian maintained that these three men put great store in novel terms that actually describe phenomena long ago known. He believed, for example, that feedback was described in physiological terms by such people as Bernard, Pavlov and Sechenov. Frolov, who described himself as the oldest pupil of Pavlov still working, similarly doubted the originality of cybernetics and said it had no philosophy of its own and could be used by people of different schools; much was currently being made of it, he continued, by neopositivists and Gestaltists. Voronin, of Moscow University, maintained that the new criticism of Pavlovianism was not so much based on new scientific facts or proof that Pavlovianism had "aged" as it was simply a revealing of positions that these critics had long wanted to take. The Young Turks were actually "Old Turks." Grashchenkov and Anokhin, he continued, were insisting on modish terms primarily in order to bring in concepts that they long ago favored but did not have cybernetic vocabularies to back up.

There was a ring of truth in this criticism. Grashchenkov had, indeed, characterized Pavlovianism in the thirties as "mechanistic" Bern-shtein shtein had called for a replacement of "reflex arc" with "reflex circle" as early as 1935; Anokhin a biographer of Pavlov, and usually very respectful of him had criticized his teaching before World War II rather sharply and had, in turn, been the subject of strictures at the 1950 Pavlov session. But to see the controversy of 1962 over Pavlov ianism entirely as a reflection of disputes of the thirties would be quite inaccurate. The new developments in neurophysiology and information theory, as exemplified by the works of such people as W. Ross Ashby and Arturo Rosenblueth, were by 1962 exercising great influence on Soviet physiologists and psychologists. These developments seemed to promise new successes in the explanation of decision-making and goal directed biological development on the basis of materialist assumptions. Since the materialist tradition in physiology was particularly strong in the Soviet Union, it was only natural that these two streams of thought would come together and had, indeed, been anticipated by certain Soviet scholars of the thirties. These older leaders spoke out in 1962 as the most prestigious members of the "cybernetic school," but they were supported by many younger workers. Although this point of view obviously did not go unopposed at the conference, the emphasis in the final resolution on the importance of cybernetics to physiology and the prominent place given to the reports of the cyberneticists indicate that they had achieved an advantageous position in their discussions with the more traditional members of the Pavlovian school.

The Definition of "Consciousness"

If the debate over the validity of the reflex concept was one in which physiologists were the most active participants, the definition of "consciousness" was the issue on which psychologists and philosophers spoke out most frequently. This debate was carried to incredibly fine degrees of detail.

In order to avoid the trap of dualism that the total separation of physiology from psychology presented, the philosophers around Rubinshtein had devised in the late fifties a formula that said that reflex activity is *both* physiological and psychological. Since at this time reflex activity was considered synonymous with psychic activity, the product of this analysis was the position that psychic (reflex) activity is studied in two different aspects by two different kinds of specialists. The physiologist studies psychic (reflex) activity in its ontological aspect, which is material, and he concerns himself with neurophysiology. The psychologist studies psychic (reflex) activity in its epistemological aspect, which is ideal, being based on the ideal forms (*obrazy*), and he concerns himself with cognition ("the refraction of external reality by internal conditions").

This formulation began to break down in the late fifties and early sixties when certain physiologists (Bernshtein and others) began to say that reflex activity is *not* synonymous with psychic activity because the reflex concept is too simple to explain psychic activity; if physiologists would go beyond the reflex approach, they could identify physiological mechanisms (the physiology of activity, the acceptor of action, and so on) that would explain many phenomena earlier thought reserved for psychologists. This approach alarmed many psychologists who feared that the more aggressive physiologists were trying to "swallow their field," as one speaker at the 1962 conference phrased it.

Furthermore, the compromise position of the late 1950's was being undermined from a different quarter. Certain philosophers and psychologists were also arguing that psychic activity was not the same as reflex activity. But while the physiologists like Bernshtein argued this so that physiology could shed its shackles to reflex theory and then proceed more successfully into the realm of psychology, Orlov had other consequences in mind: He wanted to wed physiology to "reflex activity" in order to leave "psychic activity" free for psychologists. He thought that psychic activity should be defined as the "ideal (spiritual) activity of the material brain" and should be the province of psychologists. The physiologists would study the "material brain" itself, and if they wished to describe its functions as "reflex activity," that seemed perfectly natural to Orlov, who saw the study of reflexes as in the tradition of Sechenov and Pavlov, both physiologists. So the situation in 1962 was paradoxical, and can be sharpened in the following way: Both the aggressive physiologists and the aggressive psychologists denied the premise that reflex activity is the same as psychic activity, but for contrasting reasons. The aggressive physiologists denied it because they thought that reflexes were not a sophisticated enough weapon for their continuing campaign to explain psychic activity on a physiological basis; the aggressive psychologists denied it because they looked upon the field of psychic activity as their own domain and did not wish constantly to have to assure their audiences that ontologically everything they were talking about had a material, reflex base. They were not much cheered that the aggressive physiologists no longer thought all psychic activity had a reflex base, since they felt that the intrusion of physiologists belonging to the new cybernetic tradition was no more pleasant a prospect than the intrusion of the physiologists of the old Pavlovian school.

These debates were related to worldwide discussions in physiology and psychology at this time, but they took a different tone and a somewhat different path in the Soviet Union because all concerned physiologists, psychologists, philosophers could openly call themselves only materialists. Understandably, therefore, the psychologists felt somewhat more insecure than elsewhere.

In terms of the theoretical definition of consciousness or of the psyche, the biggest difference of opinion was between F. F. Kal'sin and V. V. Orlov. The end of the spectrum toward which Kal'sin's views tended was called by his critics "vulgar materialism." The opposite end, with which Orlov's opinions were associated by those critical of him, was predictably termed "dualism."

Kal'sin, who held a position in the Gorky Pedagogical Institute, was outspoken in his defense of materialism. He believed that the leading professional philosophers had misinterpreted Lenin when they had allowed discussions of ideal forms to creep into Soviet psychology. Consciousness must be considered, he continued, not as a unity of the material and the ideal, but as a purely material phenomenon. "Thought" itself was to Kal'sin merely a particular type of movement of matter. And Kal'sin quoted several sections of Engels's writings that were usually left in decent oblivion by the professional philosophers. Engels, for example, had observed, "The movement of matter is not only crude mechanical movement, not only a change in position; it is also heat and light, electrical and magnetic potential; chemical composition and decomposition, life, and, finally, consciousness."

Kal'sin was highly critical of Kursanov and Kolbanovskii's statement "None of the founders of Marxism-Leninism looked upon thought as a form of movement of matter." How can they maintain such a position in the light of Engels's statement? he asked. And even if Engels had not expressed himself so clearly, he continued, it is obvious that a materialist must recognize thought and consciousness as material phenomena: "Everybody recognizes that thought is a process, a movement. The natural question arises, Just what is it that is moving here?" To Kal'sin, there was only one reply: matter. His simplified approach led several professional philosophers to observe that Kal'sin had not yet learned to distinguish between "a chair and the idea of a chair."

The problem of the nature of consciousness was probably the most serious and divisive issue facing Soviet philosophy of science in the 1960's. On all other questions quantum mechanics, relativity physics, genetics, and the rest coherent and defensible positions had been found, positions that gave viable theoretical statements of the problems at hand, yet allowed room for disagreement and further development of science. But the problem of consciousness seemed intractable. Soviet philosophers could not avoid the problem by calling the question of the definition of consciousness "meaningless" in the fashion of many positivistic non-Soviet scholars; they were committed to the constant improvement of an intellectual scheme that included explanations of the stages of development of all matter, and conscious man was ultimately material in that framework.

Solace could be found in the tradition of the mind-body problem as one of the most difficult of all intellectual problems, perhaps the most difficult. The issue was one for all countries and all philosophies. As Herbert Feigl, director of the Minnesota Center for Philosophy of Science, wrote in 1958:

> *The question arises inevitably: how are the raw feels related to behavioral (or neurophysiological) states? Or, if we prefer the formal mode of speech to the material mode, what are the* logical relations *of raw-feel-talk (phenomenal terms, if not phenomenal language) to the terms and statements in the language of behavior (or of neurophysiology)?*

No matter how sophisticated we may be in logical analysis or epistemology, the old perplexities center precisely around this point and they will not down. Schopenhauer rightly viewed the mind-body problem as the "Weltknoten" (world knot). It is truly a cluster of intricate puzzles some scientific, some epistemological, some syntactical, some semantical, and some pragmatic. Closely related to these are the equally sensitive and controversial issues regarding teleology, purpose, intentionality, and free will.

Within the framework of dialectical materialism it was also reasonable that the problem should be of great complexity, since dialectical materialists recognize principles of qualitative distinction and of varying difficulty on different levels of the evolution of matter. Dialectical materialists opposed the view that such phenomena as the formation of mental abstractions could ever be explained on the basis of physicochemical principles. The human nervous system, including the brain, represents the ultimate and most subtle stage of the evolution of matter; it was natural that the elaboration of its principles would be man's most difficult task.

A thorough description of the shades of opinion in Soviet writings in the late fifties and the sixties would require a book in itself. Here in broad strokes are some of the more identifiable positions.

A few Soviet authors, such as V. M. Arkhipov and I. G. Eroshkin, continued to affirm that psychic activity itself was material; they identified consciousness with nervous processes. They represented the extreme materialist wing of the authors who published on the subject. Close to them were F. F. Kal'sin, and less outspokenly, N. V. Medvedev, B. M. Kedrov, and A. N. Riakin, who characterized psychic activity and thought as a special form of the movement of matter a form of movement that is no doubt extremely complex, but nonetheless a movement of matter. The scholars so far named were criticized for leaning toward vulgar materialism by still other Soviet writers, such as M. P. Lebedev. The scholars in Rubinshtein's old circle in the Institute of Philosophy affirmed that psychic activity is both physiological and psychological, and continued to maintain that the term "ideal" is perfectly legitimate when used with reference to epistemology (the "reflected" is ideal; the "reflecting" is material). Still other writers openly denied that the term "material" can be applied to psychic activity at all. And V. V. Orlov, as we have seen, did not hesitate to speak of "the spiritual (*dukhovnyi*) activity" of the material brain. The position of the professional philosophers closest to the institutional seats of power particularly those in the Institute of Philosophy was in the middle. Their earlier compromise was breaking down, but they hesitated to insist on a new formulation in view of the current effort not to intervene in scientific discussions. But for them dialectical materialism continued to be a middle way between, on the one hand, those scholars particularly the psychologists who separated psychic activity entirely from its material substratum, and on the other hand, vulgar materialists like the behaviorists and ultracyberneticists, who questioned the very validity of the term "consciousness."

A subject of considerable concern to Soviet physiologists and psychologists in the 1960's, both at the 1962 conference and elsewhere, was what was called "the philosophical problems of the physiology of the reticular formation." The reticular formation (*formatio reticularis*) is a structure about the size of one's little finger, situated on the upper part of the spinal cord and in the central core of the brainstem. It derives its name from its appearance as that of a reticulum, or fishnet, when viewed under a microscope. It is a part of what is called the "phylogenetically primitive nervous system." In the process of phylogenetic development most of the undifferentiated masses of nerve cells –but not those, apparently, in the reticular formation became more and more organized around cellular groupings and therefore functionally rather distinct. In contrast to these parts, the reticular formation remained relatively non–specific and unorganized. Already one can see that the discovery of a part of the brain system of significant importance in mental activity and with such apparently close links to the distant evolutionary past raises questions of an interesting order.

The reticular formation seems to play a rather important role in the process of perception. Studies have indicated that the reticular formation (and some other sub-cortical systems as well, such as the basal ganglia, brainstem nuclei, and spinal motor neurons) acts as a way station for external sensory information *before* it reaches the primary cortical projection areas. Furthermore, this way station seems to play an active or "distorting" role that modifies these impulses. Such an understanding of the process of perception forces psychologists and physiologists to be even more skeptical than they previously were of defending a close correlation between the world mentally perceived by man and the world of objects external to him. But the "active" or "integrating" role of the reticular formation varies considerably with circumstances. When the sensory signal is novel, the reticular formation seems to have the least influence.

The reticular formation raises some rather significant questions, not only for Soviet scholars, but for others as well. Does the reticular formation merely act as a barrier, screening out information, or does it play a positive, contributory role? If it actually adds information, what are the sources of this information? If a phylogenetically primitive organ, one rather undifferentiated and therefore relatively unapproachable by anatomical or physiological analysis, plays a major role in perception and mental activity, can man explain it through research? What implications does the reticular formation have for the relative weights of rationality and irrationality in mental activity? Can one defend a "copy-theory of knowledge," as most dialectical materialists have done, when the full significance of the reticular formation is taken into account? Can the Pavlovian concept of reflexes, which attributed the major role in perception to the cerebral cortex, be retained?

Soviet philosophers frequently saw the reticular formation as a foundation on which physiological idealism was being rebuilt, as they thought it had once rested, at moments, on Descartes's view of the pineal gland or Müller's concept of specific nerve energies. They were therefore rather critical of those non-Soviet scholars who put particular emphasis on the reticular formation in their explanations of mental activity, such as W. Penfield, H. H. Jasper and A. E. Fessard. The "centrencephalic theory" most closely associated with Penfield was a particular object of their criticisms. In the process of advancing arguments against these attempts to explain the action of the reticular formation the Soviet authors frequently simplified the points of view of their foreign colleagues. Underneath the debate, however, there were several interesting issues. The Soviet authors did not deny the significance of the reticular formation in mental activity; they resisted, however, any interpretation of it that would give it a "dominating role," and they repeated frequently that the reticular formation had its material, evolutionary origins and should not be isolated as a mysterious source of mental activity.

Several Soviet scholars maintained that these issues came to a head in 1953 at an international conference on brain mechanisms and consciousness in Ste.-Marguerite, Canada. At that meeting the French neurophysiologist A. E. Fessard presented a summary report, a very scholarly and well-received one, in which he made several comments to which Soviet writers later took philosophic exception. One of these remarks was Fessard's opinion that sub-cortical structures "play an important if not predominant role in integrative processes that organize states of consciousness"; the other was Fessard's statement that the reticular formation possessed the property of operating on the basis of the principle of "spontaneous activities (output without corresponding input)."

The Soviet writers seemed to fear that the attribution of major or predominant influence in mental activity to subcortical organs would become the basis for idealistic and religious viewpoints; E. A. Asratian criticized Penfield's centrencephalic theory as an attempt to explain man's highest psychic functions on the basis of the most ancient and primitive parts of the brain. And a group of three other Soviet authors objected to Fessard's description of the reticular formation:

> *The reticular formation is not an isolated source of energy, functioning according to Fessard's principle of "output without input," but instead a specific kind of intermediate transformer, a "transmission station" for afferent stimuli connecting the cerebral cortex with the energy of the external and internal environment of the organism.*

The authors of the "reticular" theory of emotions have not taken into consideration that the nervous substrata of the emotions also arose in the process of evolution, in the process of adaptation of the organism to the external environment; one must not make absolute the relative dependence of emotions on this or that nervous center.

Other Soviet scholars cautioned their colleagues, however, not to press the case against men such as Penfield and Fessard too hard, since much that they wrote could be easily accommodated with a materialist interpretation of mental activity; so long as the reticular formation was not directly linked to religion, not made a new "seat of the soul" and neither Penfield nor Fessard did this it could be explained on a materialist basis. M. A. Logvin warned the scholars attending the 1962 Moscow conference not to pin labels on non-Soviet physiologists and psychologists, whose philosophical views, he continued, are quite diverse. He particularly objected to lumping together as "idealists" such materialistically oriented scientists as Penfield and H. W. Magoun with other openly idealistic scientists as Charles Sherrington and J. C. Eccles. Logvin noted that Magoun, one of the important researchers of the reticular formation, had defended the value of some Soviet work based on materialistic assumptions and had further admitted that much Western work still displays vestiges of dualism. A similar conciliatory effort was made by A. F. Makarchenko, P. G. Kostiuk, and A. E. Khil'chenko at a 1960 Kiev conference, where they observed that Penfield praised Soviet work on the reticular formation. These latter comments are illustrations, which could be amplified, that the Soviet discussion of the reticular formation and its epistemological implications remained fairly sophisticated and tentative in its assertions. The reticular formation and the question of the relative importance of the sub-cortical region to mental activity continues to be a subject of interest in Soviet philosophy of science at the present time.

Peter Kuzmich Anokhin

One of the most prominent physiologists in the Soviet Union during the past decades has been Peter Kuzmich Anokhin (1898-). In the 1920's as a student and young lecturer Anokhin worked in the laboratories of Pavlov and Bekhterev, and much of his life has been devoted to an evaluation and extension of the Pavlovian tradition. His first significant publications appeared in the early thirties; since that time he has published many works. He attended physiological congresses abroad and is well known outside the Soviet Union; his biography appears in such standard references as the *International Who's Who*. After 1955 he was head of the faculty of the First Moscow Medical Institute and since 1966 a full member of the Academy of Sciences of the U.S.S.R. His research has concerned primarily the central nervous system and embryoneurology.

Anokhin has frequently praised dialectical materialism as a philosophy of science. Throughout his career one of his primary motivations, by his own account, has been the effort to elaborate a materialist and determinist explanation of nervous activity; he has attempted to discover physiological mechanisms underlying forms of human behavior previously described by such indefinite terms as "intention," "choice," "creativity," and "decision-making."

In 1962 Anokhin stated:

> *The methodology of dialectical materialism is strong precisely because it permits one to rise to a higher level of generalizations and to direct scientific research along more effective routes leading to the most rapid solution of problems.*

Anokhin continued that dialectical materialism frequently warns a researcher against falling into interpretations that are ideologically "unacceptable for us." But he also saw a danger in this warning function: It is possible, he observed, to have a science that is philosophically correct but scientifically stagnant. The "enormous motive force hidden in the dialectical materialist methodology" would be fully revealed only if one combined the admonitory function of dialectical materialism with the "logic of scientific progress" that is, the constant checking, elimination, and confirmation of working hypotheses by experimental facts. Thus, Anokhin hoped for a synthesis of dialectical materialism and rigorous experimental science. There could be no contradiction between the two because the principles of dialectical materialism are developed *by* science. To be sure, dialectical materialism contained the a priori assumptions of materiality and lawfulness, but these were the assumptions of science itself. He commented in 1949, "Nature develops according to the laws of the materialist dialectic. These laws are an absolutely real phenomenon of the objective world."

In one of his earliest works, published in 1935, Anokhin advanced several of the ideas that were in modified forms to play an important role in his understanding of nervous activity. Future historians of neurophysiology and of cybernetic concepts in physiology will need to turn to this early source to evaluate Anokhin's claims to have anticipated such concepts as "feedback" with his "sanctioning afferentiation." Anokhin of course had no knowledge in 1935 of the mathematical foundation of information theory. Furthermore, discussions of "integrated nervous activity" were common in physiology at this time. Charles Sherrington's seminal *The Integrative Action of the Nervous System* had appeared long before, in 1906. Nonetheless, when a person now reads Anokhin's work of

1935, the vocabulary and the concepts do have a ring similar to that found in the subsequent literature of neurocybernetics. He spoke, for example, of neurophysiology in terms of "functional systems" in which the execution of functions is based largely on the set of incoming signals "that direct and correct" the process.

Throughout his life Anokhin was convinced that a physiologist should be both loyal to the Pavlovian school yet simultaneously critical of it. Always proudly calling himself a student of Pavlov, Anokhin nonetheless questioned some of his teacher's concepts. Even in his most critical moments, however, he stoutly defended the materialist assumptions underlying Pavlovianism. In the period immediately after World War II Anokhin declared that he had made errors in several of his earlier writings in which he had criticized Pavlov's method or pointed to predecessors of Pavlov in certain lines of work. Although it cannot be proved, there is considerable reason to believe that these corrections were in response to the changing political scene after the war, when efforts to establish Russian priority in science and ideological factors became more prominent than earlier. Thus, in 1949 Anokhin commented that in his survey of the history of reflex theory from Descartes to Pavlov, published in 1945, he had given too much attention to eighteenth-century materialists and had thus detracted from Pavlov's eminence. Also in 1949 he corrected a criticism of Pavlov that he had published in 1936; in the 1936 publication he had advanced the opinion that it was not correct to say that Pavlov always studied the "whole organism." This synthetic approach was more true of the early Pavlov, who studied blood circulation and digestive processes, than it was of the later Pavlov examining reflex activity. Anokhin in 1936, therefore, implied that the mature Pavlov was something of a reductionist.

In the late 1940's and early 1950's Anokhin became more orthodox while under the shadow of the criticism advanced against him at the 1950 physiology conference, but in the late fifties and sixties he returned to his earlier innovative, even speculative approach and advanced ideas concerning a new architecture of the reflex arc, the use of cybernetics in neurophysiology, and the reliance on more concepts from psychology (as compared with physiology) than earlier. In this later period he clearly believed that the Pavlovian concept of the reflex arc needed modifying, however much he continued in debt to his teacher. He commented in 1962:

> *Scientific results and theories should be judged according to whether they correspond with reality. But some people completely disregard this elementary critical approach and ask only, "Does this new thing concur with what Pavlov said?" And if it does not concur, then it is automatically proclaimed to be a "revision of Pavlov." On the basis of such comparisons we eliminate all possibility of finding anything new. I am not worrying about whether my interpretation will depart from the interpretation of my teacher Pavlov. This is quite natural; we live in a different epoch.*

Despite the slight variations in his viewpoints, Anokhin followed a fairly consistent approach to the Pavlovian tradition throughout his life.

This approach can be described as one between two extremes. In his 1949 biography of Pavlov, Anokhin saw dual dangers of opposite nature facing the followers of the great physiologist. On the one hand was the danger that the guiding ideas of Pavlov would be dissipated; on the other was the possibility of turning his teaching into dogma. Anokhin correctly predicted that the greatest danger was that of "canonization."

In 1949, before the physiological congress at which Anokhin was criticized for deviating from Pavlov's principles, Anokhin published a long article surveying what he called "The Main Problems of the Study of Higher Nervous Activity." This article, together with similar surveys published in 1955 and 1963, contains a summary of the work of Anokhin and his school.

Anokhin made it quite clear in 1949 that for twenty years he and his co-workers had attempted to modify the classical method used by Pavlov in studying conditioned reflexes. In several instances, he observed, they used "totally new approaches." They had gone beyond Pavlov's reliance on easily observable secretory glands and muscular reflexes to encephalographic investigations of conditioned reactions, to embryophysiological studies of higher nervous activity, and to morphophysiological correlations (studying in a parallel fashion both conditioned reflexes and the architectonic features of the cerebral cortex). On the bases of these new approaches they had concluded that the Pavlovian concept of conditioned reflexes was too simplified, particularly its model of the reflex arc with its three links. But Anokhin did not go so far as some other researchers, such as the Soviet academician I. S. Beritashvili in believing that psychic and nervous activity is a unique and unitary form of activity for which the conditioned reflex is not even characteristic. Beritashvili recognized the conditioned reflex as the basic law of activity of the *brain*, but not of psychic functions in general. Anokhin believed Beritashvili to be contradicting not only Pavlov but scientific evidence as well. He continued to defend the conditioned reflex as the key to psychic activity, but he wished to modify the understanding of the conditioned reflex as inherited from Pavlov.

Anokhin believed that the very approach utilized in the classical Pavlovian method blinded researchers to important processes intermediate between the conditioned stimulus and the response. He asked, "Is not the secretory indicator only an organic part of the external expression of the integrated conditioned reaction of the animal, the general form of which took shape long before the stimulation reached the effector mechanisms of the salivary gland?" In other words, Anokhin was turning attention to the internal nervous structure of the conditioned reflex and implying that it was far more complex than simply a direct transmission from a receptor (sense organ) along a nerve to the cortex, a jump along the arc to a neighboring cluster of neurons, and then a transmission of a motor stimulus to an appropriate gland or muscle. He criticized physiologists for falling into the habit of speaking of the passage of conditioned and unconditioned stimuli to the cortex of the brain even though very little was actually known of the physiology of these pathways. And as a matter of fact, it had been found by the 1940's that these stimuli pass through complex structures *before* they reach the cortex. Therefore, said Anokhin, we must be careful about speaking of the "closing" of the reflex arc when it seems so much may be happening before such a closing could occur in the cortical regions.

Pavlov had believed that the arc is closed in the cerebral cortex. He thought that the cerebral cortex contained a multitude of different centers, a "mosaic" as he called it, among which connections could be made. Any one center, made up of a group of cells, was related to a certain activity of the organism. But Anokhin noted that in recent years most neurophysiologists had moved away from this concept of localized centers. It had been demonstrated, for example, that a dog with most of its cerebral cortex removed could still form conditioned reflexes of a diverse character. To be sure, said Anokhin, the cortex plays an important function in the formation of most conditioned reflexes, but its role is much more flexible than earlier thought. The nervous system appears to be an integrated whole in which the parts have a remarkable ability to replace each other. Anokhin called for further study of the means by which this replacement occurs and for investigation of the gathering evidence that some rudimentary conditioned reflexes can be formed in dogs in the sub-cortical regions.

In the course of his endeavors to explain nervous activity on the basis of material physiological systems Anokhin utilized several terms that have come to be closely linked to his name. These included "return afferentiation"; "sanctioning afferentiation"; "acceptor of action"; and "anticipatory reflection." In recent years Anokhin has usually dropped the phrase "sanctioning afferentiation," first used by him in 1935, but still retains "return afferentiation" (*obratnaia afferentatsiia*) and "acceptor of action" (*aktseptor deistviia*). "Anticipatory reflection" (*operezhaiushchee otrazhenie*) seems to be a more recent development. Each of these phrases described a part of conditioned reflex activity, which Anokhin considered to be characteristic of all organisms of the globe, a means of "entering into temporary adaptive relations with the surrounding world."

In studying the two-way flow of nervous signals that occurs in reflex activity, Anokhin became convinced that the inward, or afferent, flow was the more crucial determinant of subsequent action. His teacher Pavlov had agreed with this observation, although it did not play so important a role in his research as in Anokhin's. In 1911 Pavlov had remarked:

> *If one divides all the central nervous system into only two halves the afferent and efferent then it seems to me that the cerebral cortex is an isolated afferent part. In this section there takes place exclusively the higher analysis and synthesis of the nervous signals that are carried to it, and from there are directed already-prepared combinations of nervous signals to the efferent section. In other words, only the afferent part is active, creative, so to speak, while the efferent is only a passive section for the execution of actions.*

Anokhin noticed that this attribution of greater significance to the afferent system than to the efferent one distinguished the Pavlovian reflex system from the Cartesian reflex arc, in which the afferent impulse played the role only of an initiator or impetus. To be sure, Descartes deserved great credit in the history of the "progress of materialistic knowledge" by first introducing the concept of the "external stimulus" and by thus opening the door to the idea of the deterministic dependence of the behavior of animals on the changes occurring in the external world. But Descartes, being a dualist, did not take notice of the question that, for a materialist, becomes pressing: Why is the response usually advantageous? Or why, if the response is not advantageous, does the organism correct its error? For Descartes, this question could be answered in terms of the "higher reasoning" of the organism. Anokhin observed in 1955 that for twenty-five years he had been attempting to answer these questions without any such appeal. Anokhin wished to penetrate "that complicated process of the fitting of reflex activity to the interests of the whole organism that is, the phenomenon that, practically speaking, deserves the description 'creative.'"

Those physiologists who followed the views of Descartes,

Anokhin continued, believed that reflex activity is adaptive or goal-directed from the very beginning of the process. Consequently, they concentrated on discovering already-prepared reflex responses. But with Pavlov's discovery of conditioned reflexes and the phenomenon of "reinforcement," it became clear that a creative, adjusting process lies at the base of reflex acivity. The inadequacy of classical reflex theory became even more clear as a result of experiments in which reflex functions were at first eliminated by vivisection, then restored by compensation. It was through such experiments that Anokhin approached these problems for the first time.

Anokhin soon came to the view that the organism could not begin the process of compensation without signals from the periphery telling of the presence of a defect. But the question still arises, How does the organism "know" that compensation is needed? Anokhin maintained that without what he called "return afferentiation" an answer to this question could not be attempted. By this term he meant "the constant correction of the process of compensation from the periphery." Schematically he represented return afferentiation in the following way:

Anokhin considered this return link in the reflex arc to be intrinsic to reflex activity: "It is difficult to imagine any kind of reflex act of an intact animal that would end with only the effector link of the 'reflex arc,' as is called for in the traditional Cartesian scheme." Every act is, instead, accompanied by an entire integral of afferentiations, greatly varied in terms of strength, localization, time of origin, and speed of transmission. These afferentiations are visual details; temperature, aural, and olfactory sensations; and kinesthetic sensations. The total variety of combinations is infinite. Together they make up one process: "In the presence of constant return afferentiation accompanying, like an echo, every reflex act, all the natural behavioral acts of an intact animal may arise, cease, and be transformed into other acts, making up as a whole an organized chain of effective adaptations to surrounding conditions."

As a simplified schematic diagram, Anokhin would represent this "organized chain" in the following way:

> *In this chain, return afferentiation serves as an "additional or fourth link of the reflex." In the final step, the desired result has been obtained, so there is no further effector action. If the process is one of compensation for a previously destroyed function (for example, by brain-slicing), the desired compensation has occurred in the final step. If the process is a more normal one, such as simply picking a glass off the table, that particular goal has also been attained in the final step.*

Anokhin guarded against his conception being understood simply as the belief that "the end of one action is the beginning of the next." Such an incorrect understanding of what Anokhin was describing would result in a different diagram, one that Anokhin rejected:

> *Anokhin meant, instead, that the end of one action is a source of return afferentiation that is transmitted to the nervous center where it is processed before it serves as the cause of a new action. It is in this central point that the "decision" as to whether the desired result has yet been obtained is made. This mechanism was called by Anokhin the "acceptor of action" and deserves a special treatment. It is the acceptor of action that controls the whole process.*

In his discussion of the acceptor of action Anokhin made an attempt to study intention and will from a physiological and deterministic standpoint. He initially asked, "How does the organism know when it has reached its goal?" And he replied, "If we stand on a strictly deterministic position, then essentially all the neurophysiological material that we have in our arsenal fails to give us an answer to this question. For the fact of the matter is that for the central nervous system of an animal, all return afferentiations, including sanctioning [that which corresponds to the desired goal] afferentiation, are only complexes of afferent impulses; from the normal point of view of causation there is no obvious reason why one of these stimulates the central nervous system to the further mobilization of reflexive, adaptive acts and another, on the contrary, halts adaptive actions."

There is only one way out, thought Anokhin, and that is the view that there exists in the organism some sort of prepared pattern of nervous impulses with which return afferentiation can be compared. This pattern had to exist before the reflex act itself occurred. If the afferent information coincides with the prepared pattern, then the desired goal has been reached. If it does not, then further effector action is necessary. The whole question then becomes, of course, What is the physiological mechanism containing this pattern, and how was the pattern originally produced?

In order to explain the way in which Anokhin attacked this problem, it is necessary to review several features of classical Pavlovian reflex theory, particularly the relationship between a conditioned reflex stimulation and an unconditioned one. It will be remembered that Pavlov believed that every conditioned reflex is formed on the basis of an unconditioned one. Thus, an unconditioned stimulus such as food in the mouth will automatically cause the flow of saliva on the first or nearly first occasion, evoking strong activity in the brain. Such an unconditioned stimulus is usually also accompanied by other stimuli that may become conditioned through training visual or olfactory sensations and so on. A "temporary" connection is formed between these points, and one can, after a little training, henceforth stimulate the secretory or motor centers of the brain merely by the conditioned stimulus. However, this temporary connection will not be maintained unless it is periodically reinforced by stimulation of the unconditioned center. That is, in the classic experiment saliva will not flow on the strength of the bell signal alone unless periodically it is followed by the presence of food in the mouth of the dog and the stimulation of the unconditioned salivary reflex on which the conditioned reflexes are based.

Anokhin now incorporated his return afferentiation into this scheme. He believed that every conditioned stimulation is sent through the sense organs to the center of the brain that *in the past* had been stimulated many times by the unconditioned stimulus, and that *shortly afterward* the center will again be stimulated by the unconditioned stimulus. Therefore, there arises the possibility of a "matching" or "mismatching" of the *representation* of an unconditioned response that the conditioned stimulus evokes and the actual conditioned response itself, following in a short period of time.

In stage one the conditioned stimulus falls upon the appropriate sense organ. In stage two it causes a conditioned reflex response based on a "representation" of an unconditioned reflex, a step that has occurred frequently in the past, but has not yet occurred in this sequence. In stage three, the unconditioned stimulus itself (food in the mouth, for instance) has occurred; the unconditioned response turns out to "match" the conditioned representation of it, and reinforcement occurs.

Anokhin now pushed farther in order to ascertain how powerful the "matching" or "controlling" mechanism was. He found that it was very powerful indeed, as one of his experiments illustrated. He conditioned a dog to go to a feeding box on the left side of a training box in response to the sound "la," and to a feeding box on the right side in response to the sound "fa." In all cases the food was sugared bread. The dog soon became thoroughly conditioned and would immediately go to the correct side. On one day (and only for one time) he introduced a change, however. He placed not bread but meat in the box on the left side and gave the appropriate signal. The dog went to the left box as was its custom, but was obviously surprised to find meat instead of bread there. It demonstrated what animal psychologists call an "orientingresearch reaction," but after this moment's hesitation, devoured the meat.

From this point onward, and for twenty days thereafter, the dog's actions were governed by this one event. No matter whether the sound "fa" or "la" was sounded, the dog would always bound to the left box. The experimenters continued to run the experiment as if the exception had never occurred, using sugared bread for food, and placing it in a box on the left side if the signal "la" was to be used and on the right if "fa" was to follow. Yet the dog long persisted in disregarding its old conditioning and searching only the left box for meat. If it found bread there (which it always did if the signal had been "la"), it refused to eat. Only after a long period of the total lack of reinforcement was the old pattern restored, so strong an impression had the one occasion made.

Anokhin believed that this experiment provided further evidence for the presence in the nervous system of a mechanism called the acceptor of action, which is based on very strong, inherited unconditioned reflexes that, in turn, can be linked to conditioned stimuli. In the case of the dog, the unconditioned reflex was the food reflex of carnivorous animals. Anokhin observed that "the acceptor of action" was an abbreviated term for the more accurate but cumbersome phrase "the acceptor of the afferent results of a completed reflex act." The word "acceptor" was a key term conveying both the ideas "to receive" and "to approve" that are contained in the Latin *acceptare*.

According to this scheme, if the nervous system of an animal is acted upon by a conditioned stimulus that has in the past been reinforced by meat, then the acceptor of action will define to what degree the received information corresponds to the earlier afferent experience of the animal. Anokhin represented schematically two cases: the one in which the match is correct and strong reinforcement occurs; the other in which there is a mismatch between the conditioned and the unconditioned stimuli.

Anokhin believed that this approach could help in an understanding of the way in which the nervous system repairs itself after it has been damaged. Let us imagine a form of reflex activity that was schematically represented by stage *A* in the above drawing, but was destroyed by the existence of some defect caused by surgery or disease. Anokhin would show in the subsequent stages the way in which the components of the appropriate nervous sub-system rearrange themselves until they arrive at an arrangement that yields the proper reinforcement as determined by the afferent signals of the past history of the organism. This reorganization effort is graphically represented by drawing the subsystem in radically different shapes, one after the other, until a proper arrangement is found, shown in the last stage.

Thus, Anokhin with his acceptor of action was trying to arrive at a physiological explanation for the goal-oriented activity of the nervous system. To be sure, he had not solved the problem of teleology, which has plagued biology throughout its history. On a rather speculative basis he had pushed the boundaries of the riddle back a little farther. Furthermore, despite his effort to identify physiological mechanisms, he had made no effort to localize the acceptor of action in the body itself. Accepting his arguments for a moment, however, one could now give something of an answer to the question, How does the organism know when it has reached its goal? He illustrated that the problems of intention and goal-seeking are amenable to physiological investigation, even if many problems remained. Many neurophysiologists of his time were beginning to think in similar terms. His concepts of reinforcement and acceptor of action bore some resemblances to much earlier research, such as Lynn Thorndike's and Lloyd Morgan's law of effect. Thomdike had shown that when he placed animals in labyrinth-pens, there occurred a curious causal phenomenon that some people incorrectly described as retroactive. It appeared that when the animal made a correct movement toward its eventual escape, it learned to repeat that movement in a similar situation later even though the success always occurred after the movement. Thorndike believed that the success, as an effect of the movement, "stamped in" the movement that had been the cause of it. Anokhin acknowledged the similarity of Thorndike's law of effect to his own viewpoint, but pointed out that Thorndike was not so interested as he in the physiological mechanisms behind the phenomenon. Thorndike identified success with pleasure or satisfaction and described the law of effect primarily in psychological or subjective terms. Anokhin attempted to describe reinforcement and goal-seeking in terms of physiology: the relationship between conditioned and unconditioned stimuli, and the means by which past experience could provide a pattern contained in the acceptor of action, against which future afferent information could be checked.

Anokhin refined his views in 1962 in a speech that he gave to the conference on the philosophic questions of higher nervous activity. In this speech, later published as a long article, he closely tied his opinions about the development of nervous systems to general principles within dialectical materialism, such as space and time as forms of existence of matter and reflection as a property of all matter.

Anokhin had long been fascinated by the related concepts of *time* and *signalization*, believing that they lie at the root of the intellectual problem of the origin of nervous systems and, ultimately, of reflex activity. Nervous systems are, in his opinion, mechanisms for the anticipation of the future. The importance of anticipation is obvious in Pavlov's classic experiments in which the dog anticipates the coming of food on the basis of signals. But in a more general, theoretical sense, one can say, with Anokhin, that anticipation of the future is the basis of most, if not all, nervous activity. The brain itself is sometimes defined as a physiological mechanism for the purpose of utilizing information from the distance receptors (eyes, ears, nose) and therefore giving the animal a time interval for preparatory reactive steps. This is, indeed, one of the standard definitions of the "brain." But if one regards the brain from the evolutionary standpoint, extending its history of development all the way back to the origin of life itself, then it becomes clear that some way of accounting for this property of anticipation must be found that is comprehensible against the background of an initial lifeless, inorganic milieu. Here again Anokhin was wrestling with the problem of teleology, but he now wished to begin, not with the end-products of evolution, but its origins.

In the section of his speech entitled *"The Conditioned Reflex as a Reflection of the Spatio-Temporal Structure of the World,"* Anokhin began with a recognition of the relevance of dialectical materialism to his approach. Not only does dialectical materialism cite space and time as the forms of existence and movement of matter, but it characterizes all matter as possessing the property of reflection. Therefore, the conditioned reflex must be described in terms of space and time, and it must be a reflection within living matter of material conditions external to living matter.

Since living matter developed in a world in which non–living matter already existed, it was, said Anokhin, "inserted" so to speak, in a world in which spatial and temporal relations were already defined. The relationship of inorganic matter to temporal parameters was entirely passive; no anticipation of events occurred. In terms of the over-all evolution of matter, qualitative new relationships, much more active ones, emerged only with the appearance of living matter. The problem at hand is the way in which these relationships emerged and their essential characteristics.

Anokhin's primary concern here, then, was not the origin of life itself, but the origin of rudimentary nervous systems. The two are obviously intimately connected Anokhin believed no life possible without the nervous function of "anticipatory reflection" in some form but the slight distinction permitted Anokhin to take a different approach from his colleague A. I. Oparin, whose ideas were, nonetheless, important to his own. Oparin, also an outspoken dialectical materialist, is discussed in Chapter VII in this study.What sort of temporal parameters of the movement of matter existed in the inorganic world at the time of the origin of life? These parameters would be the ones within which the first bit of living matter would have to survive and to which it would have to adapt. These basic characteristics of the time structure of the material world Anokhin considered to be a sort of "categorical imperative" for the development of life. Anokhin believed that the first bit of life would encounter three different types of temporal influences (different in their relationship to the bit of life, not in an absolute sense), all existing in the inorganic world around it:

1. the action of relatively constant factors;
2. the action of sequential series of repeating factors (rhythmic or a periodic);

3. the action of sequential series of factors that never repeat.

An example of the first type would be the presence or absence of nitrogen or water in the milieu. An example of the second type would be the sequence of day, evening, night, morning, or summer, autumn, winter, spring. An example of the third type would be a solar eclipse, an extremely violent storm or hurricane, or the passing nearby of a comet.

The most interesting category from the standpoint of Anokhin's analysis is the middle one, the action of sequential series of repeating factors. He considered the presence of this category an absolute necessity for the evolution of rudimentary nervous systems, for without it no method of prediction, of anticipatory reflection, could ever be developed, for there would be nothing to predict and no basis for that rhythmic interchange of matter on which life is based. To be sure, the examples already given for this category (diurnal and annual cycles) are not entirely appropriate for the first bit of life, since it is commonly supposed that life originated in deep ocean waters where these influences would be slight; nonetheless, Anokhin reasonably maintained that in such a milieu there would be repeating, perhaps a periodic, factors (flow of current, change in temperature, presence of various substances). These changing factors were absolutely essential for the development of life, and they had on the internal constitution of the first bit of living matter effects that Anokhin described in terms of dialectical materialist "reflection":

> The basis of the development of life and of its relations to the external inorganic world is the repeating influences on the organism. It is precisely these influences the result of the original properties of the spatio-temporal structure of the inorganic world that conditioned the whole anatomic organization and adaptive functions of the first living substances. In this sense the organization of living substances is in every sense of the word a reflection of the spatio-temporal parameters of the concrete milieu of their existence.

Some of these repeating factors are harmful to the existence of the organism, and some are helpful. The organism to survive must be adapted in differing ways to these contradictory factors. Anokhin then asked the question that for a physiologist becomes so important: "By means of what sort of concrete mechanisms do these repeating and vital influences cause the organism to be adapted?" Notice that the question here was not, in Anokhin's mind, incorrectly inverted. He did not wish to ask how the organism adapts itself, but how the organism is caused to be adapted by the external factors. At this most primitive level the cause must come from the outside if one is not to attribute a life principle to a closed (and very small) bit of matter itself. This, of course, would be vitalism.

Every external influence on the primitive organism will have effects inside it, said Anokhin. In some cases, these effects can be described as chains of chemical reactions making up "open systems." If we can represent a series of repeating external factors as (*ABCD*) (*ABCD*) (*ABCD*), then the series of resulting internal reactions can be designated as (*abcd*) (*abcd*) (*abcd*). As Anokhin observed: "If in the external world several specific events develop sequentially (for example, seasonal rhythms, changes in temperature, currents in the oceans), then the organism must reflect each of these in specific chemical reconstructions of its protoplasm as long as they attain a given threshold of effect."

According to Anokhin, the crucial step in the evolution of rudimentary nervous systems is now at hand. That step is the origin of mechanisms by which the time intervals separating the effects (*abcd*) (*abcd*) (*abcd*) become *shorter* than the time intervals separating the original causes (*ABCD*) (*ABCD*) (*ABCD*). This will be the first case of "anticipatory reflection." Many such cases are necessary for the evolution of life, Anokhin believed, forming a "struggle for rapidity."

Anokhin believed that this acceleration of reactions would occur on the basis of natural selection. Many of the external influences already discussed could be initiators of whole series of chemical chain reactions. The great majority of the resulting chain reactions would bear no adaptive relationship to subsequent events in the external world.

Some, however, would coincidentally result in chemical states inside the primitive organism that would have a survival value when a later external event occurred. If the external events occurred according to a repeating pattern, and if the organism reproduced itself in forms capable of diverse and rapid chain responses to these external events, then those forms producing responses that proved later to be advantageous would survive and reproduce. Thus, macrointervals of time (*ABCD*) (*ABCD*) (*ABCD*) produced changes (*abcd*) (*abcd*) (*abcd*), which were originally also separated by macrointervals of time but gradually become separated by microintervals.

Anokhin described the process as follows:

> *If in a small clump of living matter, making up an open system of many molecules, diverse and very rapid reactions become possible, then several of the sequentially repeating external influences, even those separated by large time intervals, will receive the possibility of reflecting themselves in rapid chemical transformations of this substance.*

Thus, there was created an ability of primitive organisms to reflect the external inorganic world not passively but actively, with the anticipation in their protoplasm of sequential and repetitive phenomena of the external world.

As a matter of fact, since a chain of chemical reactions was established in the protoplasm of the living organism, it is now possible for the action of only the first factor *A* to initiate the whole subsequent chain of chemical reactions.

The fact that *A* alone could serve as a "signal" for the whole subsequent chain brought Anokhin closer to his subject of reflexes. There obviously is no true first member of any series of external influences, Anokhin noted. Even *A* in our example was actually merely an intermediate member of an infinite series stretching back in time. Therefore, the possibilities of signalization are infinite, and many "temporary connections" based on such signalization are established:

> *Consequently the very fact of the appearance of "signalization" and "temporary connections" must be recognized as one of the oldest*

> *principles of the development of living matter. It is precisely in this way that one should interpret I. P. Pavlov's statement, "the temporary nervous connection is the most universal physiological phenomenon in the animal world and in us ourselves." This principle was a result of natural selection and has been driven firmly and ever more deeply into the stabilized structures of protoplasm. One really should not be surprised, then, that this function of the organism, offering it the broadest possibilities of adaptation and progress, rapidly became refined in a special substratum. And it is the nervous system that is this substratum.*

This then is the philosophical and evolutionary background against. Which Anokhin viewed the nervous system. He then proceeded to discuss the conditioned reflex from this standpoint that is, as a form on a high level of the evolutionary ladder of anticipatory reflection. While stating that dialectical materialism was of methodological assistance in developing his view, Anokhin also seemed to be suggesting an interesting addition to its principles. While dialectical materialism already contained the concept that all matter possesses the property of reflection, Anokhin was suggesting that the additional principle be added that all *living* matter possesses the property of anticipatory reflection. And the latter property was, to Anokhin, a step toward a rudimentary nervous system and conditioned reflexes themselves.

In the 1960's Anokhin's attention turned strongly toward cybernetics, which he recognized as a "revolution in thought, in our very approach to the phenomena of nature and our method of analysis." His attitude to cybernetics was, however, not untinged with the feeling that he had never been properly credited for his anticipation of its approach as early as 1935, when he began applying the term "functional system" to the nervous system. And his term "return afferentiation" was obviously another name for cybernetic feedback. In articles in 1966 and 1968 he attempted to defend the interpretation given above within the framework of cybernetics.

Suggestive as Anokhin's evolutionary interpretation of nervous systems was, it should be noticed that it suffers from the same defects in terms of definition and logic as

Oparin's view of "natural selection among prebiological systems." Apparently the only way in which it can be defended against the criticism that Mora advanced against Oparin is by saying (as Bernal did) that the origin of life or of nervous systems has a "logic of its own." This logic is one of "emergent evolution" or "qualitative leaps" in which different autonomous realms of regularities are successively attained. Yet one must admit that this concept of dialectical logic may be merely a way of dodging, or living with, the ancient problems in biology of teleology and reductionism; these problems seem to be with us still, and they appear particularly clearly in discussions of origins in biology, whether of the origin of life or of nervous systems. Anokhin has pointed to the intimate connection between life and nervous systems, and believes that systems analysis or cybernetics holds out the most fruitful approach, at the present time, to these questions within the tradition of dialectical materialism.

As this survey of philosophical issues in recent Soviet psychology and physiology draws to a close, certain limitations will be evident. The foregoing exposition and discussion have drawn heavily on the physiological aspects of psychology, a concentration deriving in large part from the characteristic stress on physiology pervading Soviet philosophical discussion of psychology. In the description of the network of issues in physiology and psychology a number of prominent Soviet authors have been neglected among them, A. R. Luria, D. N. Uznadze, B. M. Teplov, and A. N. Leontiev. A more serious limitation, perhaps, is the omission of full discussion of Soviet attitudes toward Freudian psychology. A brief note on this complicated issue will have to suffice.

Soviet psychologists have seriously underestimated the influence of the subconscious on mental activity. Freudianism, after a period of popularity in the Soviet Union in the twenties, became a prohibited subject. In recent years Soviet psychologists have begun to recognize their inadequacies in this area, although they have by no means become enthusiastic about psychoanalysis. A concern in the last ten or fifteen years among many Soviet psychologists, if one judges by the literature, has been the fear that they have surrendered to Freudianism the whole realm of the subconscious; they have wanted to make it clear that this is not so, or at least should not be so. Consequently, there have been a number of efforts in the Soviet literature to show that Freud was by no means the first person to point to the importance of the subconscious realm. This attempt was, no doubt, an attempt to relativize Freud, to make possible a turning of real attention to the phenomena usually associated with the name of Freud without appearing to embrace Freudianism after years of denying its legitimacy. They have criticized the "monopoly" of Freudianism abroad, particularly in the United States (and perhaps with good reason). They have engaged in rather detailed and controversial semantic discussions of the relative validities of the terms "non-conscious" (*neosoznatel'nyi*), "unconscious" (*bessoznatel'nyi*), and "subconscious" (*podsoznatel'nyi*), with several scholars preferring "non-conscious" to the other two terms, which they saw as closer to Freudianism. But on the whole, they have been moving more and more toward a recognition of Freud. A. M. Sviadoshch, of the Medical Institute in Karaganda, commented in 1962 to an audience of psychologists, physiologists, and philosophers:

> *Without any question S. Freud performed a service for science. He attracted attention to the problem of the "unconscious." He pointed to several concrete manifestations of the "unconscious," such as its influence on slips of the pen or of the tongue. However, he also introduced much that was improbable or fantastic to the subject of the "unconscious," such as his assertion of the sexuality of the small child. He created a mistaken psychoanalytic theory, which we deny.*

The work of the Georgian psychologist D. N. Uznadze (1886–1950) has often been cited as a Soviet alternative to Freud, although it has also been criticized in the Soviet Union as ideologically suspect. In recent years it has become less sensitive. F. V. Bassin promoted Uznadze's theory of "set" (*ustanovka*) as a basis for building a general theory of non–conscious nervous activity that he considered superior on methodological grounds to that of Freud.

Uznadze worked on his conception from the early twenties, when Freudianism attracted many Soviet thinkers, until his death. His work has been continued, in particular, by the Institute of Psychology of the Georgian Academy of Sciences. Uznadze, in his classical experiments, asked a subject to compare the size of two small balls by the sense of touch, one ball being placed in each of his hands. At first, during a long series of tests, the balls (of equal weight) were always of different volume, and the smaller ball was always placed in the same hand. Then the subject was given balls of the same volume (and weight). In reply to the question "Which ball is the larger?" the subject answered that the larger ball was in the hand that had always before received the smaller ball. Uznadze explained this illusion in terms of the "internal state" of the subject. On the basis of simple experiments such as this he built a rather elaborate theory of "set," in which the set is formed by past experience.

Several of Uznadze's defenders in the Soviet Union have differed with him on the details of the set theory. Uznadze thought the set remained non-conscious, or at least he was interested in it only when it was. Bassin, on the contrary, thought that much of the merit of Uznadze's approach consisted in the fact that it explained non-conscious factors in terms of a "functional displacement" and thus opened up this realm to objective experiment. Bassin commented, "The non-conscious set fulfills the role, in this way, of an invisible 'bridge' between definite forms of realized experiences and objective behavior, a bridge that, according to Freud, must be fulfilled by the subject's 'unconscious.'"

Even in the sixties, however, Uznadze's views met criticisms linked with older Soviet attitudes. I. I. Korotkin, of the Pavlov Physiological Institute, maintained that the set had to be interpreted either as an epiphenomenon and therefore was unacceptable in science or as a material phenomenon, which would then differ little, he thought, from the dynamic stereotype of the Pavlovian school. And M. S. Lebedinskii, from the Institute of Psychiatry of the Academy of Medical Sciences, commented that presenting Uznadze as an alternative to Freud was an unsuccessful attempt, since a Freudian analyst would have no trouble accepting Uznadze's views and still remaining a loyal Freudian.

Despite the continuing differences, Soviet psychologists and physiologists are closer at the present time to their non-Soviet colleagues in terms of their research interests and methods than they have been at any time in the last twenty years. Much of this movement has come from the Soviet side, but by no means all. In recent years there has been a quickening interest among British and American scholars in Soviet psychological and physiological research. And even the earlier hostility of non-Soviet scholars to references to dialectical materialism shows some signs of lessening, while the Soviet writers make fewer such references. Jeffrey A. Gray, of the Institute of Experimental Psychology of the University of Oxford, wrote an introductory article for a 1966 symposium on Soviet psychology that deserves to be quoted at length because of its unusual perceptivity:

The official status of Marxism-Leninism has had the result that philosophy intrudes in scientific research and writing in a way which is totally unexpected for a Western scientist. It is a shock to discover in a text ostensibly concerned with the empirical investigation of psychology that there are frequent references to Marx, Engels, Lenin and even not so long ago Stalin. One's first reaction is to dismiss this as a necessary obeisance in the direction of the political powers-that-be as no doubt, in part, it is; and, in any case, one feels that philosophy has no place in the conduct of scientific research. However, a more sympathetic consideration of the use to which these philosophers are put reveals that there is something of more importance, and perhaps even of real value, going on. In the first place, the Russian habit of making the philosophical background plain for all to see is not such a bad one; above all, it becomes clear that, with different philosophical assumptions, there would be different research and different favoured forms of expression and this connection is not to be broken simply by keeping the philosophical assumptions out of sight (and out of mind) as the Anglo-Saxon psychologist tries to do. Secondly, there is a good case to be made for the particular assumptions of Marxist philosophy as a reasonable starting point for a scientific psychology provided, of course, that we are ready to abandon them if our data suggest that other assumptions would make a better starting point. In particular, it can be argued that Marxist assumptions are more consistent with the results of recent psychological and neurophysiological research than are the assumptions contained in the extreme associationist-behaviourist point of view identified with the names of J. B. Watson and C. L. Hull; and that the recent retreat from this extreme position in Anglo-American psychology has made it possible to attempt a rapprochement between the views of human nature held in the East and the West.

It is out of a combination of the point of view advanced above by Gray and the recent effort by Soviet scholars to be less zealous in ideological criticisms of their non-Soviet colleagues that some of the most interesting, even exciting, developments in contemporary psychology and physiology may flow. And it should be added that the possibilities for a rapprochement between Soviet and non-Soviet intellectual conceptions are by no means limited to physiology and psychology; there are similar opportunities in the other fields of natural science as well.

Summary

The elaboration and refinement of the early suggestions of Engels, Plekhanov, and Lenin into a systematic interpretation of nature is the most original creation of Soviet Marxism. In the hands of its most able advocates, there is no question but that dialectical materialism is a sincere and legitimate attempt to understand and explain nature. In terms of universality and degree of development, the dialectical materialist explanation of nature has no competitors among modern systems of thought. Indeed, one would have to jump centuries, to the Aristotelian scheme of a natural order or to Cartesian mechanical philosophy, to find a system based on nature that could rival dialectical materialism in the refinement of its development and the wholeness of its fabric.

The most significant function of dialectical materialism in the Soviet Union derives from the comprehensiveness of its conception and the intimacy of its connection with current scientific theory. As a system of thought it is not of immediate utilitarian value to scientists in their work in fact, converted into dogma it has been a serious hindrance in several cases, although it may have indirectly helped in others but it does have an important educational or heuristic value. Not only professional Soviet philosophers but many scholars and students in other fields as well have a concept of a unifying principle of human knowledge, the materialist assumption that lies at the base of dialectical materialism. It is not a provable principle, but then neither is it absurd. Soviet scientists as a group have, in fact, faced more openly the implications of their philosophic assumptions than have scientists in those countries such as the United States and Great Britain where the fashion is to maintain that philosophy has nothing to do with science. Perhaps out of a meeting of the approaches of Soviet and non-Soviet scholars a way can be found to admit that philosophy does indeed influence science (and vice versa) without allowing that admission to become exaggerated or subject to political manipulation in the way in which it frequently was in the Soviet Union.

In terms of improving the intellectual position of a materialistic explanation of nature, it is clear that Soviet dialectical materialists have made genuine progress in certain fields, progress that to a degree offsets the damaging effects of their failure in genetics. Thirty or forty years ago the crucial questions that dialectical materialists faced were in the area of physics. The new ideas contained in quantum mechanics and the theory of relativity were upsetting to Soviet materialists, as they were to many other traditional thinkers. The dialectical materialists worried about the effects the new physics would have on assumptions that they had previously considered secure: belief in the existence of objective reality, the principle of causality, and the material foundation of reality. Today it is clear that this phase of anxiety has passed. No one knows what the future of physics will bring and the nature of science is to bring crises but at the present moment the philosophical problems in physics are much less difficult for dialectical materialists than they were thirty years ago. It is simply not true, as was frequently maintained several decades ago, that relativity physics and quantum mechanics "destroy materialism at its base." These areas of physics no longer contain unique threats to the assumptions of objective reality, causality, and the primary significance of matter. In terms of the new materialist interpretations of nature, relativity and quantum physics support a more satisfying explanation of natural phenomena than Newtonian physics can yield.

The works of such men as Fock, Blokhintsev, Omel'ianovskii, and Aleksandrov are important in this respect, although somewhat different in content. They pointed out that in the light of modern physics, thinkers should give up determinism in the Laplacian sense, but not causality in general. If there were no causality in nature, all possible outcomes of a given physical state would be equally probable. We know very well that according to quantum mechanics the real situation is far from such absolute indeterminism. A concept of causality, based on probability, can still be maintained. Many thinkers continue to find such a causal concept necessary for an understanding of nature. Once accustomed to it, they usually consider it vastly superior to the rigid Laplacian view.

Other Soviet scholars have turned attention to the fact that in general relativity theory the role of matter is not smaller than in classical (Newtonian) physics, but greater. The density of matter in the universe, according to general relativity, determines the configuration of spacetime. Matter therefore acquires a significance of which eighteenth and nineteenth century materialists could not have dreamed. It is true that the principle of the interchangeability of matter and energy contained in relativity theory seems to demote matter in status (why, for example, should it be considered more important than energy?), but there is another side to the coin that strengthens the convictions of dialectical materialists. Having accepted this interchangeability, they now consider matter and energy synonymous (matter-energy), and proceeding on the basis of the primacy of matter-energy, they do not encounter the ancient problem of the void faced by classical materialists. All voids apparently contain fields of some kind, at least a gravitational one, and therefore a form of matter-energy. The very concept of the void as something that can exist in the real world is therefore thrown into question.

The originality of Soviet dialectical materialism compared to other areas of Soviet thought is not only a result of the talent of a number of Soviet dialectical materialists; it also derives from the nature of classical Marxism and the breathtaking rapidity of development of science itself. Marx wrote a great deal about society, but very little on natural science. The brilliance of his original statements on society overshadowed all subsequent efforts by his supporters in political and economic theory. Before 1917, the system of historical materialism was much more developed than the system of dialectical materialism. Engels, of course, wrote extensively on philosophy of nature and in that sense launched dialectical materialism, but his efforts were rather quickly rendered inadequate by the revolutionary development of scientific theory at the turn of the nineteenth and twentieth centuries. After 1917, Soviet dialectical materialists were forced to seek new paths toward an understanding of nature, because scientific theory itself was already flowing along a new path. Deprived of an adequate Marxist explanation of nature and faced by a revolution in science, Soviet dialectical materialists during the past fifty years have made in philosophy of science an innovative effort that stands out in sharp contrast to other Soviet intellectual efforts.

Perhaps even more significant as a reason for the Soviet achievement in philosophy of nature as compared to other areas of thought in the U.S.S.R. was the system of Communist Party controls over intellectual life, a system that left more room for initiative in scientific subjects than in political ones. The best minds went into scientific subjects, and some of them, naturally enough, were attracted to the philosophic aspects of their work. In the peculiar Soviet environment the esoteric nature of the discussions of dialectical materialism has been something of a boon to writers, screening away censors. Among those authors who write on dialectical materialism, the sections of their works of the highest quality are frequently buried in rather technical discussions. Those scientists who came to dialectical materialism in the late forties for the purpose of defending their science found that interesting work was possible in the philosophy of science. The Communist Party officials continued to consider themselves experts on theories of society, and still crudely intervene today in such discussions. They learned, after several very injurious experiences, that intervention in the knotty problems of scientific interpretation is an exceedingly risky business. Their relative tolerance in this particular field in recent years has had two oddly contrasting effects. Some scientists have abandoned conscious consideration of philosophy now that they are no longer constantly forced to show the relevance of dialectical materialism to science; others have turned to it with new interest now that there is an area of wide dimensions, intellectually speaking, in Soviet philosophy where some innovation is possible.

In retrospect, it is clear that the controversies discussed in this volume reveal very different, even contradictory, aspects of Soviet society. If one is interested primarily in the way in which the Soviet system of political controls over intellectuals created a situation in which unprincipled careerists could gain extraordinary influence in a few scientific fields, then one should turn to the Lysenko affair. Here one can find abundant evidence of the damage done to science by a centralized political system in which the principle of control was extended to scientific theory itself. The Lysenko affair was one of the most flagrant denials of the right of scientists to judge the validity of scientific theory to occur in modern history. The intensely political character that science assumes in all countries is no justification for the intrusion of controls on the judgment of the adequacies of rival scientific explanations. That decision must belong to scientists.

My major goal in writing this volume was not, however, the recounting of this repressive side of the story of Soviet science; for those readers seeking this aspect, I would recommend the books on Lysenko by David Joravsky and Zhores Medvedev. Instead, I have sought to emphasize the more philosophically interesting, rather than the politically dramatic, aspects of the relation between Soviet philosophy and science. In looking through the literature of the past thirty years, I have tried to center my attention on the publications of the best intellectual quality, not the worst. I would do the same if I were studying the relation of Cartesianism to science in the seventeenth and eighteenth centuries, or of Aristotelianism to science in the Middle Ages.

In a commentary on one of my articles on Soviet science the Soviet physicist V. A. Fock wrote that I paid my "main attention to that part of the discussion which proceeded at a higher scientific level and not to articles and viewpoints rightly described by him as 'offensive parodies of intellectual investigations' (particularly numerous in the dark period from 1947 to the early 1950's)." This emphasis meant that I rarely found pertinent items of interest in newspapers, popular political journals, or textbooks of Marxism-Leninism; instead, I looked to the serious monographs and journal articles of Soviet scholars and, wherever possible, to those of natural scientists.

In the eyes of certain of my readers this approach may not seem justified. Those persons who are convinced that all references to dialectical materialism by Soviet scientists have been nothing more than responses to political pressures will doubt that anything worthwhile can be accomplished by studying the works in which these scientists have attempted to illustrate the relevance of Marxist philosophy to their work. I am convinced, however, that quite a few prominent Soviet scientists believe that dialectical materialism is a helpful approach to a study of nature. They have examined many of the same problems of the interpretation of nature that philosophers and scientists in other countries and periods have also examined, and they have slowly developed and refined a philosophy of nature that would almost certainly continue to survive and evolve even if it were no longer propped up by the Communist Party. It is a philosophy of nature that is tied very closely to science itself, and it now depends much more on science for sustenance than on Party ideology. Only by concentrating on scientific literature, rather than on political or ideological sources, can this independent, unofficial side of dialectical materialism be revealed. By recognizing the intellectual sources of much Soviet writing on dialectical materialism, one can begin to understand why there have been such wide disagreements among Soviet scientists on philosophical interpretations of such issues as the physiology of perception, the nature of the universe, and the uncertainty relation of quantum mechanics.

Before turning to a few final observations about the controversies discussed in this volume, it is appropriate to observe that contemporary dialectical materialism possesses several very serious weaknesses. Aside from the political obstacles it still faces in the Soviet Union, the most critical of these, one that seems to be a systemic failing, is the weakness of its defenses against critics standing on the position of philosophic realism. Many scientists and philosophers around the world completely agree with the position of dialectical materialism on the existence of objective reality, but simultaneously decline to consider themselves dialectical materialists. One of the best ways to illustrate their reasons for this refusal is to analyze Lenin's definition of matter:

Matter is a philosophic category for the designation of objective reality which is presented to man in his sensations, an objective reality which is copied, photographed, or reflected by our sensations but which exists independently from these sensations.

A philosophic realist can reply to Lenin that he agrees completely that objective reality exists, that there also exist "objects outside the mind," and that there is nothing supernatural in the world. But where, he can continue, did this word "matter" come from? The realist will then observe that he prefers the term "objective reality" to "matter," since it is clear to him that such phenomena as "consciousness" and "abstraction" are real, but it is far from clear to him that they are material.

The criticism of the philosophic realist reveals that the Leninist definition of matter characterizes its relation to the subject and does not contain a definition of matter itself. There is good reason to believe that Lenin realized this very well, but adopted a "relational definition" of matter nonetheless, since all alternatives to it were much more susceptible to attack. The point is that Lenin's definition of matter is both the *strength* and the *weakness* of dialectical materialism. The definition is strong because it does not depend on the level of development of natural science and thereby acquires much greater permanence. If dialectical materialists attempted to define matter itself that is, in terms of the totality of its properties this would mean that eventually the definition would become obsolescent as our knowledge of those properties changed, just as all definitions of matter given by previous materialists have become obsolescent (for example, those of the Greek atomists, who thought of matter in terms of indivisible units). Lenin considered matter to be inexhaustible in its properties and therefore undefinable in terms of them. This belief is one of the most significant differences between dialectical materialism and mechanistic materialism. The Leninist position avoided the built-in obsolescence of previous definitions of matter, but at the same time opened up dialectical materialists to the criticism that they have no clear way of demonstrating the superiority of the term "matter" to "objective reality."

Strictly speaking, of course, Lenin did not give a definition of matter. The principle of the materiality of the universe does not flow out of the Leninist position on epistemology quoted above. It constitutes, instead, a separate assumption. The assumptive character of materialism is not quite the fatal flaw that its opponents might try to make of it, since all conceptual systems contain some assumptions. Man cannot pursue his most human goal of trying to understand without paying the price of making some assumptions. It is a question of choosing one's assumptions carefully and remaining open to other possibilities. One can argue (for example, on the grounds of economy) that some assumptions are more justified than others; furthermore, one can defend materialism against realism on precisely this basis. But the assumptive character of materialism does mean that dialectical materialism does not uniquely flow out of the facts of science, as some of its defenders would maintain.

And even in the chemistry discussion prompted by a crude emulator of Lysenko there was the very real problem of the significance and meaning of models in nature. The Lysenko affair alone was totally artificial from an intellectual standpoint; the few legitimate scientific and philosophic issues raised there either were outdated or were misunderstood by the supporters of Lysenko. I think that this assessment will stand even if the inheritance of acquired characters by far not the only issue should become accepted in future biology. Lysenko was incapable of understanding the biology of his time. It would be both ahistorical and inaccurate to try to defend him in the name of future biology. Such a defense of the theory of inheritance of acquired characters would be another matter.

In the other disputes, however, we frequently found talented scientists who fully understood modern science making arguments that are in many cases at least plausible while connecting them with dialectical materialism in ways that do not seem to have been merely the results of political pressure. Some of these opinions were "wrong," as judged by contemporary science, to be sure, but they are frequently still understood by contemporary scientists as legitimate and reasonable points of view in the context of their times. Some of them are quite relevant even yet, and more are being developed.

Among the Soviet scholars' viewpoints that have been recognized as valuable, either in their time or at the present time, and in which dialectical materialism may have played a role are: V. A. Fock's and A. D. Aleksandrov's criticisms of certain interpretations of quantum mechanics and relativity; D. I. Blokhintsev's philosophical interpretation of quantum mechanics; O. Iu. Schmidt's analysis of planetary cosmogony; V. M. Ambartsumian's views of star formation and criticisms of certain cosmological theories; G. I. Naan's "quasi-closed" cosmogonical models; A. L. Zel'manov's view of a "manifold universe"; many Soviet criticisms of the concept of an absolute beginning of the universe or of a non-developmental cyclical one; A. I. Oparin's views of the origins of life and his criticism of mechanism in biology; a number of Soviet philosophers' and scientists' views on cybernetic evolution of matter; L. S. Vygotsky's opinions on thought and language; S. L. Rubinshtein's concepts of perception and consciousness; and P. K. Anokhin's revision and extension of Pavlovian physiology.

Yet one should be very careful about assuming that the scientific views of any specific one of the above Soviet scientists were, in fact, importantly influenced by dialectical materialism. In the sentence at the beginning of the above paragraph I said that dialectical materialism "may" have played a role in the intellectual development of these scientists, not that it had in any identifiable case. There is, indeed, no way of demonstrating beyond a doubt that the views of a particular scientist were importantly influenced by intellectual Marxism. Such demonstrations are not in the nature of intellectual history in general, completely aside from the question of Marxism. We can show that a scientist held idea *x*, as evidenced in his writings, and we often can show that he stated in print that there was a connection between *x* and concept *y* in Marxism as he interpreted it. But there is no way that we can prove an actual causal link between *x* and *y*. There are, in fact, many possible explanations besides that of true intellectual stimulation. The scientist may have come upon idea *x* independently and then used *y* merely as a supporting argument. He may have created such linkages for no other reason than the political pressure being exerted upon him. He may actually have used such linkages merely for purposes of his career, aware that a scientific interpretation that could be called Marxist would have a better chance of receiving official favor.

In the final analysis, the problem of causation in this study of science and Soviet Marxism is not fundamentally different from that in other areas of intellectual history and in other countries. Philosophy and politics influence scientists in all countries. The filiation of ideas is an extremely difficult process to study, but the attempt is worthwhile. Furthermore, dialectical materialism deserves particular consideration by historians and philosophers of science by virtue of its more intimate interaction with science than any other current in contemporary philosophy.

A sophisticated materialism that is open to criticism and debate, of which dialectical materialism might some day become a true form, is a philosophic point of view that can be helpful to scientists. It is most valuable to the scientist when his research approaches the outermost limit of knowledge, the area where speculation necessarily plays the greatest role, the approach to the cosmic, the infinite, or the origins or essence of forms of being. It is least valuable, and quite capable of being crudely used with harmful results, when applied to the immediate, the next stage of research. Dialectical materialism could not help a scientist with the details of laboratory work. It would never predict the result of a specific experiment. It certainly would never tell him how to raise crops or treat mental illness. But it might warn him not to fall prey to mysticism in the face of the sometimes overwhelming mystery and awe of the unknown. Through its non-reductionism it might remind him how contradictory and difficult the explication of nature is and how dangerous it is to reduce the complex phenomena on one level to combinations of simple mechanisms on a lower level. It might warn him that the sudden appearance of anomalies in research is not reason for discarding a realist epistemology or a commitment to the existence of at least some regularities in nature, whether probabilistic or strictly deterministic. It might remind him, through its emphasis on the interconnectedness of nature, of the importance of an ecological approach to the biological world and of the significance of the historical view for an understanding of the development of matter. It might encourage him to erect temporary explanatory schemes larger than any one science, but ones that do not pretend to possess final answers. At the same time, it might also assure him that the retention of commitments to epistemological realism and natural order is by no means a renunciation of art or mystery in nature. Nothing is more baffling than the ingenuity of man and his creations and the beauty of nature, of which man is a part. A sophisticated materialism could handle such considerations equally as well as a sophisticated idealism, and would start from assumptions that are more consonant with the naturalism implicit in most science.

8

Logical Philosophy and its Relationship

Introduction

In this broad sense, logic in the twentieth century has come to include, not only theories of formal entailment, but informal logic, probability theory, confirmation theory, decision theory, game theory and theories of computability and epistemic modelling as well. As a result, over the course of the century the study of logic has benefited, not only from advances in traditional fields such as philosophy and mathematics, but also from advances in other fields as diverse as computer science and economics. Through Frege and others late in the nineteenth century, mathematics helped transform logic from a merely formal discipline to a mathematical one as well, making available to it the resources of contemporary mathematics. In turn, logic opened up new avenues of investigation concerning reasoning in mathematics, thereby helping to develop new branches of mathematical research–such as set theory and category theory–relevant to the foundations of mathematics itself. Similarly, much of twentieth century philosophy–including advances in metaphysics, epistemology, the philosophy of mathematics, the philosophy of science, the philosophy of language and formal semantics–closely parallels this century's logical developments. These advances have led in turn to a broadening of logic and to a deeper appreciation of its application and extent. Finally, logic has provided many of the underlying theoretical results which have motivated the advent of the computing era, learning as much from the systematic application of these ideas as it has from any other source.

This chapter is divided into four sections. The first, 'The Close of the Nineteenth Century', summarizes the logical work of Boole, Frege and others prior to 1900. The second, 'From Russell to Gödel', discusses advances made in formal logic from 1901, the year in which Russell discovered his famous paradox, to 1931, the year in which Gödel's seminal incompleteness results appeared. The third section, 'From Gödel to Friedman', discusses developments in formal logic made during the fifty years following Gödel's remarkable achievement. Finally, the fourth section, 'The Expansion of Logic', discusses logic in the broader sense as it has flourished throughout the latter half of the twentieth century.

Close of the Nineteenth Century

'Logic is an old subject, and since 1879 it has been a great one.' This judgment appears as the opening sentence of W.V. Quine's 1950 *Methods of Logic.* The sentence is justly famous–even if it has about it an air of exaggeration–for nothing less than a revolution had occurred in logic by the end of the nineteenth century.

Several important factors led to this revolution, but without doubt the most important of these concerned the mathematization of logic. Since the time of Aristotle, logic had taken as its subject matter formal patterns of inference, both inside and outside mathematics. Aristotle's *Organon* had been intended as nothing less than a tool or canon governing correct inference. However, it was not until the mid-nineteenth century that logic came to be viewed as a subject which could be developed mathematically, alongside other branches of mathematics. The leaders in this movement–George Boole (1815–64), Augustus DeMorgan (1806–71), William Stanley Jevons (1835–82), Ernst Schröder (1841–1902), and Charles Sanders Peirce (1839–1914)–all saw the potential for developing what was to be called an 'algebra of logic', a mathematical means of modelling the abstract laws governing formal inference.

A important factor concerning the advancement of logic during this period had to do, not with the mathematization of logic, but with the logicizing of mathematics. The idea of reducing mathematics to logic had been advocated first by Gottfried Wilhelm Leibniz (1646–1716) and later by Richard Dedekind (1831–1916). In general, it amounted to a two-part proposal: first, that the concepts of (some or all branches of) mathematics were to be defined in terms of purely logical concepts and, second, that the theorems of (these same branches of) mathematics were in turn to be deduced from purely logical axioms. However, it was not until the late nineteenth century that the logical tools necessary for attempting such a project were discovered.

Not accidentally, the attempt to logicize mathematics coincided with a process of systematizing and rigorizing mathematics generally. Many commentators had called for such work to be done. Earlier discoveries by Gerolamo Saccheri (1667–1733), Karl Gauss (1777–1855), Nikolai Lobachevski (1793–1856), Janos Bolyai (1802–60) and Bernhard Riemann (1826–66) in the development of non-Euclidean geometry had led to a new sensitivity about axiomatics and about foundations generally. Thus, by the late 1800s, the critical movement–which had begun in the 1820s–had eliminated many of the contradictions and much of the vagueness contained in many early nineteenth century mathematical theories. Bernard Bolzano (1781–1848), Niels Abel (1802–29), Louis Cauchy (1789–1857) and Karl Weierstrass (1815–97) had successfully taken up the challenge of rigorizing the calculus. Weierstrass, Dedekind and Georg Cantor (1845–1918) had all independently developed methods for founding the irrationals in terms of the rationals and, as early as 1837, William Rowan Hamilton (1805–65) had introduced ordered couples of reals as the first step in supplying a logical basis for the complex numbers. By 1888, Dedekind had also developed a consistent postulate set for axiomatizing the set of natural (or counting) numbers, *N*. Building upon these results, as well as others by H.G. Grassmann (1809–77), Guiseppe Peano (1858–1932) was then able to develop systematically, not just a theory of arithmetic and of the rationals, but a reasonably detailed theory of real limits as well. The results appeared in Peano's 1889 *Arithmetics Principia*.

Beginning with four axioms governing his underlying logic, and Crucial to further advancement was the introduction of the quantifiers and the development of the predicate calculus. The introduction of the quantifiers resulted from the independent work of several authors, including Peirce and Schröder, but predominantly it came about through the work of Gottlob Frege (1848–1925). It is possible that Peirce was the first to arrive at the notion of a quantifier, although explicit mention of either the universal or the existential quantifier does not appear in his published writings until 1885. Frege, in contrast, first published his account in 1879 in his now famous *Begriffsschrift* (meaning literally 'concept writing'). Hence the date in Quine's famous aphorism. Subtitled 'a formula language, modeled upon that of arithmetic, for pure thought', the *Begriffsschrift* took as its goal nothing less than a rigorization of proof itself. Said Frege,

> *My intention was not to represent an abstract logic in formulas, but to express a content through written signs in a more precise and clear way than it is possible to do through words. In fact, what I wanted to create was not a mere* calculus ratiocinator *but a* lingua characterica *in Leibniz's sense.*

The result was the introduction of a very general symbolic language suitable for expressing the type of formal inferences used in mathematical proofs. By combining expressions representing individuals and predicates (properties and relations) with the prepositional connectives ('and', 'or', 'not', etc.) and quantifiers ('all', 'some'), Frege succeeded in producing a language powerful enough to express even the most complicated of mathematical statements. Frege immediately put his language to work, applying it as he did to his project of logicizing arithmetic. By the time his *Die Grundlagen der Arithmetik* appeared in 1884, Frege had arrived at the appropriate logical definitions for the necessary arithmetical terms and had begun work on the essential derivations. The derivations themselves appeared in his two-volume *Grundgesetze der Arithmetik* in 1893 and 1903. Thus, it is Frege, along with Boole, who is generally credited as being one of the two most important founders of modern formal logic.

The *Begriffsschrift* was Frege's first work in logic, but it is a landmark one, for in it he develops the truth-functional prepositional calculus, the theory of quantification (or predicate calculus), an analysis of propositions in terms of functions and arguments (instead of in terms of subject and predicate), a definition of the notion of mathematical sequence and the notion of a purely formal system of derivation or inference. Of these contributions, Frege's unique insight with regard to the predicate calculus was to note the inadequacy of the traditional subject/predicate distinction, and to replace it with one from mathematics, that of function and argument. For Frege, once arguments have been substituted into the free variables of a prepositional function, a judgement is obtained. In short, a word in a statement which can be replaced by other words is itself an argument, while the remainder of the sentence is the function. The fact that the *Begriffsschrift* additionally contains the first formally adequate notation for quantification, together with the first successful formalization of first-order logic, has guaranteed it a position of unique importance in the history of logic.

Unfortunately, for much of his life, Frege's work was met with indifference or hostility on the part of his contemporaries. Response to the *Begriffsschrift* was typified by a rather caustic review by Cantor, who had not even bothered to read the book. Frege's axioms for the propositional calculus (which use negation and material implication as primitive connectives) turned out not to be independent and, in addition to his one stated rule of inference, detachment (also called *modus ponens,* the rule that given well-formed formulas of the form p and p ? q, one can infer a well-formed formula of the form q), an unstated rule of substitution was also used. However, the most important reason for the *Begriffsschrift's* poor reception was the difficulty of working with Frege's rather idiosyncratic and cumbersome logical notation, which has not survived. Thus, it was not until others were able to replace Frege's notation, and further rigorize his logic, that Frege's discoveries began to receive the prominence they deserved. None the less, it is Frege's analysis of propositions into functions and arguments, along with his introduction of the quantifiers, which to this day remain at the heart of the development of modern logic.

Yet a third factor contributing to the revolutionary advancement of logic during the late nineteenth century was Cantor's discovery of set theory. Intuitively, a set can be thought of as any collection of well-defined distinct objects. In Cantor's words, by the idea of a set 'we are to understand any collection into a whole M of definite and separate objects m of our intuition or our thought'. The objects which determine each set are called the elements or members of that set.

With little more than these beginnings, Cantor was able to show that the cardinality of the set of all subsets of any given set (i.e. the power set of that set) is always greater than that of the set itself, thus introducing the modern hierarchy of sets. In addition, he showed that the set of real numbers is non-denumerable (or, equivalently, that the cardinality of the continuum of reals, R, is greater than that of N). Cantor proved both results in 1891 by means of a diagonal argument, an argument designed to construct objects, on the basis of other objects, and in such a way that the new objects are guaranteed to differ from the old. Cantor's diagonal argument therefore provides an important example of a modern impossibility proof since, by using it, he proved the non-denumerability of the reals by showing that a one-to-one correspondence between the natural numbers and the reals is impossible. It also followed from Cantor's work that infinite sets could consistently be placed in one-to-one correspondence with proper subsets of themselves, thus disproving Euclid's general axiom that the whole is necessarily greater than the part. Such unintuitive results meant that set theory, like non-Euclidean geometry before it, would soon lead to further questions about the nature of proof. The question of when to rely upon axioms which, up until this point had still regularly been based upon 'clear and distinct ideas', therefore became crucial. The fact that there was also an intuitive identity between the extension of a predicate and its corresponding set meant that developments in set theory were bound to affect logic.

From Russell to Godel

If it is fair to say that by the end of the nineteenth century logic had become invigorated as a result of its connections with mathematics, it is also fair to say that mathematics was becoming similarly invigorated as a result of its connections with logic. In fact, it was the interplay between logic and mathematics which led a host of figures in both disciplines–including most famously David Hilbert (1862–1943), L.E. J. Brouwer (1881–1966), Arend Heyting (1898–1980), A.N. Whitehead (1861–1947), Bertrand Russell (1872–1970), Ernst Zermelo (1871–1953), Kurt Gödel (1906–78), Alfred Tarski (1902–83), Alonzo Church (b. 1903) and W.V. Quine (b. 1908)–to concentrate their efforts upon foundational issues in logic and mathematics. Many were prompted to do so as a result of a problem which affected both logic and mathematics equally, the problem of the antinomies.

In 1900 many of the world's best philosophers met to attend the Third International Congress of Philosophy, held in Paris from 1 to 5 August. Following the Congress, a significant number of philosophers remained in Paris for the Second International Congress of Mathematicians, which was being held immediately afterwards, from 6 to 12

August. It was at these later meetings that Hilbert delivered his famous keynote address welcoming the mathematical world to Paris. Conscious of his place in history, Hilbert took the occasion to remind his audience of the challenges facing them as they entered a new century:

> *If we would obtain an idea of the probable development of mathematical knowledge in the immediate future, we must let the unsettled questions pass before our minds and look over the problems which the science of to-day sets and whose solution we expect from the future. To such a review of problems the present day, lying at the meeting of the centuries, seems to be well adapted. However unapproachable these problems may seem to us and however helpless we stand before them, we have, nevertheless, the firm conviction that their solution must follow by a finite number of purely logical processes. This conviction of the solvability of every mathematical problem is a powerful incentive to the worker. We hear within us the perpetual call: There is the problem. Seek its solution. You can find it by reason, for in mathematics there is no* ignorabimus.

To emphasize his challenge, Hilbert presented his now famous list of twenty-three major unsolved logical and mathematical problems. First on Hilbert's list was Cantor's continuum problem, the problem of determining whether there is a set with cardinality greater than that of the natural numbers but less than that of the continuum. Second on the list, but equally important from the point of view of the logician, was the problem of proving an axiom set consistent or, as Hilbert put it, the problem of the 'compatibility of the arithmetical axioms'. Although primarily of theoretical interest in 1900, it was a problem which would, within the year, take on new urgency.

It was also at these meetings that Russell met Peano. By all accounts the meeting was a congenial one. As Russell reports:

> *The Congress was a turning point in my intellectual life, because I there met Peano. I already knew him by name and had seen some of his work, but had not taken the trouble to master his notation. In discussions at the Congress I observed that he was always more precise than anyone else, and that he invariably got the better of any argument upon which*

he embarked. As the days went by, I decided that this must be owing to his mathematical logic.

Thus, impressed by Peano and his logic, Russell returned to England, motivated to begin work on his *Principles of Mathematics* and confident that any problem he might set for himself would quickly be solved. As one might guess, Russell's *Principles* was to be heavily influenced, not only by Peano's *Arithmetices Principia,* but also by Frege's *Begriffsschrift* and *Grundlagen.* Russell finished the first draft of the manuscript, as he tells us, 'on the last day of the century', 31 December, 1900.

Five months later, in May 1901, Russell discovered his now famous paradox. The paradox comes from considering the set of all sets which are not members of themselves, since this set must be a member of itself if and only if it is not a member of itself. As a result, one must attempt to find a principled way of denying the existence of such a set. Cesare Burali Forti (1861–1931), an assistant to Peano, had discovered a similar antinomy in 1897 when he had observed that since the set of ordinals is well-ordered, it therefore must have an ordinal. However, this ordinal must be both an element of the set of ordinals and yet greater than any ordinal in the set.

After worrying about the difficulty for over a year, Russell wrote to Frege with news of the paradox on 16 June, 1902. The antinomy was a crucial one, since Frege claimed that an expression such as *f(a)* could be considered to be both a function of the argument *f* and a function of the argument *a*. In effect, it was this ambiguity which allowed Russell to construct his paradox within Frege's logic.

Let *w* be the predicate: to be a predicate that cannot be predicated of itself. Can *w* be predicated of itself? From each answer its opposite follows. Therefore we must conclude that *w* is not a predicate. Likewise there is no class (as a totality) of those classes which, each taken as a totality, do not belong to themselves. From this I conclude that under certain circumstances a definable collection does not form a totality.

Russell's letter to Frege, in effect telling him that his axioms were inconsistent, arrived just as the second volume of his *Grundgesetze* was in press. Immediately appreciating the difficulty, Frege attempted to revise his work, adding an appendix to the *Grundgesetze* which discussed Russell's discovery. Nevertheless, he eventually felt forced to abandon his logicism. A projected third volume of the *Grundgesetze* which had been planned for geometry never appeared. Frege's later writings show that Russell's discovery had convinced him of the falsehood of logicism, and that he had opted instead for the view that all of mathematics, including number theory and analysis, was reducible only to geometry.

As I think about acts of integrity and grace, I realize that there is nothing in my knowledge to compare with Frege's dedication to truth. His entire life's work was on the verge of completion, much of his work had been ignored to the benefit of men infinitely less capable, and upon finding that his fundamental assumption was in error, he responded with intellectual pleasure clearly submerging any feelings of personal disappointment. It was almost superhuman and a telling indication of that which men are capable if their dedication is to creative work and knowledge instead of cruder efforts to dominate and be known.

With the appearance of Russell's paradox, Hilbert's problem of proving consistency took on new urgency. After all, since (in classical logic) all sentences follow from a contradiction, no mathematical proof could be trusted once it was discovered that the underlying logic was contradictory. Important responses came not only from Hilbert and Russell, but from Brouwer and Zermelo as well.

The seeds of Hilbert's response were contained in his 1904 address to the Third International Congress of Mathematicians. In this address, Hilbert presented his first attempt at proving the consistency of arithmetic. Hilbert recognized that an attempt to avoid the contradictions by formalizing one's metalanguage would only lead to a vicious regress. As a result, he opted instead for an informal logic and metalanguage whose principles would be universally acceptable. His basic idea was to allow the use of only finite, well-defined and constructible objects, together with rules of inference which were deemed to be absolutely certain. Controversial principles such as the axiom of choice were to be explicitly excluded. The programme variously became known as the finitary method, formalism, proof theory, metamathematics and Hilbert's programme. Together with Wilhelm Ackermann (1896–1962), Hilbert went on to publish *Grundzüge der Theoretischen Logik (Principles of Mathematical Logic)* in 1928 and, together with Paul Bernays (1888–1977), the monumental two-volume work *Die Grundlagen der Mathematik (Foundations of Mathematics),* in 1934 and 1939, respectively. This latter work recorded the results of the formalist school up until 1938, including work done following the publication of Gödel's 1931 incompleteness theorems, after which it was concluded that the original finitistic methods of the programme had to be expanded.

Hilbert's finitary method had similarities to a second important response to the antinomies, that of Brouwer and the intuitionists. Like Hilbert, Brouwer held that one cannot assert the existence of a mathematical object unless one can also indicate how to go about constructing it. However, Brouwer argued, in addition, that the principles of formal logic were to be abstracted from purely mental mathematical intuitions. Thus, since logic finds its basis in mathematics, it could not itself serve as a foundation for mathematics. For similar reasons, Brouwer rejected the actual infinite, accepting only mathematical objects capable of effective construction by means of the natural numbers, together with methods of finite constructibility. On this view theoretical consistency would be guaranteed as a result of reformed mathematical practice.

Russell's own response to the paradox was contained in his aptly named theory of types. Russell's basic idea was that by ordering the sentences of a language or theory into a hierarchy (beginning with sentences about individuals at the lowest level, sentences about sets of individuals at the next lowest level, sentences about sets of sets of individuals at the next lowest level, etc.), one could avoid reference to sets such as the set of all sets, since there would be no level at which reference to such a set appears. It is then possible to refer to all things for which a given condition (or predicate) holds only if they are all at the same level or of the same 'type'. The theory itself admitted of two versions.

According to the simple theory of types, it is the universe of discourse (of the relevant language) which is to be viewed as forming a hierarchy. Within this hierarchy, individuals form the lowest type; sets of individuals form the next lowest type; sets of sets of individuals form the next lowest type; and so on. Individual variables are then indexed (using subscripts) to indicate the type of object over which they range, and the language's formation rules are restricted to allow only sentences such as 'a*n*

$\in$

b*m*' (where $m=n+1$) to be counted among the (well-formed) formulas of the language. Such restrictions mean that strings such as 'x*n*

The ramified theory of types goes further than the simple theory. It does so by describing a hierarchy, not only of objects, but of closed and open sentences (propositions and prepositional functions, respectively) as well. The theory then adds the condition that no proposition or prepositional function may contain quantifiers ranging over propositions or prepositional functions of any order except those lower than itself. Intuitively, this means that no proposition or prepositional function can refer to, or be about, any member of the hierarchy other than those which are defined in a logically prior manner. Since, for Russell, sets are to be understood as logical constructs based upon prepositional functions, it follows that the simple theory of types can be viewed as a special case of the ramified theory.

Russell first introduced his theory in 1903 in his *The Principles of Mathematics*. Later, in 1905, he abandoned the theory in order to consider three potential alternatives: the 'zigzag theory', in which only 'simple' propositional functions determine sets; the 'theory of limitation of size', in which the purported set of all entities is disallowed; and the 'no-classes theory', in which sets are outlawed, being replaced instead by sentences of certain kinds. Nevertheless, by 1908 Russell was to abandon all three of these suggestions in order to return to the theory of types, which he develops in detail in his article 'Mathematical Logic as Based on the Theory of Types'. The theory finds its mature expression in Whitehead and Russell's monumental work defending logicism, *Principia Mathematica,* the first volume of which appeared in 1910.

In order to justify both his simple and ramified theories, Russell introduced the principle that 'Whatever involves *all* of a collection must not be one of the collection'. Following Henri Poincaré (1854–1912), Russell called this principle the 'Vicious Circle Principle' (or VCP). Once the VCP is accepted, it follows that the claim-first championed by Peano and later by Frank Ramsey (1903–30)–that there is an important theoretical distinction between the set-theoretic and the semantic paradoxes, is mistaken. The reason is that in both cases the VCP provides a philosophical justification for outlawing self-reference.

Yet a fourth response to the paradoxes was Zermelo's axiomatization of set theory. In 1904 Zermelo had solved one of Hilbert's twenty-three problems of the 1900 Congress by proving that every set can be well ordered. Four years later, in 1908, he developed the first standard axiomatization, Z, of set theory, improving upon both Dedekind's and Cantor's original, more fragmentary treatments.

Zermelo's axioms were designed to resolve Russell's paradox by restricting Cantor's naive principle of abstraction-the principle that from each and every predicate expression a set could be formed. Specifically, Zermelo's axiom of replacement would disallow the construction of paradoxical sets (such as the set of all sets that are not members of themselves), but would still allow the construction of other sets necessary for the development of mathematics. *ZF,* the axiomatization generally used today, is a modification of Zermelo's theory developed primarily by Abraham Fraenkel (1891–1965).

As a result of the work of many others, including Thoralf Skolem (1887–1963), Leopold Löwenheim (1878–1957), and

John von Neumann (1903–57), it is recognized that many additional axioms variously used to formalize set theory may be derived from the above list of axioms. For example, the separation axioms are derivable from the axiom schema of replacement; the pairing axiom is derivable from the power set axiom together with the axiom schema of replacement; and the union axiom is derivable from the axiom of extensionality, the sum With propositional and predicate logic formalized, the systematic metalogical study of formal properties of logical systems, and the investigation of informal, philosophical problems resulting from such results, began to develop. Among the most important metalogical results to be proved at the level of propositional logic are the completeness and soundness results (which show, respectively, that all valid formulas are theorems of the system and that all theorems of the system are valid formulas), the deduction theorem (which states that, if there is a proof from '*s1*, s2, ..., sn' to '*sn+1*', then there is also a proof from '*s1*, s2, ..., sn-1' to 'if *sn* then *sn+1*'), and the decidability result (which shows that there is an effective, mechanical decision procedure–such as truth tables-for determining the validity of any arbitrary formula of the system).

Among the metatheorems proved for first-order logic are similar completeness and soundness results, Tarski's theorem concerning the undefinability of (arithmetical) truth, and the now famous Löwenheim-Skolem theorems. These are a family of metatheoretical results proved by Löwenheim in 1915, and extended by Skolem in 1920 and 1922, to the effect that if there is an interpretation in which an enumerable set of sentences is satisfiable in an enumerably infinite domain then the set is also satisfiable in every infinite domain and, similarly, that if there is an interpretation in which a set of sentences is satisfiable in a non-empty domain then it is also satisfiable in an enumerably infinite domain. This latter theorem gives rise to the so-called 'Skolem's paradox', the unintuitive (but ultimately non-contradictory) result that systems for which Cantor's theorem is provable, and hence which must contain non-denumerable sets, nevertheless must be satisfiable in a (smaller) enumerably infinite domain.

Despite such impressive results, Hilbert's goal of discovering a consistency proof for arithmetic remained elusive. The explanation came with a paper published by Gödel in 1931. Today Gödel is remembered for several important results, any one of which would have given him a position of importance in the history of logic. Most famously, these included his proofs of the completeness and compactness of first-order logic in 1930, and the incompleteness of arithmetic in 1931. They also include results in constructive logic in 1932 and computation theory in 1933, as well as his 1938 proof that the continuum hypothesis is consistent with the Zermelo-Fraenkel axioms of set theory–in other words, that the usual axioms of set theory could never prove the continuum hypothesis false.

Of these results, several require further comment. It was in Gödel's 1930 doctoral dissertation at the University of Vienna that the completeness of the first-order predicate calculus was proved for the first time. Completeness is the property that every valid formula of the system is provable within the system. This turns out to be equivalent to the claim that every formula is either refutable or satisfiable. Building on the results of Löwenheim and Skolem, Gödel succeeded in proving a result slightly stronger than this, a result which also entails the Löwenheim-Skolem theorem. Gödel then generalized his result to first-order logic with identity and to infinite sets of formulas. At the same time Gödel proved the compactness theorem for first-order logic, which states that any collection of well-formed.

By introducing his famous system of 'Gödel numbering', Gödel assigns natural numbers to sequences of signs and sequences of sequences of signs within the system. This is done in such a way that, given any sequence, the number assigned to it can be effectively calculated, and given any number, it can be effectively determined whether a sequence is assigned to it and, if so, what it is. Metamathematical predicates used to describe the system in this way can be correlated with number-theoretic predicates. For example, the metamathematical notion of being an axiom can be expressed using the predicate *Ax(x)*, which in turn corresponds to exactly those Gödel numbers, *x*, which are correlated to the axioms of the system. Referring to the set of axioms can then be accomplished simply by referring to this set of Gödel numbers.

Theorem VI of Gödel's 1931 paper states that in a formal system of the specified kind there exists an undecidable proposition (that is, a proposition such that neither it nor its negation is provable within the system). It is this theorem which is commonly referred to as Gödel's first incompleteness theorem (G1). In addition to proving G1, Gödel also proves several other corollaries: for example, that among the propositions whose validity is undecidable are both arithmetic propositions (Theorem VIII) and formulas of pure first-order logic (Theorem IX), and that a statement which expresses the consistency of the system and which can be written as a formula in the system is itself among those formulas not provable in the system (Theorem XI). It is this last theorem, Theorem XI, which is commonly referred to as Gödel's second incompleteness theorem (G2). Both of Gödel's incompleteness theorems required that logicians and mathematicians view formal systems of arithmetic in a new light.

From Godel to Friedman

Hilbert's programme in effect had two main goals concerning the foundations of mathematics. The first was descriptive, the second justificatory. The descriptive goal was to be achieved by means of the complete formalization of mathematics. The justificatory goal was to be achieved by means of a finitary (and hence epistemologically acceptable) proof of the reliability of those essential but non-finitary (and hence epistemologically more suspect) parts of mathematics. Work by both formalists and logicists during the first two decades of the century had effectively accomplished the former of these two goals. Ideally a finitary consistency proof would accomplish the latter.

Gödel's second incompleteness theorem, G2, is often thought relevant to Hilbert's programme for just this reason. Specifically, it is often claimed that G2 implies three philosophically significant corollaries: first, that any consistency proof for a theory, *T*, of which G2 holds will have to rely upon methods logically more powerful than those of theory *T* itself; second, that (in any significant case) a consistency proof for theory *T* can yield no epistemological gain and so cannot provide a satisfactory answer to the sceptic regarding *T*'s consistency; and third, that as a result of this, G2, if not strictly implying the outright failure of Hilbert's programme, at the very least indicates that significant modifications to it are needed. All three of these 'corollaries' are controversial. In fact, Gödel himself explicitly makes the point that the truth of G2 should not be viewed, by itself, as sufficient reason for abandoning Hilbert's goal of discovering a finitary consistency proof.

I wish to note expressly that Theorem XI (and the corresponding results for *M* and *A*) do not contradict Hilbert's formalistic viewpoint. For this viewpoint presupposes only the existence of Other consequences also followed. Among them was a renewed interest in the problem of decidability, the problem of finding an effective, finite, mechanical, decision procedure (or algorithm) for determining whether an arbitrary, well-formed formula of a system is in fact a theorem of the system. A positive solution to such a problem is a proof that an effective procedure exists. A negative solution is a proof that an effective procedure does not exist. By 1931 it was already well-known that a positive solution existed in the case of prepositional logic, that a decision procedure existed in the form of truth tables for determining whether an arbitrary formula was a tautology. However, in 1936 Church proved that there could be no such decision procedure for validity (or equivalently, for semantic entailment) in first-order logic.

Church proved that there does exist a decision procedure for determining that valid first-order sentences are valid.

However, at the same time he also proved that there is no corresponding procedure for showing of sentences which are not valid that they are not valid. In other words, although there is an effective, finite, mechanical, positive test for first-order validity, there is, and can be, no such negative test. Thus, given an arbitrary first-order sentence, there can be no effective, finite, mechanical decision procedure for determining whether or not it is valid.

Central to Church's theorem was the idea of computability. Intuitively, a computable function may be said to be any function for which there exists an effective decision procedure or algorithm for calculating a solution. Several suggestions were offered as a means of making this notion more precise, not only by Gödel and Church, but by Emil Post (1897–1954) and Alan Turing (1912–54) as well. One such precise suggestion was to identify effective computability with that given by a Turing machine; another was to identify it with a series of functions identifiable in the lambda calculus; yet a third was to identify it with that of a general recursive function. Perhaps surprisingly, all three notions turned out to be equivalent. As a result, the thesis of identifying computability with the mathematically precise notion of general recursiveness became known alternatively as Church's thesis or the Church-Turing thesis.

Intuitively, a Turing machine can be thought of as a computer which manipulates information contained on a linear tape (which is infinite in both directions) according to a series of instructions. More formally, the machine can be thought of as a set of ordered quintuples, (*qi*, *Si*, Sj, Ii, qj), where *qi* is the current state of the machine, *Si* is the symbol currently being read on the tape, *Sj* is the symbol with which the machine replaces *Si*, *Ii* is an instruction to move the tape one unit to the right, to the left, or to remain where it is, and *qj* is the machine's next state. From this rather impoverished set of operations it turns out that a wide range of functions are computable. It also turns out that the decision problem for first-order validity is capable of being modelled via Turing machines since, by assigning unique numbers to first-order sentences, there will be a function which returns 1 if given a number representing a valid sentence and 0 otherwise. The question of whether this function is finitely and mechanically calculable turns out to be equivalent to the so-called halting problem, the problem of discovering an effective procedure for determining whether the appropriate Turing machine will ever halt, given arbitrary input. Church's theorem is equivalent to the result that such a function is not finitely and mechanically calculable.

The latter may be defined as a set of functions whose members are said to be either primitive recursive or general recursive, and which are themselves constructed from a set of more fundamental functions by a series of fixed procedures.

Specifically, a *constant* function is a function that yields the same value for all arguments. A *successor* function is a function that yields as its value the successor of its argument, for example, $s(1)=2$, $s(35)=36$. An *identity* function of n arguments is a function that yields as its value the ith of its n arguments. Together, the constant, successor and identity functions are called the *fundamental* functions is being defined by composition and g, $h1$, ...hm are previously defined functions, then $f(x1,\ldots, xn)=g(h1(x1, \ldots, xn), \ldots, hm(x1, \ldots, xn))$.

Similarly, a new function, f, of n arguments is defined by *primitive recursion* as follows: first, if a designated argument is 0, then f is defined in terms of a previously defined function, g, of n-1 arguments whose arguments are taken to be exactly those of f except for the designated argument, 0. In other words, if f is being defined by recursion and g is a previously defined function, then $f(x1, \ldots, xn-1,0)=g(x1, \ldots, xn)$. Second, if the designated argument is not 0, and is instead the successor, c+1, of some number, c, then f is defined in terms of a previously defined function, g, of n+1 arguments whose arguments are taken to be exactly those of f except for the designated argument, c+1, together with c and the value of f when its arguments are exactly those arguments already given for *g*. In other words, if *f* is being defined by recursion and *g* is a previously defined function, then $f(x1, ..., xn\text{-}1, c+1) = g(x1, ..., xn\text{-}1, c, f(x, ..., xn\text{-}1, c))$.

Any function which is either a fundamental function or can be obtained from the fundamental functions by a finite number of applications of composition and primitive recursion is then said to be a *primitive recursive* function.

Next, a new function, *f* of *n* arguments is defined by *minimization* such that its value (whenever it exists) is the least *c* such that, given a previously defined function, *g*, whose arguments are exactly the arguments of *f* together with *c*, *g* has the value 0. If there is no such *c*, then *f* remains undefined for those arguments. In other words, if *f* is being defined by minimization and *g* is a previously defined function, then $f(x1, ..., xn)$=the least *c* such that $g(x1, ..., xn, c)=0$, provided that there exists some *c*; otherwise *f* is undefined.

Any function which is either a fundamental function or can be obtained from the fundamental functions by a finite number of applications of composition, primitive recursion, and minimization is then said to be a *general recursive* (or simply *recursive*) function. The proved identification of the class of recursive functions with the other two classes of functions thought to express the intuitive concept of computability lent strong support to the Church-Turing thesis.

Having observed both the incompleteness of first-order theories of arithmetic and the undecidability of first-order validity, questions naturally turned to other issues relating to computability. One such issue is that of computational complexity. Another concerns the extent and nature of arithmetical incompleteness. Recently, advances have been made in both of these areas.

The computational complexity of a problem is a measure of the computational resources required to solve the problem. In this context, the distinction is made between problems solvable by polynomial-time algorithms and problems which, if solvable, have solutions which are testable in polynomial time but which, if not solvable, do not. Problems of the former kind are said to be members of the class of problems P, while problems of the latter kind are members of the class NP. Those problems in NP which are measurably the hardest to solve are said to be NP-complete. Today the problem of whether P=NP remains open. Nevertheless, in 1971 Stephen Cook proved that the problem of satisfiability (the problem of determining, given an arbitrary set of sentences, whether it is possible for all of the sentences to be jointly true) is at least as difficult to solve as is any NP-complete problem [1.87]. The result is important since it unifies the class of NP-complete problems in a way that was unappreciated prior to 1971.

Similarly, advances have been made concerning the extent and nature of arithmetical incompleteness. Chief among these are the 1981 independence results of Harvey Friedman. Friedman's contributions include the discovery of a series of mathematically natural propositions (concerning Borel functions of several variables) that are undecidable, not just in *ZFC,* but in the much stronger system of *ZFC* together with the axiom of constructibility, *V=L,* as well. This is important not simply because such propositions are of a level of abstraction significantly lower than previous results, but because virtually all propositions previously thought to have been undecidable have been made decidable by the addition of the axiom of constructibility to *ZFC.* Importantly, Friedman has also shown that some of these propositions, although undecidable in *ZFC,* are decidable in a competing theory, Morse-Kelly class theory with choice. Such results are important since they indicate in what ways competing axiomatizations of set theory in fact differ.

The Expansion of Logic

As long ago as Galen (*c.* 129-*c.* 199) it was recognized that some sound arguments could not adequately be analysed in terms of either Aristotelian or Stoic logic. For example, neither the argument 'if Sophroniscus is father to Socrates, then Socrates is son to Sophroniscus', nor the argument 'if Theon has twice as much as Dio, and Philo twice as much as Theon, then Philo has four times as much as Dio' is provably valid in such systems. Thus, modern first-order logic can be viewed as a means of extending the ancient idea of formal validity, since it successfully displays many formally valid inferences which ancient logic fails to capture.

One way of understanding the advent of contemporary non-classical and informal logics is to view them in much the same way. Such logics regularly attempt to describe, in a systematic way, additional types of reliable inference not captured in classical first-order logic. They do so in two ways: first, *extensions* of classical logic attempt to exhibit reliable forms of inference in addition to those displayed in first-order logic much as first-order logic exhibits reliable forms of inference in addition to those displayed in ancient logic or in modern propositional logic. Second, *competitors* to classical logic advocate alternative ways of understanding the idea of valid inference itself, rejecting in one way or another the concept of validity as it is described in first-order logic.

Among the most philosophically interesting of the competitors to classical logic are intuitionistic logics, relevance logics and paraconsistent logics. Of these, intuitionistic logics were the first to appear. Motivated by the intuitionistic idea that satisfactory proofs must refer only to entities which can be successfully constructed or discovered, intuitionist logic requires that we find examples, or that we find algorithms for finding examples, of each object or set of objects referred to in a proof. Formalized by Heyting, intuitionistic logic therefore abandons those forms of classical proof (including indirect proof) which do not contain the appropriate constructions. A standard axiomatization consists of rules for substitution and detachment, together with the following axioms:

1. $p \rightarrow (pp)$
2. $(pq) \rightarrow (qp)$

3. $(P \to q) \to ((Pr) \to (qr))$
4. $((P \to q)\ (q \to r)) \to (p \to r)$
5. $(p\ (p \to q)) \to q$
6. $p \to (pq)$
7. $(pq) \to (qp)$
8. $((P \to r)\ (q \to r)) \to ((pq) \to r)$
9. $\neg P \to (p \to q)$
10. $((p \to q)(p \to \neg q)) \to \neg p.$

Like intuitionistic logic, relevance logic is a competitor to classical logic which emphasizes a non-classical consequence relation. Like intuitionistic logic, too, relevance logic results from a dissatisfaction with the classical consequence relation. Developed by Alan Ross Anderson (1925-73), Nuel Belnap (b. 1930) and others, the logic stresses entailments which involve connections of relevance between premisses and conclusions, rather than simple classical derivability conditions. It is intended that the relevance consequence relation thereby avoids both the paradoxes of material implication and the paradoxes of strict implication. (The former involve the unintuitive but, strictly speaking, non-contradictory results to the effect that whenever the antecedent is false or the consequent is true in a material implication, the resulting implication will be true, regardless of content; the latter involve the unintuitive, but likewise non-contradictory, results that a necessary proposition is strictly implied by any proposition and that an impossible proposition strictly implies all propositions, regardless of content. Following Lewis's 1912 definition, one sentence, *p*, is said to strictly imply a second sentence, *q*, if and only if it is not possible that both *p* and ~*q*.)

As it is normally formalized, relevance logic turns out to be a type of paraconsistent logic. Such logics tolerate, but do not encourage, inconsistencies. They do so in the sense that a contradiction (the joint assertion of a proposition and its denial) may be contained within the system; at the same time they are consistent in the sense that not every well-formed formula is a theorem.

Other competing logics include combinatory logic (a variable-free branch of logic which contains functions capable of playing the role of variables in ordinary logic); free logics (logics in which it is not assumed that names successfully refer, hence logics which do not make the kind of existence assumptions normally associated with classical logic); many-valued logics (logics which countenance more than the two possible classical truth values, truth and falsity; historically, such logics have been motivated by the problem of future contingents, the problem first raised by Aristotle but popularized by £ukasiewicz of determining whether contingent statements concerning future states have truth values prior to the time to which they refer); and quantum logics (logics designed to take account of the unusual entailment relations between propositions in theories of contemporary quantum physics; hence, a logic in which the law of distributivity fails).

In contrast to the above logics, extensions of first-order logic typically have as their goal the construction of a broader, more inclusive type of consequence relation than that found in classical logic. Formal extensions, such as modal extensions of both prepositional and predicate logic, do so by expanding the concept of formal entailment to include a class of formally valid arguments in addition to those of first-order logic. In contrast, informal extensions do so by expanding the concept of validity to include informal (or material) validity in addition to formal validity. Finally, inductive or non-monotonic extensions do so by expanding the concept of consequence to include, not just entailments, but implications, corroborations, and confirmations as well. Thus it is that logic (in the broad sense) has come to include, not just theories of formal entailment relations, but probability theory, confirmation theory, decision theory, game theory and theories of epistemic modelling as well.

It may be obtained from classical prepositional or predicate logic by the addition of axioms and rules of inference governing operators such as *K* and *B* in *'Kp'* ('it is known that *p*') and *'Bp'* ('it is believed that *p*'). Similarly, deontic logics (logics emphasizing inferential relations and entailments which result from deontic properties of sentences) may be obtained by the addition of axioms and rules of inference governing operators such as *O* and *P* in *'Op'* ('it ought to be the case that *p*') and *'Pp'* ('it is permissible that *p*').

Other extensions include counterfactual logics (logics which are primarily concerned with conditional sentences containing false antecedents); erotetic or interrogative logics (logics emphasizing inferential relations and entailments pertaining to questions and answers); fuzzy logics (logics concerned with imprecise information, such as information conveyed through vague predicates or information associated with so-called fuzzy sets, sets in which membership is a matter of degree); imperative logics (logics emphasizing inferential relations and entailments which result from imperatives); mereology (the formal study of inferences and entailments which result from the relationship of whole and part); theories of multigrade connectives (logics whose connectives fail to take a fixed number of arguments); plurality, pleonotetic or plurative logics (logics emphasizing inferential relations and entailments pertaining to relations of quantity and involving plurality quantifiers such as 'most' and 'few'); preference logics (logics emphasizing inferential relations and entailments which result from preferences); second-order and higher-order logics (logics carried out in higher-order languages in which quantifiers and functions are allowed to range over properties and functions as well as over individual i.e. individuals); and tense or temporal logics (logics designed to be sensitive to the tense of sentences and to the changing truth values of sentences over time).

In contrast to the above logics, informal logic is the study of arguments whose validity (or inductive strength) depends upon the material content, rather than the form or structure, of their component statements or propositions. (The logical form of a sentence or argument is obtained by making explicit the expression's logical constants and then by substituting free variables for its non-logical constants; logical form is thus typically contrasted with the material content–or subject matter–of the non-logical constants for which the free variables are substituted.) Such logics are extensions of formal logic in that they recognize the significance of formal validity, but also recognize the existence of valid arguments which are not instances of valid argument forms. Thus the argument from 'If Hume is a male parent then Hume is a father' and 'Hume is a father' to 'Hume is a male parent', although valid, is not formally so. In fact, far from being formally valid, it is an instance of the invalid argument form of affirming the consequent.

Since all so-called invalid argument forms have valid arguments as uniform substitution instances, the claim is made that formal characterizations alone will never be able to capture completely the concepts of either validity or invalidity.

Finally, mention should be made of formal systems which expand the traditional consequence relation by weakening it to a relation which is less than that of a valid inference. Such systems include inductive and non-monotonic logics, as well as theories of probability and confirmation. All such systems are concerned with so-called ampliative arguments, arguments whose conclusions in some important sense go beyond the information contained in their premisses. Such arguments are defeasible in the sense that their premisses fail to provide conclusive evidence for their conclusions and, hence, allow for the later overturning or revision of a conclusion. Ampliative arguments include both inductive inferences and inferences to the best explanation. They may be either acceptable or unacceptable depending upon their (inductive or probabilistic) strength or weakness. Similarly, confirmation theories evaluate the degree to which evidence supports (or confirms) a given hypothesis, emphasizing the rational degree of confidence that a cognitive agent should have in favour of a hypothesis given some body of evidence.

Such theories are typically (but not always) based upon probability theory, the mathematical theory of the acceptability of a statement or proposition, or of its likelihood. The standard account, first axiomatized in 1933 by Andrej Kolmogorov (b. 1903) [1.160], can be summarized as follows:

Given sentences *s* and *t*, probability is a real-valued function, *Pr,* such that

1. *Pr(s)*=0,
2. *Pr(t)*=1, if *t* is a tautology
3. *Pr(st)*=*Pr(s)*+*Pr(t)*, provided that *s* and *t* are mutually exclusive (i.e. ~(*s* & *t*))
4. *Pr*(*s*¦*t*)=*Pr(s* & *t)*/ *Pr(s)*, provided that *Pr(s)*?0.

Default logics, which permit the acceptance or rejection of certain types of default inferences in the absence of information to the contrary, provide one type of non-monotonic alternative to probabilistic theories.

Defeasible theories often give rise to so-called 'applied logics', including theories of belief revision and theories of practical rationality. Theories of belief revision (for example in [1.123]) are typically designed in such a way as to model changes in one's belief set which come about both as a result of the acceptance of new beliefs and the revision of old beliefs.

Theories of practical rationality (for example in [1.159]) typically include both decision theory (the theory of choice selection under various conditions of risk and uncertainty, given that each option has associated with it an expected probability distribution of outcomes, gains and losses), and game theory (the theory of choice selection by two or more agents or players when the outcome is a function, not just of one's own choice or strategy, but of the choices or strategies of other agents as well). Such theories may be either bounded or not, depending upon whether they take account of possible cognitive limitations of the decision-makers.

Philosophical logic deals with formal descriptions of natural language. Most philosophers assume that the bulk of "normal" proper reasoning can be captured by logic, if one can find the right method for translating ordinary language into that logic. Philosophical logic is essentially a continuation of the traditional discipline that was called "Logic" before the invention of mathematical logic. Philosophical logic has a much greater concern with the connection between natural language and logic. As a result, philosophical logicians have contributed a great deal to the development of non-standard logics as well as various extensions of classical logic (e.g., modal logics), and non-standard semantics for such logics.

Logic and the philosophy of language are closely related. Philosophy of language has to do with the study of how our language engages and interacts with our thinking. Logic has an immediate impact on other areas of study. Studying logic and the relationship between logic and ordinary speech can help a person better structure his own arguments and critique the arguments of others. Many popular arguments are filled with errors because so many people are untrained in logic and unaware of how to formulate an argument correctly.

9

Logical Positivism and Philosophy of Twentieth Century

Introduction

The 'death' of the movement was due, however, not only to the dispersal of its members, but also to a widespread recognition of the defects of its ideas. Now in this sense, probably most of the philosophy studied in our universities is dead, for most of it is open to more-or-less fatal criticisms; and criticism is regarded as one of the main approaches to the great philosophers and movements of the past. However, what hastened the widespread rejection of logical positivism was not merely the (unsurprising) discovery that its doctrines were open to criticism, but the aggressive and even arrogant way in which those doctrines were propounded to the world. Chief among these was the 'elimination of metaphysics'. It was claimed by members of the movement that they had noticed something about existing and traditional philosophy, which would completely overturn it and render it largely otiose. There appeared articles with such titles as 'The Elimination of Metaphysics through the Logical Analysis of Language' and

'The Turning Point in Philosophy'. Carnap posed the question: 'Can it be that so many men, of various times and nations, outstanding minds among them, have devoted so much effort, and indeed fervour, to metaphysics, when this consists of nothing more than words strung together without sense?' International conferences were called with a view to disseminating the new 'insights', and a grandiose project, *The Encyclopedia of Unified Science,* was launched to give definitive expression to the new 'scientific' era in which philosophical and other discourse would become part of the discourse of science. In these circumstances it was not surprising that critics of the new ideas were more than usually prompt, forthright and thorough in their criticisms.

Nevertheless, logical positivism has an established place in the history and continuing development of philosophy. At least three reasons might be given for this. One is purely historical, regarding the considerable impact and influence of the movement in its heyday. A second lies in the intrinsic interest of its ideas, which I hope to bring out in what follows. A third lies in the fact that even if no one today would call himself a logical positivist, some of its main positions, such as verificationism, and emotivism in ethics, are still referred to as parameters within which discussions of particular topics, such as ethics or the philosophy of religion or of science, are to be conducted. Again, it can be argued that even if the parent plant is dead, many of its seeds are alive and active in one form or another. In an interview in 1979, A.J. Ayer, a leading philosopher of our time, who had been an advocate of logical positivism in the 1930s, was asked what he now saw as its main defects. He replied: 'I suppose the most important was that nearly all of it was false'. Yet this did not prevent him from admitting, shortly afterwards that he still believed in 'the same general approach'.

In a number of ways 'the same general approach' is still widespread today, and indeed was so long before the advent of logical positivism. Empiricism, in one sense or another, is a major thread running through Western philosophy since the seventeenth century, including logical positivism and much of the philosophy of today. The same is true of 'reductionism', and especially the assumption that mental phenomena can be reduced, in some sense, to the vocabulary of the material or physical. Another idea, which was central to logical positivism and remains of central importance today, is that philosophical questions are largely questions of language, and that theories of meaning are therefore of central importance.

The movement originated in the 1920s among philosophers and scientists of the 'Vienna Circle', under the leadership of Moritz Schlick, professor at the University of Vienna. After some years of writing and discussion, the Circle organized its first international congress in 1929, attracting sympathizers from many countries. In 1930 it took over a journal, renamed *Erkenntnis,* for the publication of its ideas. The British philosopher A.J. Ayer attended its meetings in 1933 and made its ideas widely known in the English-speaking world with the publication of *Language, Truth and Logic* in 1936.

By that time the dispersal of the Circle was already under way, a number of members having emigrated, mainly to the United States and Britain. But there were also fundamental intellectual differences within the Circle.

An important influence was that of Wittgenstein, though he was not a member of the Circle. Having retired from philosophy after the publication of his *Tractatus Logico-Philosophicus* in 1922, he was coaxed back into the subject by Schlick in 1927 and had regular meetings with Schlick and Waismann, another member of the Circle. But from 1929 he declined to meet with other members of the Circle, whose views he found unsympathetic–a sentiment that was reciprocated by Otto Neurath, for one. Nevertheless, the *Tractatus* was regarded by the Circle as a classic statement of the new outlook in philosophy, and the work was read out and discussed sentence by sentence in the period 1924-6. It was, moreover, Wittgenstein who first formulated the 'verification principle' by which the new philosophy became known. Nevertheless it was he who made the most decisive break from these ideas when embarking on his 'later' philosophy in the early thirties.

The philosophy of the Circle became known as 'logical positivism' or 'logical empiricism'. The former name is more usual, but the latter, preferred by Schlick, seems to me to be more appropriate. It has the advantage of indicating the affinity of the Circle's ideas with those of the empiricist tradition begun by Locke in the seventeenth century, and later represented by such thinkers as Mill and Russell. It is also readily connected with the Circle interest in empirical science. Hence, although this article is entitled 'logical positivism', I shall prefer to use the term 'logical empiricism'.

The term 'logical' indicates a primary interest in language and meaning, as opposed to knowledge. The main questions for such philosophers as Locke and Descartes had been about the sources and extent of knowledge, and the empiricist Locke claimed that sense experience was the only source. In the new empiricism, by contrast, the primary question was not 'How do we know that *p*?', but 'What does "p" mean?'

The new approach may be illustrated by the problem of 'other minds'. How can I know that other people really have thoughts and feelings, when I can only observe their bodily movements and the sounds they utter? In the new philosophy this problem of knowledge is transformed into one about meaning. What does it *mean* to say that another person has such and such a feeling? According to the new philosophy, it can mean *no more* than what is observable. Any statement about feelings as distinct from what is observable will be meaningless. 'It is not false, be it noted, but meaningless: we have no idea what it is supposed to signify'.

The logical empiricists recognized two, and only two, kinds of meaningful statements. They are, first and mainly, empirical statements, verifiable by observation. Second, there are statements, such as those of logic and mathematics, whose truth can be known a priori; but these were regarded as not presenting 'new' knowledge, but merely an analysis of what was known already. Any other statements were to be dismissed as meaningless 'pseudo-statements'.

The Verification Principle

'The meaning of a proposition is the method of its verification.' So ran the principle as formulated by Wittgenstein and Schlick. (It was first formulated by Wittgenstein, but its most frequent use occurs in the writings of Schlick.)

In this sentence we have an answer to a long-standing question of philosophy, namely, What does meaning–the kind of meaning that language possesses–consist in? This question has sometimes been answered in terms of words and sometimes in terms of sentences or speech-acts. Locke answered it by reference to mental entities corresponding to words, claiming that 'words signify nothing but the Ideas that are in the mind of the speaker'. Wittgenstein, in the *Tractatus,* had postulated 'names' as the fundamental units of meaning, these 'names' being correlated with fundamental 'objects' in the world; with a further 'picturing' relation between propositions and corresponding 'states of affairs' in the world. The verificationist answer, as I have said, was in terms of the method of verification of a given proposition. Here was an apparently simple principle which could provide a focus for both adherents and critics of the new movement.

As well as providing an answer to the question 'What is meaning?', the principle was intended to provide a criterion to distinguish what is meaningful from what is not. Thus, if there is *no* method of verification–no way of verifying the proposition–then it must be meaningless.

It seems obvious that there is something right about the verification principle; that there is, at least, an important connection between meaning and verification. Thus, if we are unsure what someone means by his words, we can often find out by asking how one would verify what he said. And sometimes, at least, the admission that there is no conceivable method of verification will lead us to conclude that what was said is meaningless.

On the other hand, there are a number of difficulties with the principle, which may be divided into three, corresponding to the terms 'proposition', 'is' and 'method of verification'.

The original German word for 'proposition' is *Satz,* and the straightforward translation of this is 'sentence'. There is a difficulty, however, about treating sentences as objects of verification. Such a sentence as 'It is raining' cannot be regarded as true or false in itself. It is only when someone *uses* the sentence on a particular occasion that *what he says* is true or false. In answer to this and other difficulties philosophers have used the term 'proposition' to mean, roughly, what is asserted by means of a declarative sentence. It is, according to this usage, propositions that are true or false, and not the sentences by means of which they are asserted.

But now another difficulty may arise. 'Proposition' is sometimes defined as meaning an entity that is necessarily true or false. But if this is so, then the question of meaningfulness has already been decided in using the term 'proposition', for only what is meaningful can be described as true or false. In other words, it would be self-contradictory to speak of a meaningless proposition.

One way of overcoming this would be to put 'putative' before 'proposition'; another, which I shall adopt, is to use the word 'statement'. In ordinary English we can ask of a statement, made by someone, both whether it means anything and, if so, whether it is true. Some philosophers have also defined 'statement' to mean what is necessarily true or false (and hence meaningful), but there is no need for a verificationist to follow this usage. In this article I shall prefer the word 'statement', but will sometimes follow the writers under discussion in using 'sentence' or 'proposition', as the case may be. This seems the least confusing way of proceeding.

My next difficulty was about the word 'is', in the claim that meaning 'is' a method of verification. How are we to understand this identification? 'Meaning' and 'method' are concepts of different types. 'One can sensibly talk about *using* a method, but [not] "using a meaning"'. A method may be easy or difficult to carry out, it may take a long or a short time, etc.; but these things cannot be said about the meaning of a statement.

Nevertheless, it was thought essential to account for the meaning of verbal expressions by reference to something other than verbal expressions. Otherwise what would be the connection between language and reality? This need was expressed as follows by Schlick:

> *In order to arrive at the meaning of a sentence or proposition we must go beyond propositions. For we cannot hope to explain the meaning of a proposition merely by presenting another proposition. I could always go on asking 'But what does this new proposition mean?'. The discovery of the meaning of any proposition must ultimately be achieved by some act, some immediate procedure.*

A similar thought was sometimes expressed with reference to words as distinct from propositions, and here the notion of 'ostensive definition' (as distinct from verification) was invoked, as when we point to an object to explain the meaning of a corresponding word, such as the word 'red'. Wittgenstein expressed this thought as follows: 'The verbal definition, as it takes us from one verbal expression to another, in a sense gets us no further. In the ostensive definition however we seem to take a much more real step towards learning the meaning'. This passage, however, was written after Wittgenstein's break from verificationism, and in the ensuing pages he argued that such appeals to 'reality' as distinct from language could never supply the desired detatchment from language. He imagined someone trying to explain the word 'tove' by pointing to a pencil and pointed out that, in the absence of all verbal information, this act might be taken to mean all sorts of different things.

This brings us to the third difficulty; about 'method of verification'. What would be the relevant method in the case of, say, the statement 'It is raining'? I might verify this by putting my hand out of the window. But this act might serve to verify all sorts of statements; and, on the other hand, all sorts of methods might be used to verify the statement.

I began with a difficulty about identifying meaning with a method, as indicated by the word 'is' in the verification principle. Suppose now that we removed this word and spoke instead of a 'correspondence' between meaning and method. This would still leave us with the difficulty just mentioned. It is hard to see how Schlick's requirement of 'going beyond propositions'–breaking out of the circle of language–could ever be satisfied.

The Criterion of Verifiability

Not every verificationist was concerned, or mainly concerned, about the question of what meaning consists in. One of the main aims of the movement, as I said, was to distinguish what is meaningful from what is meaningless, with the special aim of showing statements of metaphysics to belong to the latter class. Now such a criterion can easily be deduced from the verification principle. If the meaning of a statement is the method of its verification, then it will follow that if it lacks such a method–if it is not verifiable-then it will, likewise, lack meaning. It is possible, however, to advocate this criterion independently, without deduction from the verification principle. Thus, one might claim that unverifiable statements are meaningless, without putting forward an account of what meaning consists in; and this was the position of A.J. Ayer. He expressed the *criterion of verifiability,* as I shall call it, as follows:

> *We say that a sentence is factually significant to any person if, and only if, he knows how to verify the proposition that it purports to express-that is, if he knows what observations would lead him, under certain conditions, to accept the proposition as being true, or reject is as being false.*

It should be noted that the word 'verify' is used here in the sense of 'verify whether...' and not 'verify that...' The latter would presuppose that the proposition in question is *true.* But a proposition that is known to be true is, by the same token, known to be meaningful; so that the criterion would be redundant. The relevant sense of 'verify' is that in which this is *not* known, so that, as Ayer implies, we do not yet know whether the proposition will turn out true or false. Moreover, either of these results would satisfy the criterion:

what is at stake is not the truth of the proposition, but whether it has meaning. Unverifiable propositions, being meaningless according to the criterion, would be neither true nor false.

By means of such a criterion it was hoped to proceed immediately to the 'elimination of metaphysics', without getting involved in questions about what meaning consists in. The discovery that the propositions in question are meaningless would explain why philosophers who had wrestled with them through the ages had never, apparently, succeeded in getting anywhere, while, at the same time, providing the key to a resolution of their problems.

It soon appeared, however, that the criterion was beset with difficulties. First, there is simply the question of acceptance. In a broadcast debate with F.C. Copleston, A.J. Ayer introduced the word 'drogulus' to stand for 'a disembodied thing' whose presence could not be verified in any way. He put it to Copleston: 'Does that make sense?' But Copleston replied that it did make sense. He claimed that he could form an idea of such a thing and that this was enough to give it meaning.

Second, it proved difficult to formulate the criterion in such a way as to yield the desired results–to exclude the statements of metaphysics while admitting those of 'science' (including everyday empirical statements). The formulation I quoted from Ayer was too weak, because all sorts of observations might 'lead' someone to regard a proposition as true or as false. In a further formulation he introduced the more rigorous notion of 'deduction'. A statement is meaningful, he held, if 'some experiential propositions can be deduced from it', the latter being defined as propositions 'which record an actual or possible observation'. A typical example would be ' This is white', which might result from the fact that it is snowing. But while it is clear that this fact and this statement are related, the relation does not seem to be one of deduction. There is, for example, no logical deduction from the statement 'It is snowing' and, say, the fact that I am looking out of the window, to the conclusion that 'this is white', or the conclusion that I am seeing something white.

Suppose, however, that this difficulty could be resolved. In that case 'It is snowing' would be vindicated because 'This is white' is deducible from it. But clearly there is more than that to the meaning of 'It is snowing'. And might not the remainder be merely pseudomeaning, for all that Ayer's criterion has shown? This difficulty was illustrated in a striking way by Carl Hempel, who supposed that some straightforward empirical statement, such as 'It is snowing', had been *conjoined* with a piece of 'metaphysical nonsense', such as 'The absolute is perfect'. This conjunction would yield the same deductions as the empirical component by itself; so that the conjunction as a whole would have to be declared meaningful.

Various formulations were attempted by Ayer and others to escape these and other difficulties, but it seems that what is required is not merely deduction, but *analysis*. There must be a way of showing that the *whole* meaning of a statement is, somehow, accounted for by observations and the corresponding observation statements. Moreover, as we shall see, this affects all kinds of ordinary statements and not just the rather fanciful example constructed by Hempel.

Analysis

According to Waismann, ordinary empirical statements were to be analysed into 'elementary propositions', whose whole meaning would consist in corresponding verificatory experiences.

To analyse a proposition means to consider how it is to be verified. Language *touches* reality with elementary propositions. It is clear that assertions about bodies (tables, chairs) are not elementary propositions. What elementary propositions describe are: phenomena (experiences).

There is a difference between this and verification in the ordinary sense. The latter is an activity of some kind, and hence it made sense to speak of a *method* of verification. But what seems to be meant in the passage just quoted is that elementary propositions are to be verified by *having* the corresponding 'experiences', as distinct from any activity.

But how is such analysis to proceed? From the true statement that there is a table in my room, it would not follow that anyone is having a relevant experience, since there may not be anyone in the room. The view adopted, known as 'phenomenalism', allowed for this possibility. According to it, a crucial role is played by *hypothetical* statements, such as 'If someone were in the room, he would have such and such experiences'. But in what sense are such statements entailed by the statement under scrutiny? Even if I am in the room, and endowed with normal eyesight, I may fail to see the table. 'You can't miss it' is notoriously unreliable.

There was also a difficulty about entailment in the other direction. From the fact that I am having the experience of seeing something brown, etc., it would not follow that there is a table in the room; and neither would it follow from my having the experience of seeing *a table,* for I might have these and other experiences in the course of a dream or hallucination. Just how far this kind of scepticism may be taken is a matter for debate, but the sceptical view is encouraged by the empiricist reliance on 'experience', conceived as something that occurs in us, is 'imprinted by the senses', etc. This problem has been recognized from the beginnings of empiricism, and in the case of logical empiricism it led to the view that statements about tables and chairs are not 'conclusively' verifiable. Wittgenstein spoke of them as 'hypotheses' which could not be 'definitively verified'.

A similar development took place with regard to general statements, such as 'All men are mortal'. The 'all men' in this statement is not analysable into any finite conjunction of names, and the truth of the statement would not follow from any finite number of verificatory experiences. The same is true of scientific laws, such as 'Water expands below 4°C', whose meaning is not confined to any finite number of observations.

The discovery that the meaning of scientific laws, in particular, went beyond any finite verification, was especially serious for a philosophy which regarded scientific statements as paradigms of meaningful discourse. One solution, advocated by Schlick, was to deny that a scientific law is a statement: it is really, he maintained, 'an instruction for the forming of statements'. A genuine statement must be 'conclusively verifiable', and this would be true only of the particular experiential statements which would be produced under that 'instruction'. A statement, he insisted, 'has a meaning only in so far as it can be verified; it only *signifies* what is verified and absolutely *nothing* beyond this'; there cannot be a 'surplus of meaning' beyond that.

Another approach was used to deal with statements whose verification is impossible for technical reasons. Consider a statement about the far side of the moon. When Schlick and Ayer considered this example, verification was impossible and, for all they knew, might always remain so. The same is true, for us, about speculations concerning life on other planets. But it would seem absurd to claim that whether such questions have meaning depends on the present technology of space exploration. The answer was to describe the relevant statements as 'verifiable in principle'. 'I know what observations would decide it for me, if, as is theoretically conceivable, I were once in a position to make them'.

But what should we say about scientific statements or theories whose meaning seems to go far beyond their verificatory content, even 'in principle'? Consider the statement that the universe is expanding, and assume that it is based on the observation of a 'red shift' in the light emitted from remote galaxies. It seems clear that the statement is not merely about red shifts. Yet such a 'reduction' of meaning seems to be required by the verificationist analysis. In this connection verificationists enlisted P.W. Bridgman's idea of 'operationism'. On this view, the meaning of statements about distant parts of the universe, for example, would indeed correspond to the relevant scientific 'operations';

there would be no more to it than that. This meant, as Bridgman pointed out, that ordinary words might change their meaning when used in a scientific context. The meaning of 'length', he claimed, 'has changed completely in character' in the context of astronomy. 'Strictly speaking, length when measured by light beams should be called by another name, since the operations are different'.

A further difficulty arose about the analysis of statements about the past. The statement 'It rained yesterday' might be verified, in the ordinary sense, by present evidence including, perhaps, asking other people. But the statement obviously does not *mean* any of these present things. Ayer offered a variety of analyses in attempting to meet the difficulty, including the bold claim that the tense of a statement is not part of its meaning, so that there would be no difference of meaning between 'George VI was crowned in 1937', 'George VI is being crowned in 1937', and 'George VI will be crowned in 1937'.

The Elimination of Experience

In one way or another, the analysis of different kinds of statements would lead to the ultimate 'elementary propositions', variously described as 'experiental propositions', 'observation-statements' etc., whereby, as Waismann put it, 'language *touches* reality'. At this stage the speaker or hearer passes from linguistic activity to the occurrence of a suitable experience or sensation which is supposed to give meaning to the words. But here arose a problem which produced a serious split in the ranks of the Circle. Experience and sensation are personal and in some sense private; must not the same be true, then, of meaning? On this view, the sentence 'I am thirsty', as Carnap argued, 'though composed of the same sounds, would have different senses when uttered by [different people]'. But what, in that case, becomes of the claims of science, and of language itself, to be communal activities? There was, thought Carnap, a ready way of disposing of such awkward questions. 'These pseudo-questions', he declared, 'are automatically eliminated by using the formal mode.' By the 'formal mode' he meant a discourse that confined itself to statements and did not try to go beyond these, to what he called 'the material mode'. There was, he held, no need to talk about 'the content of experience', 'sensations of colour' and the like; we should instead refer to the corresponding *statements,* which he called 'protocol statements'. These, and not the corresponding experiences, would occupy the fundamental role in the system–that of 'needing no justification and serving as the foundation for all the remaining statements of science'.

Carnap was uncertain about the form that these protocol statements should take, proposing such expressions as 'Joy now', 'Here now blue' and 'A red cube is on the table'. But Otto Neurath argued that such expressions could not be fitted into the system of science, unless the reference of 'now' and 'here', and the identity of the speaker, were known to others. He gave the following as a suitable example: 'Otto's protocol at 3:17 o'clock. This example, however, is still not sufficiently purged of personal elements. We who read it today would not know where in the system to place '3:17' or 'the room'; and the same would be true of 'Otto', were it not for independent knowledge of who Otto was. But perhaps Neurath was indicating the way towards a still more complex kind of statement, which would be wholly independent of reflexive reference.

Now it is clear that Neurath's example is meant as a protocol statement, because the term 'protocol' is used in it. But, now that the connection with experience has been cut, what entitles it to this of knowledge. But in the new *logical* empiricism the matter could not be passed over so lightly.

A further development, beyond the scope of this article, was the celebrated 'private language' argument of Wittgenstein's later work, in which he made a decisive break from empiricist ideas about language, arguing that the alleged 'private' meaning would not be meaning, in the required sense, at all.

The Unity of Science

Another common feature that the new empiricism shared with the old was its 'reductionism'. Locke, for example, had insisted that the various kinds and aspects of knowledge could all be reduced to a single type and source, namely the 'sensations' with which our sense organs furnish us in 'experience'. In the case of the new empiricism a similar but linguistic reductionism led to a grandiose project known as the *International Encyclopedia of Unified Science.* In this work it was hoped to show that all the different sciences, including the physical, biological and human sciences, could be expressed in a fundamental common vocabulary. Carnap's proposal for this purpose was what he called the 'thing-language'–that which we use 'in speaking about properties of the observable (inorganic) things surrounding us': such words as 'hot', 'cold', 'heavy', 'light', 'red', 'small', 'thick' etc.

One aspect of the 'unity of science' that is of particular interest is its application to human beings. It was thought that descriptions of human feelings, for example, could be reduced to statements about observed behaviour ('behaviourism'), or other physical occurrences such as those in the brain. (Here we see the beginnings of the 'physicalism' which, in various forms, is prominent in the philosophical literature of today.) The difficulties of behaviourism can be brought out by their effect on Wittgenstein in his 1930–33 lectures as recorded by G.E. Moore. 'When we say "He has toothache"', asked Wittgenstein, 'is it correct to say that his toothache is only his behaviour, whereas when I talk about my toothache I am not talking about my behaviour?' This cannot be so, because 'when we pity a man for having toothache, we are not pitying him for putting his hand to his cheek'. Again, 'is another person's toothache "toothache" in the same sense as mine?'. He now saw that according to the verification principle, the meanings, following the difference in methods of verification, must indeed be utterly different. Indeed, the difference was not merely *between* methods of verification, since there is no verification at all in case of the first person:

> *'there is no such thing as verification for "I have", since the question "How do you know you have toothache?" is nonsensical'.*

Carnap tried to accommodate such difficulties in a number of writings. In one of these he admitted that a person N1, 'can confirm more directly than N2 a sentence concerning N1's feelings, thoughts etc.'; but, he went on, 'we now believe, on the basis of physicalism, that the difference is only a matter of degree'.

The 'Elimination of Metaphysics'

One of the main objectives of logical empiricism was to provide a way of demarcating the meaningful statements of science and ordinary life from the 'pseudo-statements' of metaphysics. Now the word 'metaphysics' may mean various things. The verificationists used such examples as Heidegger's statement 'The nothing nothings', and F.H. Bradley's talk about 'the Absolute', as in the statement: 'The Absolute enters into, but is itself incapable of, evolution and progress'. This, said Ayer, was 'a remark [he had] taken at random' from Bradley's *Apperance and Reality.* It was, he claimed, 'not even in principle verifiable', and therefore nothing more than a 'metaphysical pseudo-proposition'.

Now such a sentence, plucked 'at random' out of its context, might well strike the reader as meaningless. But is this so because it is unverifiable? Perhaps, if we read Bradley's argument, we would find there the means of assessing the truth of his statement. *That* would be the appropriate method of verification in this case. In taking the statement out of its context, Ayer had merely denied us access to the relevant method of verification.

Perhaps what Ayer had in mind was that the method of verification would not be *empirical.* Now this might well be true; but what would it show? It might show that the statement itself is not empirical; but perhaps it was never intended to be so. Merely to classify it as non-empirical is not to show that it is a 'pseudo-proposition', nor even that it is unverifiable.

Another class of statements to which verificationists turned their attention were those about God. Such statements, it was argued, were not necessarily meaningless, but the meaning ascribed to them should not exceed their verificatory content. Carnap spoke of an early phase of the concept of God, in which He was conceived as a corporeal being dwelling, say, on Mount Olympus. Such statements would satisfy the verificationist criterion, but this would not be true of the metaphysical accretions of later phases of the concept. Ayer put the matter thus:

> *If the sentence 'God exists' entails no more than that certain types of phenomena occur in certain sequences, then to assert the existence of a god will be simply equivalent to asserting that there is the requisite regularity in nature.*

Here again the requirement is that of sense experience, of the observation of phenomena by means of the senses. But is this the only kind of experience? In a further passage Ayer spoke of 'mystical intuition'. He would not, he said, deny that 'the mystic might be able to discover truths by his own special methods'. But, he went on, the mystic's statements, like others, 'must be subject to the test of actual experience'. But is not the mystic's experience itself a kind of 'actual experience'? Some further argument would be needed to show that such experiences cannot count as verificatory.

This difficulty is part of a fundamental problem about the whole empiricist programme. How is their preference for empirical statements, and empirical methods of verification, itself to be justified? John Locke, the father of empiricism, posed the question: 'Whence has [the mind] all the materials of reason and knowledge?', to which he replied: ' in one word, from *experience:* in that, all our knowledge is founded; and from that it ultimately derives itself. But if this were so, how could this knowledge itself have been obtained? The claim that all knowledge comes from experience cannot itself be derived from experience.

A similar difficulty arises if we turn the verification

principle or the criterion of verifiability on themselves. They are not themselves empirical statements: must they not suffer the same fate as other non-empirical statements? As Bradley observed, 'the man who is ready to prove that metaphysics is wholly impossible is a brother metaphysician with a rival theory of first principles'.

This difficulty was recognized by verificationists, who made various proposals to overcome it. Schlick claimed that the verification principle was 'nothing but a simple statement of the way in which meaning is *actually* assigned to propositions, both in everyday life and science'; while Ayer said that his criterion of verifiability was to be regarded 'not as an empirical hypothesis, but as a definition'. But what reason would there be for accepting this definition or Schlick's claim? As we have seen, they are not confirmed by the various ways in which the word 'meaning' is actually used.

Another idea was to describe the principle or the criterion as a 'proposal' or 'methodological principle'. This would exempt them from self-application, since a proposal cannot be described as true or false, verified or unverified. But what would it mean to adopt the proposal in question? According to it, I am to describe certain statements as meaningless. But how can I do that unless I *believe* them to be meaningless?

Leaving this difficulty aside, we may ask what would be gained if such a proposal were to be adopted. One of the motives behind it, as we have seen, was that of reductionism and the 'unity of science'. It was thought, and hoped, that the multifarious jungle of human discourse could all be reduced to a single type. Wittgenstein, in his later writings, spoke disparagingly of such aspirations as due to a 'craving for generality'. He now maintained that the uses of language–'language-games', as he called them–are irreducibly various, and that the philosopher's task was to notice and expound the differences, resisting any temptation to impose an artificial uniformity.

The Accommodation of Ethics

How are ethical statements to be accommodated under

verificationist criteria? Should they be accommodated at all? Carnap, at one stage, declared: 'we assign them to the realm of metaphysics'. But while it might be thought that metaphysical discourse can safely be set aside as unnecessary for the conduct of human life, this could hardly be so in the case of moral discourse. Could the latter be regarded, perhaps, as a kind of empirical discourse?

According to Schlick, this was the proper way of regarding it, as is clear from the first sentence of his book on the subject: 'If there are ethical questions which have meaning, and are therefore capable of being answered, then ethics is a science'. He went on to maintain that all of these conditions are fulfilled in the case of ethics. Words such as 'good', he claimed, are used to express desires; and these belong to the science of psychology. The 'proper task of ethics' was to examine the causal processes, social and psychological, which would explain why people have the desires they have.

But cannot something be described as morally good or desirable even if people do not desire it? According to Schlick, this would make no sense. 'If I assert that a thing is desirable simply in itself, I cannot say what I mean by this statement; it is not verifiable and is therefore meaningless'. There is no place in Schlick's account for the aspect of morality on which Kant laid so much emphasis: the conflict between desire and duty, which is such a familiar aspect of moral life. He rejected Kant's account of moral discourse, accusing him of being out of step with the ordinary meaning of 'I ought'. Yet it makes good sense for a person to say, for example, that he ought to do X because he promised, even though it is contrary to his desire.

A more common approach among logical empiricists towards moral statements was to deny that they are really statements. In a paper published in 1949, Ayer referred to his earlier view, 'which I still wish to hold, that what are called ethical statements are not really statements at all, that they are not descriptive of anything, that they cannot be either true or false'. This view, he now admitted, 'is in an obvious sense incorrect', since in ordinary English 'it is by no means improper' to speak of ethical statements as statements or descriptions, or to describe these as true or false. Nevertheless, he continued, 'when one considers how these ethical statements are actually used, it may be found that they function...very differently from other statements'. Yet, after all, 'if someone still wishes to say [they] are statements of fact, only it is a queer sort of fact, he is welcome to do so'. Here again is the craving for uniformity–a wish to deny that ethical facts are facts, because they do not conform to a preferred model.

To support his denial Ayer resorted to the existence of moral disagreement. 'Let us assume that two observers agree about all the circumstances of a case, but that they disagree in their evaluation of it.' In that case, he claimed, 'neither of them is contradicting himself. Now in such a case we might indeed conclude that there is 'no fact of the matter'. But there are many other cases in which this is not so. If I have said I will do you a favour, then I *would* be contradicting myself if I denied responsibility for doing what I said. In that case it would be true, and a fact, that I am under an obligation to do what I said. And, as Ayer recognized, the word 'true' is freely used in moral discourse, in various contexts.

As we have seen, Schlick regarded moral statements as factual and verifiable, while Carnap and Ayer tried to dispose of them in other ways. Another writer tried to analyse them into factual and non-factual components. This was the moral philosopher C.L. Stevenson, whose works were cited with approval by logical empiricists. According to the first of Stevenson's 'working models', the statement 'This is wrong' means 'I disapprove of this; do so as well'. (He dealt similarly with the words 'ought' and 'good'.) The first part of this, he pointed out, is amenable to verification, while the second, being an imperative, is not.

A difficulty with which Stevenson wrestled was about applying this account to someone who is *asking* himself whether X is wrong. This would not be a factual psychological question, about whether he does in fact approve of X; the question for him would be whether he *ought* to approve of it.

Another difficulty is that of making sense of the imperative 'do so as well'. A person can be requested to do something only if he can choose to do it; but this is not the case with approval. If you give me suitable reasons, I may *come to see* that X is wrong; but I cannot do so at will, in response to an imperative.

The connection with reasons was, however, denied by Stevenson. He admitted that 'a man's willingness to say that X is good, and hence to express his approval, will depend partly on his beliefs', but pointed out that 'his reasons do not "entail" his expression of approval'. Now this is true enough; indeed, it is not clear in what sense an expression can be 'entailed'. But it remains the case that on the basis of suitable reasons I may come to see (recognize, know it to be a fact) that X is wrong and ought not to be done. If such facts do not conform to the reductionist programme of logical empiricism, then it may be the programme that should be questioned, rather than the status of moral facts.

In the early 20th century, logical positivism a descendant of Comte's basic thesis but an independent movement sprang up in Vienna and grew to become one of the dominant schools in Anglo-American philosophy and the analytic tradition. Logical positivists (or 'neopositivists') reject metaphysical speculation and attempt to reduce statements and propositions to pure logic. Critiques of this approach by philosophers such as Karl Popper and Thomas Kuhn have been highly influential, and led to the development of postpositivism. In psychology, the positivist movement was influential in the development of behavioralism and operationalism. In economics, practising researchers tend to emulate the methodological assumptions of classical positivism, but only in a de-facto fashion: the majority of economists do not explicitly concern themselves with matters of epistemology. In jurisprudence, "legal positivism" essentially refers to the rejection of natural law.

In contemporary social science, strong accounts of positivism have long since fallen out of favour. Practitioners of positivism today acknowledge in far greater detail observer bias and structural limitations. Modern positivists generally eschew metaphysical concerns in favor of methodological debates concerning clarity, replicability, reliability and validity. This positivism is generally equated with "quantitative research" and thus carries no explicit theoretical or philosophical commitments. The institutionalization of this kind of sociology is often credited to Paul Lazarsfeld, who pioneered large-scale survey studies and developed statistical techniques for analyzing them. This approach lends itself to what Robert K. Merton called middle-range theory: abstract statements that generalize from segregated hypotheses and empirical regularities rather than starting with an abstract idea of a social whole. Other new movements, such as critical realism, have emerged to reconcile the overarching aims of social science with various so-called 'postmodern' critiques.

10

Cybernetics in Philosophy

Introduction

The need for a forum of this sort arose first in connection with problems being studied by Wiener and Bigelow in the design of mechanisms for controlling artillery directed against fast-moving aircraft, which Rosenbleuth saw to be similar to problems he had been studying in the erratic control of goal-directed motor behaviour in human patients. The topic of feedback processes on which the group subsequently focused attention thus arose from a comparative study of biological and artifactual control systems. Invariably associated with problems in the design of effective control systems are problems of communicating data to the system on which its corrective responses can be based, and of communicating these responses to the appropriate effector mechanisms. In the biological organism these communication functions are served by the afferent and the efferent nervous systems, respectively, parallel to the radar and the aiming mechanisms of the anti-aircraft battery. This joint emphasis upon communication and control systems accounts for the somewhat inelegant subtitle of

Wiener's seminal book: *Cybernetics, or Control and Communication in the Animal and the Machine.*

The name 'cybernetics', chosen by Wiener for this new field of study, derives from the Greek *kubernçtçs,* meaning steersman or pilot. Inasmuch as 'governor' derives (via the Latin *gubernare*) from the same root, cybernetics was provided with a ready-made technological ancestry, beginning with James Watt's invention in the late eighteenth century of devices (called 'governors') regulating the rotational speed of steam engines. Political antecedents can be traced back to Plato, who several times in the *Republic* and the *Statesman* likened the leader of a well-run political order to the *kubernçtçs* of a ship. A recognizably philosophic lineage also goes back to Plato, with his reference in the *Phaedrus* (247C7-8) to reason as the *kubernçtçs* of the soul.

Wiener's interest in technical problems of communication had been anticipated by H. Nyquist and R. Hartley, progenitors of the statistical concept of information. Despite Wiener's having numbered C. Shannon among his original group, contributions to this area after publication of the latter's article 'The Mathematical Theory of Communication' tended to be categorized under the title 'communication theory' (or 'information theory') rather than 'cybernetics'. Another emerging field of research in which cybernetics was initially involved, but soon suffered loss of name-recognition, was the theory of computing machinery. Although Wiener was a key contributor, along with V. Bush (MIT), H. Aiken (Harvard) and J.von Neumann (Princeton), to planning sessions leading to the construction of the first electronic digital computers, he remained more interested in possible neurological parallels with these mechanisms than in their logical design. Subsequent contributions to the field of digital computation owes relatively little to its broadly cybernetic origin.

Cybernetics' early emphasis upon functional parallels between biological and mechanical control systems soon led, in industrial circles, to its identification with factory automation and other forms of robotics. In academic circles, by contrast, cybernetics came to be associated with the then arcane field of artificial intelligence (AI), which took its start as part of an effort to reduce human involvement in the large scale computer-based air-defence systems under development at MIT's Lincoln Laboratory in the 1950s. Key contributors to AI at this early stage were O. Selfridge, a younger member of Wiener's original group, along with M. Minsky and S. Papert, all of MIT. Philosophers who became involved with AI through their association with MIT in the 1950s included H. Dreyfus, who was generally critical of the enterprise, and K. Sayre, who saw potential in AI as a new approach to traditional problems in the philosophy of mind. The first institutional centre for combined research in philosophy and AI was established by the latter in the early 1960s at the University of Notre Dame.

Despite its history of involvement with technological developments, the major spokesmen of cybernetics from Wiener onward have been explicitly concerned with its broader philosophic implications. The following discussion treats both its philosophic antecedents and its potential for further philosophic contributions. Remarks on the current interaction between AI and cybernetics are reserved for the final section below.

Basic Concepts

Philosophy before Plato was marked by a series of attempts to find a small set of basic principles in terms of which the manifest variety of the observable world could be understood as coherently integrated. Cybernetics returns to this task with a set of explanatory concepts based upon the presence of variety itself. Primary among these are the concepts of feedback, of entropy and of information.

Feedback

Any operating system functions within a variable environment with which it interacts through its input and output couplings. Feedback occurs whenever variation at the output works upon the environment in such a fashion as to produce a corresponding variation at the input of the system as well. Of primary concern to cybernetics are two kinds of feedback pertaining specifically to deviation from a stable state of the operating system. Positive feedback occurs when deviation from a stable state produces outputs that lead to yet further deviation. An example is an increase in membership of an interbreeding population which produces an even greater increase in subsequent generations. Feedback of this sort is labelled 'positive' because it tends to increase deviation from stability of the system in which it occurs. Negative feedback occurs, by contrast, when momentary deviation from a stable state leads to inputs that counteract further deviation. Common examples of negative feedback are the operations of a thermostat to maintain a steady temperature within an enclosure, and the subtle shifts in bodily posture by which a skier maintains balance on a downhill run.

Types of negative feedback may be further differentiated with respect to the bearing of the regulatory mechanisms that maintain a system in a stable mode of operation. Homeostasis is a type of feedback by which deviation from stability is counteracted by adjustments internal to the system itself. Among familiar forms of homeostasis in biological organisms are mechanisms regulating body temperature and chemical composition of the blood. Another type of negative feedback is exemplified by the heat-seeking missile that changes direction with a manoeuvring target, and by the daisy that aims its blossom to catch the full light of the sun. Feedback of this latter sort has been labelled 'heterotelic', for its role in adjusting the system's external relations to factors in its operating environment.

All organic and most mechanical systems incorporate variables that must be restricted to a narrow range of values if the system is to remain operational. A mammal will soon die, for example, if the oxygen content of its blood falls below a certain level, just as a reciprocating engine will soon freeze if it loses oil pressure. Negative feedback may be conceived generally as a type of regulatory restraint by which the values of a system's crucial variables are maintained within a range compatible with sustained operation. The central role of negative feedback in the economy of an operating system, to paraphrase Ashby, thus is to block the transmission of excessive variety to a system's protected variables. An important result of feedback regulation is to maintain the system at a state of low entropy relative to its operating environment.

Entropy

As originally defined by Clausius, entropy measured the proportion of total energy within an isolated system that is available for doing useful work. If part of a system is significantly hotter than another part (e.g., a steam chamber), then work can be done as heat passes from the hotter to the colder part (e.g., the pistons of a steam engine). If all parts of the system are at approximately the same temperature, however, the heat energy within the system is incapable of accomplishing useful work. According to the first law of thermodynamics, the total energy within a closed system remains unchanged through time. But as its energy available for work becomes expended through irreversible physical processes (e.g., discharge of steam into the pistons), its capacity to produce additional work progressively decreases. This means that the entropy of the system, in Clausius's sense, progressively increases. The second law of thermodynamics, originally stating that the energy available for useful work within a closed system tends to decrease with time, thus received concise restatement to the effect that the entropy within a closed system tends always to increase.

The concept of entropy was provided with a statistical basis through the work of Boltzmann and Planck, beginning with Planck's definition of 'complexion' as a specific configuration of its components on the microlevel of a physical system. Boltzmann developed techniques for quantifying the proportion of a system's possible complexions correlated with each of its distinguishable macrostates. Under the assumption that all complexions of a system are equiprobable, he then established the a priori probability of its existing in a given macrostate as equal to the proportion of complexions associated with that macrostate to the total number of possible complexions of the system overall. In this treatment, the greater the proportion of complexions associated with a given macrostate (i.e. the greater its probability), the less highly organized the system will be when existing in that macrostate, and the less capable accordingly of producing useful work. This enabled a redefinition of entropy in terms of probabilities. If P is the probability of a system's existing in a given macrostate, and k is the quantity known as Boltzmann's constant, then the entropy S of the system in that macrostate is given by the equation '$S=k \log P$'. (Logarithms were used in this function to make entropy additive.)

As a result of Boltzmann's treatment, an increase in entropy came to be understood not only as a decrease in energy available for useful work, but also as a decrease in organization (structure, order) of the system concerned. Yet another formulation of the second law of thermodynamics now became appropriate: closed systems tend to become configured into increasingly more probable macrostates, which is to say macrostates exhibiting increasingly less order. What this means in practical terms is familiar to anyone responsible for cleaning house, or for keeping weeds out of a vegetable garden.

Further development of the concept of entropy came with its extension into the mathematical theory of communication. The relevance of this extension appears with the reflection that our information about the specific microstructure of a given system derives largely from observation of its macrostates, and that the more complexions there are that might possibly underlie a given macrostate the less information we have about its actual microstructure in particular. The situation is analogous to that of a detective with a general description only of a wanted person: the more people there are who fit the description, the less information at hand about the actual culprit. When entropy is taken as a measure of the variety of microstates that might underlie an observable macrostate of a system, as in Boltzmann's application, then it also provides a measure of our lack of information about the structure of the system on the microlevel.

Information

Communication theory, pioneered by Nyquist and Hartley in the 1920s and formulated systematically by Shannon in 1948, is the study of the efficient transmission of messages through a communication channel. In its most general form, a communication channel consists of an input ensemble (A) of symbols (al, a2 ...ar) and an output ensemble (B) of symbols (b1 ; b2, ...bs), statistically interrelated by a set of conditional probabilities (P(bj /ai) specifying for each output bj the probability of its occurring in association with each input ai. For purposes of formal descriptions, both A and B are assumed to include a variety of symbol events, one and only one of which occurs at a significant moment of system operation. A simple illustration is a telegraph circuit, where A and B are comprised by events at the key and sounder, respectively, with conditional probabilities P(bj/ai) determined by the physical characteristics of transmitting medium.

Because of the variety of symbol events at the input, there is uncertainty in advance (a priori probability less than 1) about what event (E) will actually occur there at a given moment of operation. This uncertainty is removed when E actually occurs (with an a posteriori probability of 1). The removal of this uncertainty is designated 'information'. Information varies in quantity with the amount of uncertainty removed by E's occurrence, according to the formula 'I(E)=log 1/ P(E)'. (As in the equation defining thermodynamic entropy S, logarithms are employed here to achieve additivity. Logarithms to the base 2 are commonly used for convenience in application to digital computers.) If E has an a priori probability, say, of 50 per cent (think of a flip of an unbiased coin), then the information provided by its occurrence measures log 1/0.5 (=log 2), which amount to 1 bit (for '*bi*nary unit') of information. In general, the amounts of information provided by the occurrence of a given event is identical to the number of times (e.g., 1.74 for an event 30 per cent probable) its a priori probability must be doubled to equal unity.

The average information (H(A)) available at A is the sum of the quantities of information provided by its individual events, each multiplied by the event's probability of occurrence. It is easily shown mathematically that H(A) increases both with number of independent events at A and with the approach of these events to randomness, i.e., to equiprobability. Because of this (and because of the similarity of the mathematical definition of H(A) to that of thermodynamic entropy S), the quantity H(A) is often called the 'entropy' of A.A related measure is the 'equivocation' (H(A/B)) of A with respect to B, which is the average amount of uncertainty remaining about events occurring at input A after the occurrence of associated events at output B. This quantity is given by the sum of the conditional probabilities of each event a; given each event bj in turn, each probability being multiplied by the logarithm of its reciprocal. This quantity approaches zero as events at B increase in reliability as indicators of events at A, which makes H(A/B) a negative indicator of a channel's reliability as a conveyor of information. The capacity of a channel overall for the conveyance of information is directly proportional to the information available at its input and indirectly proportional to its equivocation. A channel's capacity in this regard is referred to as its 'mutual information' (I(A;B)), and accordingly is measured by the quantity H(A)-H(A/B).

While communication theorists typically are concerned with the design of channels for technological applications, communication channels play prominent roles in many natural processes as well. Communication of information in a natural setting often involves complex sets of channels known as 'cascades'. A cascade of channels consists of a sequence of individual channels so arranged that the output of the first serves as the input of the second, and so on *seriatim.* A perspicuous illustration is the cascade beginning at the cornea, and proceeding serially through the lens, the several layers of the retina, the optic chiasma, and eventually to the optical cortex. Each individual channel along the way has an integral part to play in the information-processing that constitutes visual perception.

A fact stressed by Shannon and other pioneers of communication theory, but too often unheeded by cognitive theorists employing an 'information-processing' vocabulary, is that information in this technical sense (information(t)) has very little to do with meaning or cognitive content. The symbol events with which communication theory deals might receive meaning under interpretation by human users, but by themselves have no semantical characteristics whatever. One of the major challenges of cybernetics is to gain insight into how information(t) can be converted by processes in the nervous system into information with cognitive significance (information(s))-into information in the sense of knowledge or intelligence. It is no service to clarity to assume, as is common in cognitive theory today, that information(s) appears 'ready-made' at the inputs of the central nervous system.

Negentropy

As entropy is a measure of a relative lack of information(t), so information(t) corresponds to a lack of entropy. The expression 'negative entropy' was used by Wiener in describing information(t) as the absence of entropy, and was later shortened by Brilluoin to 'negentropy'. Other forms of negentropy are structure (departure from random arrangement of a system's components) and productive energy (capacity within a system for useful work). These three forms of negentropy are mutually convertible.

Energy is converted into structure when water is pumped into an elevated reservoir. And structure is converted to energy in turn when water falls to drive the turbines of a generator. Energy is converted into information(t) with the detection of a signal on a modulated radio wave. Structure is convertible to the same effect in the operation of a computer by a coded punch card. Inasmuch as information(t) is basically a statistical quantity, its conversion into structure and energy is harder to illustrate; but an intuitive sense is provided by the thought-experiment known as 'Maxwell's demon'. The 'demon' in question is situated along a passageway connecting two containers of gas, and operates a trapdoor controlling access of individual molecules to either chamber. Initially this (closed) system is in a state of maximum disorder (maximum entropy), with gas molecules distributed randomly from moment to moment. The 'demon', however, is capable of receiving information(t), distinguishing slow-from fast-moving molecules, and of admitting only fast into one chamber and only slow into the other. As an eventual result of the trapdoor's operation, the system reaches a state of maximum structure (segregation of molecules by rate of motion) and of maximum usable energy (temperature difference between chambers due to molecular impact), both purchased by the information(t) that enables the 'demon' to discriminate rates of motion. Entropy re-enters the picture with the observation that actual transformations of this sort among forms of negentropy generally involve some loss of usable energy, in accord with the second law of thermodynamics.

An important principle of communication theory (Shannon's tenth theorem) states that a system can correct all but an arbitrarily small fraction of errors at its input if its equivocation H(A/B) is no greater than the mutual information I(A;B) of its correction channel. An equivalent formulation is Ashby's law of requisite variety, to the effect that a device's capacity to serve as a regulator (to block variations producing instability) cannot exceed its capacity as a communication channel (marshalling variation for the communication of information (t)).

Explanatory Principles and Methodology

Cybernetics has been conceived from the start as a unifying discipline, providing continuity across the borderlines of more specialized sciences. It has drawn freely upon the explanatory resources of other disciplines, but always with an eye towards applications beyond the boundaries of their original employment. As the biological concept of homeostasis was extended by Ashby to the design of machines with adaptive capacities *(Design for a Brain)*, for instance, so the physical concept of entropy was extended into biology with the work of Schrodinger. Observing that all natural processes produce increases of entropy in their general vicinity, Schrodinger characterized life as the capacity of an organism to resist progressive entropy in the form of structural loss by 'continually sucking orderliness from its environment'. The 'essential thing in metabolism' he remarked, 'is that the organism succeeds in freeing itself from the entropy it cannot help producing while alive', which it accomplishes by 'feeding' on negentropy at the expense of an accelerated progression of entropy in its immediate locale. This characterization highlights the remarkable ability of living organisms to receive energy from foodstuffs existing at lower energy levels than themselves, comparable in effect to a toaster being warmed by a cold piece of bread. It is by reversing otherwise natural flows of energy in this manner that a living system maintains itself, as Wiener puts it, 'as a local enclave in the general stream of increasing entropy'.

Another biological concept with broad application in cybernetics is that of adaptation. Homeostasis itself is an adaptive process, as are other forms of negative feedback. Among organisms functioning in variable environments, moreover, there is a tendency to adjust their feedback capacities in response to pervasive environmental change, which amounts to an adaptation of adaptive capacities. Species evolution itself provides examples of this higher-level adaptation, as in the development of spiny leaf structures by plants adjusting their moisture-conserving procedures to increasing levels of infrared radiation. While the interest of evolutionary biologists in such adjustments is likely to focus upon the underlying genetic mechanisms, however, interest from the cybernetic point of view will lie more with their effect upon the interchange of negentropy between organism and environment. The primary role of such adaptive processes, from this point of view, is to maintain an efficient coupling between organisms and environment by which the organism can gain the negentropy needed to sustain its vital processes. By way of augmenting Schrödinger's characterization, it may be said that a living organism not only is capable of receiving negentropy from its proximate environment, but also belongs to a reproductive group that tends to preserve this capacity throughout its membership by adaptive changes in the feedback mechanisms involved.

While relying upon other disciplines for certain of its explanatory principles, cybernetics also has developed explanatory resources of its own which would not fit comfortably into the specialized sciences. One is the principle of mutual convertability among energy, structure and information, cited above as being due largely to Brillouin. Another is the aforementioned law of requisite variety, in which Ashby reformulated Shannon's tenth theorem of communication theory in terms directly germane to the feedback capacities of living organisms. According to this law, the range of environmental variation to which an organism can adapt is limited by the capacities of its information(t)-processing systems. A consequence is that a large amount of the negentropy received by highly adaptive organisms like the human being must come in the form of information(t), and that a large portion of their physiological structure must be devoted to processing this information(t). This result is basic to a cybernetic analysis of human mental capacities.

These considerations make clear that the sense in which cybernetics is a unifying discipline has little in common with the 'unity of science' projected by logical positivism in the early twentieth century. While this latter was to have been achieved by way of reduction to physics, cybernetic theorists from the beginning have been explicit in denying primacy to the physical sciences. The biological principles upon which cybernetics relies are no more reducible to physics than its physical principles are reducible to biology; and neither science can deal with information(t) as a form of negentropy. The manner of unification offered by cybernetics is rather that of a context in which comparable phenomena from diverse disciplines can be studied in common terms, without loss of autonomy of the part of the disciplines concerned.

Because of its essentially interdisciplinary character, there is no single method or set of methods by which cybernetics can be distinguished from related fields of enquiry. The experimental techniques of neurophysiology and control engineering were important to cybernetics at its inception, along with the more formal methods of mathematical logic, calculus and the theory of computation. In recent years, cybernetic studies have employed techniques of systems analysis, organizational theory and computer modelling as well. From this it follows not that anyone employing these techniques of investigation *ipso facto* is engaged in cybernetics, nor that being engaged in cybernetics requires employing one or more of these techniques, but only that someone might employ any of these methods in cybernetic inquiry.

What is methodologically distinctive about cybernetics is rather the way in which it undertakes to bring the diverse resources of allied fields of study into mutual relevance. As an integrative discipline, cybernetics is more philosophy than science. In his quasi-autobiographical account of how the discipline came into being, Wiener cites Leibniz time and again (at one point naming him 'patron saint' of the field and mentions the influence of Royce and Russell as former teachers. Among other philosophers favourably noted are Pascal, Locke, Hume and Bergson. While acknowledging debt to these thinkers in various respects, Wiener seems particularly sympathetic with their interest in metascientific issues, and with their sense of how philosophy informs the foundations of science. The way in which cybernetics was pursued by Wiener is not unlike the way philosophy has been pursued by its major exponents throughout the centuries. Its manner of proceeding, in most general terms, was to adopt a few basic concepts and principles of elucidation, and in terms of these to elaborate a coherent picture of the world at large. For Wiener, the basic concept was that of system or organization, and the principles of elucidation those of feedback and communication. Due perhaps to the predominantly scientific orientation of other members of his original group, however, there seems to have been little incentive to work out the ramifications of this methodological perspective in detail.

A specific mode of enquiry that has proven serviceable to cybernetics in its more recent development is patterned after a familiar method of philosophic analysis. In its typical philosophic application, the method begins with a necessary feature of the concept or kind being analysed, and proceeds by adding other features until a combination is found that characterizes all and only instances of the thing in question. Successfully completed, the procedure yields a characterization unique to the thing being analysed, in terms of its necessary and sufficient conditions. In its specifically cybernetic application, the procedure is concerned instead with complex systems and their modes of behaviour, and the techniques of analysis might be experimental (e.g., in mechanical or computer simulations) as well as conceptual; but in other respects the procedure is similar. Beginning with a component of the complex structure under investigation, it proceeds by combining other components in some appropriate order until the structure in question has been exhibited (mechanically or conceptually) as a combination of parts. What marks the procedure as specifically cybernetic is the character of the components with which it deals, and (in application to biological systems especially) the order in which these components are appropriately combined. The components will have to do typically with the regulation of the system, and with its management of information(t) and other forms of negentropy. In biological applications, moreover, and in any other where the structures being studied show signs of having evolved from less complex substructures, the order of combination should follow lines of plausible evolutionary development.

This procedure overall might be characterized as a form of analysis by synthesis, or what in recent philosophy of mind has been called 'bottom-up' (in contrast with 'top-down') analysis. The synthetic or integrative bearing of the method is philosophical in character, and is directed towards understanding a complex structure in terms of its functional components. The components themselves are understood, as indicated above, according to the explanatory resources of the specialized sciences. The methodology of cybernetics thus is both philosophic and scientific, and might be pursued in a study as well as in a laboratory.

Analysis of Goal-Directed Behaviour

Among the earliest indications of the interdisciplinary and more broadly philosophic implications of the concept of negative feedback was the paper 'Behavior, Purpose, and Teleology' published by Rosenblueth, Wiener and Bigelow in 1943. The seminal idea of this paper, in Wiener's estimation, is that the central nervous system does not function as a self-contained organ, taking inputs from the senses and issuing outputs into the musculature, but that it characteristically acts instead as part of a negative feedback loop circling from the effector muscles out through the environment and back again through the sensory system. The neurological research that led to this insight had been concerned with goal-directed behaviour like picking up a pencil, and the authors saw fit accordingly to illustrate the type of feedback involved by the target-seeking missiles being developed as part of the current war effort. Weapons of this sort are guided by some sort of communication link (sounding-echoing, magnetic, thermal, etc.) with the intended object, through which an error-correcting mechanism operates to maintain the missile on a course leading to contact with the target. In their initial enthusiasm for this analogy between human and overtly mechanical feedback operations, the authors proposed target-seeking behaviour of this sort as a model for goal-directed (purposive, teleological) behaviour generally.

A telling difficulty of this model raised in subsequent criticism is that target seeking of this sort requires the physical presence of the intended goal, whereas human purposive behaviour (e.g., searching for a lost earring) is often directed towards goals that are absent. What has been taken as teleological activity in biological processes (e.g., growth of an oak from an acorn), similarly, appears to be directed toward goal states (e.g., the morphology of a mature tree) that are present, if ever, only late in the process. A model of goal-directed behaviour better suited to this wide range of examples is based upon the concept of equilibrium or stability, and stresses the manner of direction rather than the (external) goal itself. The missile is guided to its target by feedback mechanisms designed to maintain a stable correspondence between its actual motion and the path leading to impact with the target; from the guidance system's standpoint (although not the designer's) the resulting impact is coincidental. When loss of an earring disrupts a person's morning toilet, the search-behaviour ensuing is guided by feedback procedures (search patterns) directed towards restoring equilibrium to the person's dressing routines. And when an acorn begins to sprout and sends down its roots, its subsequent growth is guided by genetically established feedback processes towards a state of stable homeostasis (that of a fully leaved tree) within its living environment. Goal-directed behaviour in each case is governed by negative feedback, and is aimed at establishing or maintaining some facet of the operating system in a state of equilibrium.

Sharing this general form of goal-directed behaviour does not relegate either human purpose or biological teleology to the mechanical status of target-seeking missiles. Artifacts like guided missiles typically are engineered to perform certain predetermined operations when functioning properly under certain conditions, such that failure to perform accordingly under those conditions would be an indication of system malfunction. Human acts performed on purpose, however, are to some extent discretionary, which means *inter alia* that failure to perform when conditions warrant does not indicate a breakdown of the feedback systems involved. Human purpose in this sense is not deterministic, and thus is in accord with the general restriction in the cybernetic framework against deterministic explanations of natural processes at large (*see* below). Teleological growth is distinguished from target-guided behaviour, in turn, not only by being directed toward an absent goal, but also in that the goal-directed activity involved takes the form of morphological change rather than change in vectored motion. At any stage of its morphological development, an organism must relate to its immediate environment with sufficient stability to acquire the negentropy it needs to remain alive. Only at a relatively advanced stage of growth, however, does an organism achieve homeostasis in a form that can be maintained without further morphological change (exfoliation of leafy structure, growth of teeth, etc.). The sense in which biological growth is ideological is that it is directed toward a goal state which arrives literally only towards the end of the growth process itself, which is a state of stable homeostasis within the living environment. In view of the fact that such growth is guided by feedback mechanisms set in place by the organism's genetic structure, however, there is no call to interpret biological teleology as a causal process in which cause follows effect.

Forms of Adaptation

Morphological development is guided primarily by the genetic mechanisms of the individual organism. Another form of change to which organisms submit is adaptation, which occurs primarily in response to environmental variations. Among modes of adaptation that had been studied systematically before the advent of cybernetics are evolution and natural selection among biological species, and the behavioural conditioning of individual organisms. Cybernetic analysis of the feedback characteristics shared by these modes of adaptation led to the discovery of another mode pertaining to the formation of perceptual patterns, which appears significant for our understanding of cognitive processes.

Evolution and Natural Selection

A biological species is a group of interbreeding individuals with traits transmitted genetically to succeeding generations. Membership of a local subgroup (deme) of a given species fluctuates with changes in local conditions, as more or fewer members live to maturity in response to variations in food supply, predation, etc. Adjustments to short-term environmental variations of this sort generally occur without significant alteration of the group's genetic pool, and hence without alteration of the traits typically shared by its membership. Adjustment to more pervasive environmental changes like geophysical upheaval or shift in climate, on the other hand, may require alteration of a group's specific traits, enabling its members to take advantage of new food sources (e.g., thicker beaks for cracking seed shells) or new means of protection (e.g., lighter colouration for a snow-covered habitat). Long-term adaptation of this sort may be initiated by a shift in reproductive dominance to individuals within the group already approximating the features in question, which thereby become proliferated among members of subsequent generations. An eventual result may be the emergence of a reproductive group based on a genetic pool sufficiently altered to constitute a new species. Species evolution thus may be viewed as the product of a homeostatic process that enables reproductive groups to maintain stability under changing environmental circumstances. The mechanism of adjustment is alteration of the genetically controlled traits affecting the group's viability within its immediate environment.

Whereas species evolution is an adaptive process operating on the level of the reproductive group, natural selection in turn is an adaptive process on the level of the biota or ecosystem. A biota is a system of species interacting within a shared environment, each occupying a distinctive role or niche in which it can maintain stable relationships with its companion species. Niches are distinguished by the kinds of living space (trees, meadows, etc.) and food source (seeds, insects, etc.) they provide. The normal state of a biota is to provide all the niches needed to keep its community in balance, and at the same time to keep its existing niches full with as many individuals as the negentropic resources of its locale can sustain. Momentary decreases in population within a given niche may be countered by increased reproduction on the part of its occupying species, or by the immigration of competing species from adjacent locales. Excessive increases in population, on the other hand, will be countered by decreased reproduction, perhaps in the form of the extinction (or severe depletion) or one or another competing group. When competition for the limited resources of a niche puts a newly emerged species at hazard, the eventual result will be either its irradication by a more competitive species or its establishment as part of a well-balanced ecosystem. Natural selection thus may be viewed as a homeostatic process by which an ecosystem maintains integrity in a changing environment by changes in relationship among its constituent species.

Learning

Failure of a newly emerged species in its original biota does not preclude success in some other locale to which it may have migrated. Characteristics that enable a group to perform competitively in various locales (part of what above was termed 'negentropic flexibility') thus provide multiple opportunities for success, and are likely to become part of the group's genetic endowment. A major provision of this sort appeared in the course of species evolution with the ability of individual organisms to adapt their behaviour to local changes in their immediate environment. Whereas adaptation of primitive life-forms like bacteria and protozoa depends upon genetic mutations affecting the species at large, and hence requires a period of several generations, adaptation of individual behaviour to immediate contingencies can occur repeatedly within the lifetime of a single organism. This ability is known in behavioural science as 'conditioning' or 'learning'.

Any operating system produces outputs that are conditional to some extent upon its inputs. A system is capable of learning when it can adjust this conditional association between input and output to its own benefit. While a biological system in an entirely fixed environment would gain nothing by such an adjustment, a complex organism in a variable environment will generally benefit from some behaviour patterns and be harmed by others. Organisms capable of learning have been genetically disposed to avoid adversive (painful or 'punishing') stimuli and to seek stimuli they find agreeable (pleasant or 'reinforcing'). Inasmuch as natural selection will favour species whose members are reinforced under beneficial circumstances, and are punished by harmful, the effect of this disposition in an enduring species is to support survival of the individuals in which it operates. Learning then boils down to a process of shaping the behaviour of organisms to elicit predominantly agreeable stimuli from their current environments, and to minimize the occurrence of adversive stimuli, with the long-term result of enhancing the survival probabilities of the species involved.

In favourably conditioned organisms, stimuli indicative of beneficial circumstances will tend to elicit behaviour likely to secure those benefits, while stimuli indicative of harmful circumstances will tend to elicit avoidance behaviour. But when an organism begins to find adversive stimuli associated with previously beneficial circumstances (e.g., a once clear stream showing signs of contamination), or vice versa, the organism will undergo significant changes in the probabilities of its behavioural outputs conditional upon inputs signalling the presence of those circumstances. Inputs previously prompting the organism to take advantage of the circumstances they signify will now tend to elicit avoidance behaviour instead, or perhaps will simply lose their power to elicit any distinctive behaviour whatever. And *vice versa* for inputs previously leading to avoidance behaviour. By thus adjusting its conditional probabilities between sensory inputs and behavioural output in response to changing 'contingencies of reinforcement' (the phrase comes from Skinner), the organism adapts its behaviour to a changing environment. In its most general cybernetic description, learning is a feedback process in which the environmental effects of an organism's behaviour are channelled back through its sense receptors, and used to shape that behaviour in patterns conducive to the organism's advantage under current environmental circumstances.

Perceptual Patterning

According to Ashby's law of requisite variety, the range of environmental variation to which an organism can adapt its behaviour is limited by its capacities as an information(t)-processing system. A straightforward consequence is that the variety of adaptive responses an organism can make to a fluctuating environment is limited by the variety of circumstances distinguishable within its afferent nervous system. Cybernetically inspired experiments at MIT in the 1950s indicated that frogs, for example, are capable of distinguishing only five or six different patterns of stimulation in their visual environment, one being a spot the size of a fly moving a frog's tongue-length away in front of its eyes. The frog's behaviour in its biotic niche is confined to responses to these distinctive patterns (and a few others like them pertaining to other sense modalities), each of which is communicated through a set of nerve fibres dedicated to that pattern specifically. With this method of information(t)-processing requiring a one-to-one correspondence between message channel and message type, a radical extension in the number of patterns distinguishable by the animal's afferent system would require additional bulk that could impair its mobility. Perceptual patterning of this sort might be described as 'hard-wired', rather than adaptive in the manner of more advanced visual systems.

A major advance in adaptive capacity came with the evolution of organisms able to extend radically their range of discriminable circumstances without corresponding increase in bulk of their afferent nervous systems. The manner of information(t)-processing making this possible permits different stimulus patterns shaped in response to different environmental circumstances to pass through a single integrated network of afferent channels. This expedient of adaptive pattern-formation is similar in its feedback characteristics both to evolution and natural selection (adaptation on the species level) and to behavioural conditioning (adaptation on the level of the individual organism), and might be conceived alternatively as a very rapid evolution of afferent neuronal structures or as a much accelerated learning of the sensory system.

By 'perceptual pattern' here is meant a more or less specific set of neuronal events that occur interactively in response to a more or less specific configuration of external events in the perceptual environment. Between the external configuration and the neuronal pattern will extend a cascade of information (t)-channels (e.g., external object to cornea to lens to retina to optic chiasma, etc.), each serving both to convey relevant information(t) into the upper reaches of the afferent system and to help forge the features of the resulting pattern along the way. The key function of the cascade overall is to fashion a configuration of neuronal events that stands in a relation of high mutual information with the configuration in the external environment. If the mutual information between external and neuronal configuration is sufficiently high (i.e., they are sufficiently alike in information(t)-structure), then the latter will serve the organism as an effective guide of behaviour it undertakes with respect to the former. The latter in this respect is an adequate 'representation' of the former.

A corollary of Ashby's law cited above is that organisms under selective pressure to increase their variety of adaptive behaviours will be under pressure as well to employ their information(t)-processing channels as efficiently as possible. Efficiency might be served by various 'noise reductions', 'boundary tracing' and other information(t)-processing techniques of sorts well studied by communications engineers, performed at various stages along the cascade (e.g., at the retina or optic chiasma). The result is a neuronal representation of largely 'schematic' character, incorporating less detail than might be had at earlier stages in the cascade. If more detail proves necessary for successful guidance of behaviour undertaken with regard to what is represented, or if different representations are required to guide other behavioural projects, adjustments are made throughout the cascade to produce patterns at the upper afferent levels with informational(t)-features adequate to the task at hand.

Pattern-formation procedures of this general sort join with the efferent faculties of the behaving organism to constitute a homeostatic system, the normal state of which is a series of afferent patterns providing perceptual guidance for the organism's ongoing behaviour. Deviation from the norm is indicated by incipient loss of perceptual control, and the system regains stability by restructuring the representations by which this behaviour is currently being guided. Recent theoretical analysis suggests that perceptual pattern-formation of this sort provides a basis for certain cognitive processes typical of the human organism specifically.

Higher Cognitive Functions

Similarities in feedback characteristics among the processes of natural selection, learning and perceptual patterning, have been extensively explored in cybernetic literature. Parallels in natural selection and learning were pointed out by Wiener and were further developed by Skinner. An empirical basis for the account of perceptual patterning outlined above was proposed by Sayre. Extensions of this line of analysis to higher cognitive functions like language use and reason remain more conjectural. The brief discussion following is an indication of potential rather than actual accomplishment.

In a perceptual environment providing regularly recurring stimulus configurations, the afferent system of a perceptually adaptive organism may develop standard representations which find employment time and again in the pursuit of its perceptually guided projects. Such representations, which might be labelled 'percepts', are normally activated by stimulation of the external sense organs, and in this sense are controlled by the external environment. Percepts available to a given organism would typically include representations of familiar plants, animals, etc. Individuals of species capable of language will also form percepts responding to symbol configurations issued by other members of their linguistic community. To learn the language of one's community is to learn to associate percepts representative of familiar objects with other percepts representing the standard symbols of the language (at first vocal, later written, etc.), in such a fashion that the former percepts are capable of being activated by the latter. Percepts that in this fashion have been brought under the control of linguistic symbols, as well as of the objects they standardly represent, may be referred to as 'meanings'.

Meanings in this sense are not abstract entities, but rather well-entrenched neuronal patterns capable of functioning in the actual information(t)-processing activities of linguistically competent organisms. Since the essential feature of a given meaning structure is its relation of mutual information with the object it represents, and since physically different structures can share in this relation with a single given object, the same meanings can be present in different organisms. Linguistic communities emerge as many individuals learn to activate the same meanings upon presentation of the same symbol configurations, and to associate those symbols with the same objects in their shared environment.

Meaning B may be said to be redundant relative to meaning A when all features of the world represented by B are represented by A as well. The meaning 'ripe' applied to grapefruit, for instance, renders the meaning 'yellow' redundant. Conversely, applicability of 'yellow' to a given grapefruit is required for 'ripe' to be correctly applicable. The former controls the latter in this connection by restricting the circumstances of its correct application. Concepts may be conceived as meanings that have been removed from exclusive control of perceptual configurations (either linguistic symbols or objective circumstances), and brought under the control in this fashion of other meanings. While the percept 'yellow' is activated only by yellow objects, and the meaning 'yellow' either by objects or their linguistic symbols, the concept 'yellow' can be activated not only by objects and symbols, but in certain applications also by the meaning 'ripe'. Understood in this fashion, percepts, meanings and concepts are all stable structures of neuronal activity, distinguished with regard to their manner of control.

Concepts may be said to participate in a shared linkage if they are mutually relevant to each other's application, perhaps conditional upon the applicability of other concepts within the same linkage. The concepts 'ripe', 'yellow' and 'soft' thus share linkage with the concept 'grapefruit', reflecting the coincidence of colour and tactual properties discovered through the experience of a linguistic group with palatable grapefruit. If a variety of grapefruit were encountered in which a colour other than yellow (say, pink) turned out to be a more reliable sign of palatability, however, the relevant conceptual linkages of the individuals undertaking to eat this fruit would soon adapt to this novel set of circumstances. Due to the public nature of the language from which these concepts are derived, it would not be necessary for other individuals to undergo the same experiences themselves for their conceptual linkages to be appropriately modified. They can be modified through conversation with the individuals first affected. Conceptual linkages thus serve as facilities of information storage, subject to homeostatic adjustment and augmentation through the experiences of subgroups within the linguistic community. Such linkages in effect are neuronal mappings of regularly associated objective circumstances, subject to adaptation by continued encounters with a shared living environment. By using these maps to chart the course of anticipated behaviour, rational agents can explore alternatives before committing themselves to action.

Relation to Alternative Paradigms

Cybernetics has been guided from the outset by the conviction that a wide range of human mental functions can be reproduced mechanically. Among Wiener's original associates in the 1940s were several figures (e.g., O.G. Selfridge, W.H. Pitts, W.S. McCulloch) who subsequently became known for contributions to AI. Spokesmen for cybernetics up through the late 1970s, (e.g., F.H. George, K. Gunderson, K.M. Sayre) still considered AI to be an integral part of that discipline. The original ties between cybernetics and AI were effectively severed during the 1980s, however, to the extent that early contributors to machine intelligence who had remained closely identified with the former movement (e.g. W.R. Ashby, D.M. MacKay, F.H. George, Wiener himself) are seldom cited in current histories of the latter. This divorce appears to have been due largely to the recent takeover of AI by the computational paradigm, and to an ideological slide towards materialism on the part of its advocates.

Materialism, in its bare essentials, is the doctrine that everything in the universe comes under the purview of physical science. The primacy of physics in this regard was challenged in Wiener's original manifesto and was explicitly rejected in Ashby. Wiener's disavowal was based in part upon his realization that the determinism implicit in classical physics is incompatible with the variety inherent in biological processes. A theoretical basis for rejecting determinism in the natural world generally lies in the principle (a version of the second law of thermodynamics) that all irreversible processes tend to involve a loss of negentropy, which entails that causes generally tend to be more highly structured (involve less variety) than their effects. A consequence, as Wiener puts it, is that even the most 'complete collection of data for the present and the past is not sufficient to predict the future more than statistically'. Ashby's repudiation of materialism was more direct, pointing out that the materiality of the systems studied by cybernetics is simply irrelevant. Some organized systems (e.g., computers) are strictly physical, while others (e.g., linguistic communities) quite probably are not; and the fact that systems of both sorts can perform comparable feedback and information(t)-processing functions is no indication that they share the same ontological status. It should be noted at the same time, however, that the likely non-physical status of some cybernetic systems does not convert automatically into evidence for dualism. Contrary to current dogma in some quarters that materialism and dualism are the only ontological options on the horizon, a more plausible alternative from the cybernetic point of view is some version of neutral monism (as in Spinoza or early Russell). Sayre attempts to articulate a monism in which neither information(t)-functions of cognitive activity nor probabilistic functions at the quantum level of matter are further reducible to mental or physical features, making mathematical (statistical) structures more basic ontologically than either mind or matter.

The computational paradigm in recent AI and cognitive science rests upon the thesis that cognitive processes are computations performed upon representations, where the computations in question are of the sort typified by a standard digital computer. Barring the arbitrary introduction of randomizing elements, the computations performed by a properly operating digital computer are deterministic in outcome, which means that the same input invariably produces the same output. Even when the machine is computing probabilities, its procedures of computation are deterministic in this fashion. This makes mechanical computation an inappropriate model for indeterministic natural processes in general, and especially so for the highly negentropy-intensive (entropy producing) cognitive functions of human organisms. This difficulty, coupled with the well-known conceptual problems computationalists have encountered trying to account for the semantic properties of mental representations, should dispose anyone approaching cognition within a cybernetic framework towards disfavour of the computational model.

A more sympathetic reception may be accorded the connectionist paradigm that emerged during the late 1980s, which portrays various forms of cognitive activity as informational exchanges within a network of interconnected nodes with varying weights and excitation levels. Description of these networks is often couched in terms of conditional probabilities, and hence could be recast without distortion in the technical terminology of communication theory.

Connectionist researchers have already begun to study certain feedback characteristics of such systems, along with certain ways in which they might function in the control of motor behaviour. If attention were directed as well towards how these networks might adjust homeostatically in response to changes in a cognitively stimulating environment, connectionism might produce significant insight into the cybernetic workings of our cognitive faculties.

Bibliography

Achinstein, Peter (1967) The Concepts of Evidence, Baltimore, Johns Hopkins University Press. Achinstein, Peter (1983) The Nature of Explanation, Oxford, Oxford University Press.

Achinstein, Peter (1988) "The Illocutionary Theory of Explanation", in Joseph Pitt (ed.), Theories of Explanation, Oxford, Oxford University Press.

Allen, C, Bekoff, M., and Lauder, G. (1998) Nature's Purposes, Cambridge, Mass., MIT Press.

Ayer, A.J. (1961) "What is a Law of Nature", in The Concept of a Person, London, Macmillan.

Barnes, Barry (1974) Scientific Knowledge and Social Theory, London, Routledge.

Barnes, Barry, Bloor, David, and Henry, John (1996) Scientific Knowledge: A Sociological Analysis, Chicago, University of Chicago Press.

Beauchamp, Tom L., and Rosenberg, Alex (1981) Hume and the Problem of Causation, Oxford, Oxford University Press.

Berkeley, George (1710) Principles of Human Knowledge.

Bloor, David (1974) Knowledge and Social Imagery, London, Routledge.

Boyd, B., Gaspar, R, and Trout, J.D. (1991) The Philosophy of Science, Cambridge, Mass., MIT Press.

Braithwaite, Richard B. (1953) Scientific Explanation, Cambridge, Cambridge University Press.

Burtt, Edwin A. (1926) The Metaphysical Foundations of Modern Physical Science, London, Routledge.

Butterfield, Herbert (1965) The Origins of Modern Science, New York, Free Press.

Carnap, Rudolph (1952) The Continuum of Inductive Methods, Chicago, University of Chicago Press.

Cartwright, Nancy (1983) How the Laws of Physics Lie, Oxford, Oxford University Press.

Churchland, Paul, and Hooker, Clifford (eds) (1985) Images of Science: Essays on Realism and Empiricism, Chicago, University of Chicago Press.

Cohen, I. Bernard (1985) The Birth of a New Physics, New York, Norton.

Conant, James B. (1957) Harvard Case Histories in the Experimental Sciences, Cambridge, Mass., Harvard University Press.

Curd, Martin, and Cover, Jan A. (1997) Philosophy of Science: The Central Issues, New York, Norton.

Darwin, Charles (1979) On the Origin of Species, New York, Avenel.

Dawkins, Richard (1986) The Blind Watchmaker, New York, Norton.

Duhem, Pierre (1954) The Aim and Structure of Physical Theory, New York, Doubleday.

Feyerabend, Paul (1975) Against Method, London, Verso.

Feynman, R. (1984) QED: The Strange Story of Light and Matter, Princeton, Princeton University Press.

Fine, Arthur (1986) "The Natural Ontological Attitude", in The Shakey Game, Chicago, University of Chicago Press.

Glymour, Clark (1980) Theory and Evidence, Princeton, Princeton University Press.

Goodman, Nelson (1973) (first published 1948) Fact, Fiction and Forecast, 3rd edn, Indianapolis, Bobbs-Merrill.

Gutting, Gary (1980) Paradigms and Revolutions, Notre Dame, University of Notre Dame Press.

Hempel, Carl G. (1965) Aspects of Scientific Explanation, New York, Free Press.

Hempel, Carl G. (1988) "Provisos", in A. Grunbaum and W. Salmon (eds), The Limitations of Deductivism, Berkeley, University of Chicago Press.

Hoefer, C., and Rosenberg, A. (1994) "Empirical Equivalence, Underdetermination and Systems of the World", Philosophy of Science 61: 592-607.

Horwich, Paul (1982) Probability and Evidence, Cambridge, Cambridge University Press.

Horwich, Paul (1993) World Changes: Thomas Kuhn and the Nature of Science, Cambridge, Mass., MIT Press.

Hume, David (1888) A Treatise of Human Nature, Oxford, Oxford University Press.

Hume, David (1974) Inquiry Concerning Human Understanding, Indianapolis, Hackett.

Jeffrey, Richard (1983) The Logic of Decision, Chicago, University of Chicago Press.

Kant, Immanuel (1961) The Critique of Pure Reason, London, Macmillan.

Kitcher, Philip (1995) The Advancement of Science, Oxford, Oxford University Press.

Kneale, William (1950) Probability and Induction, Oxford, Oxford University Press.

Kuhn, Thomas S. (1957) The Copernican Revolution, Cambridge, Mass., Harvard University Press.

Kuhn, Thomas S. (1977) The Essential Tension, Chicago, University of Chicago Press.

Kuhn, Thomas S. (1996) The Structure of Scientific Revolutions, 3rd edn, Chicago, University of Chicago Press.

Latour, Bruno, and Woolgar, Steve (1979) Laboratory Life, London, Routledge.

Laudan, Larry (1977) Progress and its Problems, Berkeley, University of California Press.

Leibniz, G.W. (1981) New Essays on Human Understanding, Cambridge, Cambridge University Press.

Leplin, J., and Laudan, L. (1991) "Empirical Equivalence and Underdetermination", Journal of Philosophy 88:449-72.

Leplin, Jarrett (1988) A Novel Argument for Scientific Realism, Oxford, Oxford University Press.

Leplin, Jarrett (ed.) (1984) Scientific Realism, Berkeley, University of California Press.

Lewis, David (1974) Counterfactuals, Oxford, Blackwell.

Lewis, David (1984) "Causation", in Philosophical Papers, vol. 2, Oxford, Oxford University Press.

Lloyd, Elizabeth (1987) The Structure of Evolutionary Theory, Princeton, Princeton University Press.

Locke, John (1690) Essay on Human Understanding.

Longino, Helen (1990) Science as Social Knowledge: Values and Objectivity in Scientific Inquiry, Princeton, Princeton University Press.

Mach, Ernst (1906) The Analysis of Sensation.

Mackie, John L. (1973) Truth, Probability and Paradox, Oxford, Oxford University Press.

Mackie, John L. (1974) The Cement of the Universe, Oxford, Oxford University Press.

Merton, Robert K. (1973) The Sociology of Science, Chicago, University of Chicago Press.

Mill, John S. (1843) A System of Logic.

Miller, Richard (1987) Fact and Method: Explanation, Confirmation and Reality in the Natural Sciences, Princeton, Princeton University Press.

Nagel, Ernest (1977) Teleology Revisited, New York, Columbia University Press.

Nagel, Ernest (1979) The Structure of Science, Indianapolis, Hackett.

Nagel, Ernest, and Newman, James R. (1954) Gödel's Proof, New York, State University of New York Press.

Newton-Smith, William (1981) The Rationality of Science, London, Routledge.

Pickering, Andrew (1984) Constructing Quarks, Chicago, University of Chicago Press.

Pitt, Joseph (ed.) (1988) Theories of Explanation, Oxford, Oxford University Press.

Popper, Karl R. (1959) Logic of Scientific Discovery, New York, Basic Books. (First published in German, 1935.)

Popper, Karl R. (1984) Objective Knowledge, New York, Harper and Row.

Quine, Willard.V.O. (1951) From a Logical Point of View, Cambridge, Mass., Harvard University Press.

Quine, Willard.V.O. (1961) Word and Object, Cambridge, Mass., MIT Press.

Railton, Peter (1988) "A Deductive-Nomological Model of Probabilistic Explanation", in Joseph Pitt (ed.), Theories of Explanation, Oxford, Oxford University Press.

Reichenbach, Hans (1938) Experience and Prediction, Chicago, University of Chicago Press.

Reichenbach, Hans (1951) The Rise of Scientific Philosophy, Berkeley, University of California Press.

Rosenberg, Alex (1985) The Structure of Biological Science, Cambridge, Cambridge University Press.

Rosenberg, Alex (1992) Philosophy of Social Science, Boulder, Westview.

Salmon, Wesley (1966) The Foundations of Scientific Inference, Pittsburgh, University of Pittsburgh Press.

Salmon, Wesley (1984) Scientific Explanation and the Causal Structure of the World, Princeton, Princeton University Press.

Salmon, Wesley (1988) "Statistical Explanation and Causality", in Joseph Pitt (ed.), Theories of Explanation, Oxford, Oxford University Press.

Salmon, Wesley (1989) Four Decades of Scientific Explanation, in Wesley Salmon and Philip Kitcher, Scientific Explanation, Minnesota Studies in the Philosophy of Science 13, Minneapolis, University of Minnesota Press.

Salmon, Wesley, and Kitcher, Philip (1989) Scientific Explanation, Minnesota Studies in the Philosophy of Science 13, Minneapolis, University of Minnesota Press.

Savage, Leonard (1972) Foundations of Statistics, New York, Dover. Shapere, Dudley (1964) "Review of Structure of Scientific Revolutions", Philosophical Review 73:383-94.

Shapin, Steven (1996) The Scientific Revolution, Chicago, University of Chicago Press.

Sheffler, Israel (1976) Science and Subjectivity, Indianapolis, Bobbs-Merrill.

Smart, J.J.C. (1968) Between Science and Philosophy, London, Routledge.

Spector, Marshall (1968) Concepts of Reduction in Physical Science, Philadelphia, Tempel University Press.

Stove, David C. (1967) Hume, Probability and Induction, Oxford, Oxford University Press.

Suppe, Fredrick (1977) The Structure of Scientific Theories, Urbana, University of Illinois Press.

Thompson, Paul (1989) The Structure of Biological Theories, Albany, State University of New York Press.

Van Fraassen, Bas (1980) The Scientific Image, Oxford, Oxford University Press.

Van Fraassen, Bas (1988) "The Pragmatic Theory of Explanation", in Joseph Pitt (ed.), Theories of Explanation, Oxford, Oxford University Press.

Westfall, Richard (1977) The Construction of Modern Science, Cambridge, Cambridge University Press.

Wright, Larry (1976) Teleological Explanation, Berkeley, University of California Press.

Index

H

I

K

L

M

N

O

P

Q

R

U

V

W

Y